# DESIGN OF SLURRY TRANSPORT SYSTEMS

# ACKNOWLEDGEMENTS

Much of the material for this book has been compiled from work carried out over a period of time by BHR Group engineers. The author wishes to acknowledge this contribution from his colleagues, past and present. They are: P. J. Baker, R. J. Galka, A. L. Headford, T. J. M. Moore, R. S. Silvester and G. F. Truscott.

The author and publishers wish to thank the following for the use of illustrations:

Bredel b.v., Ch. 2 Fig. 12;
Butterworth and Co. (Publishers) Ltd, Ch. 4 Fig. 9;
Cameron Iron Works Inc., Ch. 6 Fig. 1;
GEC Mechanical Handling, Ch. 4 Fig. 3;
Krebs Engineers USA, Ch. 4 Fig. 6;
McGraw Hill International Publications Co., Ch. 4 Fig. 2;
Mono Pumps Ltd, Ch. 2 Fig. 10
National Supply Co. (UK) Ltd, Ch. 2 Fig. 9;
Putzmeister Ltd, Ch. 2 Fig. 14;
Schwing GmbH, Ch. 2 Fig. 13;
Serck Audco Valves, Ch. 6 Fig. 6;
Simon Warman Ltd, Ch. 2 Figs 1 and 2, Ch. 3 Fig. 16.

The publishers have made every effort to trace the copyright holders, but if they have inadvertently overlooked any, they will be pleased to make the necessary arrangements at the first opportunity.

# DESIGN OF
# SLURRY TRANSPORT
# SYSTEMS

## B. E. A. JACOBS

(B H R Group)

## CRC Press
Taylor & Francis Group
Boca Raton  London  New York

CRC Press is an imprint of the
Taylor & Francis Group, an **informa** business

CRC Press
Taylor & Francis Group
6000 Broken Sound Parkway NW, Suite 300
Boca Raton, FL 33487-2742

First issued in paperback 2019

ISBN-13: 978-1-85166-634-8 (hbk)
ISBN-13: 978-0-367-86393-7 (pbk)

**British Library Cataloguing in Publication Data**

Jacobs, B. E. A.
   Design of slurry transport systems.
   1. Slurry. Hydraulic transport in pipelines
   I. Title
   621.8672

   ISBN 1-85166-634-6

   Library of Congress CIP data applied for

# CONTENTS

# PREFACE

Interest in practical hydraulic transport systems has ebbed and flowed over the years although study of the fundamentals has continued steadily. A review of the literature shows there was a flurry of interest in the 1940s. In the 1950s significant technical progress was made in several countries through a strong research effort. In the UK, experimental work was conducted particularly on the handling of coarse coal slurries. This work was mainly carried out by the British Hydromechanics Research Association in conjunction with the National Coal Board. In France, Durand and Condolios carried out a large amount of work on the hydraulic transport of aggregates. During the 1960s several countries became involved in developing hydraulic transport for mining and a number of coal mine haulage systems were installed. A new surge of interest became evident after the oil crisis in the mid 1970s.

At the present time there are many organisations throughout the world carrying out research and development in the field of slurry transport, although there are few long-distance pipelines either under construction or recently commissioned. It is understandable that the greatest interest is shown in these major lines because of their substantial engineering content. It must be noted, however, that there are many small pipelines being designed and built particularly in the mining, chemical and food processing industries, for which the details remain unpublished.

Important issues which have significant implications on the growth of slurry pipeline systems, particularly for coal and power generation, include the price of oil in relation to that of coal. The level of investment put into the energy industry by government and industrial organisations will influence growth. For example, the fall in oil prices from their peak has led to a decline in investment in other forms of energy.

Environmental pressures brought to bear by the general public will also be significant. Requirements are now placed on public utilities and mining companies to incorporate effective means of waste disposal into their future plans. Slurry systems are used in Flue Gas Desulphurisation and in waste material transport.

The use of hydraulic transport for feeding gasifiers and liquefaction plants is long term and apart from experimental facilities, the widespread commercial use of such equipment is still in the future.

The purpose of this book is to benefit users, manufacturers and engineers by drawing together an overall view of the technology. It attempts to give the reader an appreciation of the extent to which slurry transport is presently employed, the theoretical basis for pipeline design, the practicalities of design and new developments.

Each chapter is self-contained, thus the reader requiring information on a particular topic will find it principally in the appropriate section with only a minimum of cross-referencing. For this reason references are given at the end of each chapter.

The book is structured so that the reader is led through the sequence in which a slurry pipeline transport system could be designed. First, the type and size distribution of the materials to be transported and the flow rate and pipe diameter define the flow regime. The various types of flow regime are described and the methods used to predict the pressure gradients discussed. The information generated leads to the type of pump required and materials of construction. Slurry pumps are described along with the effect of wear on both pumps and pipelines.

At this stage, it would be possible to carry out a rough economic analysis to assess the likely cost advantages compared with other forms of transport. On the assumption that the economics are satisfactory, the designer would then consider the further engineering and cost implications of slurry preparation and dewatering. The necessary instrumentation and control functions are also considered. Further aspects such as start-up, shut-down and particle degradation are discussed, together with applications additional to that of simple bulk materials transport. A section dealing with carrier fluids, other than water, is included as well as some information on three-phase mixtures.

The book finishes with a chapter on existing applications including some cost information where available. Costs for proposed pipelines need to be generated for each application but a methodology to generate approximate costs is described.

The reader will discover, if he is not already aware, that there is no simple way to design a solids handling pipeline. For large schemes complex feasibility studies must be undertaken including practical development work. Research is also required for the new techniques being considered.

# 1. PREDICTION OF FLOW PARAMETERS

## 1. INTRODUCTION

Theoretical aspects of slurry pipelining are reviewed, with the objective of identifying the current limits of knowledge. It has been said that there is nothing quite so practical as a good theory. Indeed, the ultimate test of any theory is its agreement with reliable experimental results. Nevertheless, such results cannot be known to be reliable without some prior theoretical understanding. For a practical subject like slurry pipelining, such abstractions may seem a little irrelevant, if the objective is seen simply in terms of getting the product from A to B. However, if we consider that A and B may be a long way apart and that a few per cent error in operating conditions may have critical cost implications, the need for good theory becomes apparent. In the practical hydrotransport context, a further requirement exists that the theory be accessible to and usable by the designers of slurry systems, not just the academic colleagues of the theoretician.

Literature in the field of hydraulic transport is continually being reviewed over the whole spectrum of subject area. Whenever a new theory is presented, a collection of prior experimental data is usually produced to support it. The publication of new test data is, conversely, often set against the background of existing theories. On another level, reviews have been written with the objective of producing a design guide, or simply to gather together existing work and make some sense of it. With this in mind, this section begins with a guide to the various sources of information from which the literature has been drawn. There follows a discussion of pipeline design in terms of the design process, which is frequently subject to constraints that prevent it from passing rigorously from hypothesis to conclusion. The designer is often saddled with insufficient information, and time or cost restrictions that preclude him from obtaining what he needs.

## 1.1 SOURCES OF INFORMATION

The 'Hydrotransport' series of conferences, organised by BHRA in conjunction with other interested organisations, have proven a productive source of information. Coupling these Proceedings with those of the Slurry Transport Association, now entitled the Coal and Slurry Technology Association, has helped towards a balance between theory and practice. Many other sources of information exist, however, ranging from the very academic to the very practical. Reviews tend to mix theory and application. Stern (1), for instance,

1

considered slurry transport technology in the context of its possibilities for South Africa. He highlighted the basically different approaches needed for short and long distance slurry lines. He referred to existing systems in other countries, making the useful point that a material must not lose its value when crushed or wetted before it can be considered for slurry transport. In this context, iron ore is an ideal material as it makes excellent pellets from granulated fines, easing its passage through the steel mill. Referring to South Africa's shortage of water, Stern recommended looking to high solids concentrations. If this were to be combined with coarse material transport, the entire spectrum of slurry technology, from non-Newtonian to settling suspensions, would be invoked in the one system. Alternatively, as suggested by Parkes and Lindsay (2) for lifting coal from mines, the coarse material could be lifted mechanically and the fines hoisted hydraulically.

Rigby (3) took a wide historical view concerning slurry transport, referring to its early development in the American gold rush of the mid-nineteenth century. Since then, the Ohio Cadiz coal line, built in 1957 but later moth-balled (1963) for economic reasons, pioneered the large scale (147 km long $\times$ 254 mm diameter) transport of material at high throughputs (1.5 Mtpa). It was later overtaken by the Black Mesa, 439 km $\times$ 457 mm $\times$ 5 Mpta, supplying coal to the Mohave power station in southern Nevada. The first iron ore concentrate line (Savage River) was built in Tasmania (1967), with conservative slope specifications to cope with the solids specific gravity. It too has been overtaken in scale by the Brazilian Samarco line, carrying 7 Mtpa of iron ore concentrate over 400 km. Other materials have included limestone (UK), gold slime (Australia, South Africa), phosphate (Canada, South Africa), copper concentrate (Papua New Guinea), copper tailings (Chile), and zinc sulphites (Japan), to name but a few. In many cases, grinding for slurry transportation is consistent with other process requirements, such as ore concentration by flotation. Hydraulic transport of waste material, such as bauxite residue, coal mine tailings and sewage, is also a widely accepted and practised application.

In an overview of slurry transport technology, Lee (4) looked at various means of transporting coal, ranging from the very fine yet concentrated mixtures for direct combustion, to dilute transport of run-of-mine (ROM) coal. ROM coal usually reaches a maximum size of 50 mm (2 inches), and two opposite approaches to its transportation have arisen. One has been to pump it in low concentrations with devices such as jet pumps. The application for such systems include ship loading, where no shortage of water exists and the pumping distances are short. Where there is a shortage of water, high concentration coarse coal suspensions, stabilised both by the presence of fines and the high total solids, come into their own. The difficulty here is that of finding a high pressure pump capable of handling coarse solids. For this, the most promising candidate appears to be the Boyle rotary ram pump, capable of shearing lumps of coal obstructing the valve gear. Another kind of concentrated suspension, the 70 per cent coal-water-mixture (CWM) avoids the dewatering step by offering itself for directly fired combustion.

Other reviews have focused more particularly upon the theory. Shook (5), for example, considered hydrotransport in the context of general two-phase flow. He compared capsule and slurry transport, commenting that capsule transport was, at that time, further advanced technically despite being a later development. He mentioned the possibility of turbulence suppression in channel flow and some benefit to be gained from the use of drag-reducing additives. Some difficulties were highlighted concerning the prediction of critical velocity for settling slurries, and scale-up from model to prototype. Turbulence suppression in non-Newtonian and fibrous suspensions was noted, with the observation that most of the equations for the flow of such media were empirically based. In another comparison between capsule and suspended sediment transport, Lazarus (6) presented a break-even curve for the two systems. For a 300 m line, capsules were favoured at throughputs of sand-weight material above 300 kg$^{-1}$. Capsules had the advantage that they could be made neutrally buoyant by not filling them completely, whilst slurries were found to be favoured by their independence of a fabricated vehicle.

Thomas and Flint (7) reviewed pressure drop prediction in slurry lines, considering low and high concentrations of both settling and stable suspensions. They refer to the use of an effective turbulent viscosity for the Reynolds number determination of a turbulently flowing non-Newtonian slurry, and to the practice of splitting a slurry of mixed particle sizes into homogeneous and heterogeneous portions. Kazanskij (8) presented an especially comprehensive review, tabulating thirty-seven different correlations for pressure drop in heterogeneous flow alone, between the years of 1954 and 1976. They concluded that a scale-up criterion based purely on theory was disallowed by a lack of fundamental knowledge concerning the physics of suspension. To some degree, that situation has changed as explained in sections 3.1 and 3.2. The change is, however, recent and not yet complete. Duckworth (9), reviewing the field at the same time as Kazanskij (8), noted a large number of empirical expressions in more common use than any based on theory alone. One difficulty, of course, with the latest theoretical developments is their complexity, requiring them to be tied to a powerful computer, perhaps inaccessible to the pipeline system design team.

## 1.2 PIPELINE DESIGN

The designer is faced with a number of problems. We can assume that he starts with a material to be transported at a known rate. We can also take it that there is a well-specified destination, perhaps a port, for shipping the material abroad, perhaps a central depot for inland distribution, or a down stream reactor in a chemical production plant. He must now select a means of transport. Where applicable, road and rail offer a certain flexibility at high cost in running expenses, but little capital expenditure is required unless new roads or rail lines have to be laid. Should he choose a slurry line for its low running costs, he must ensure that it will pay for itself well within the life of the project. Short payback periods are the rule if finance is to be sought for the venture. Delays due to opposition by existing transport contractors, government planning requirements and unforeseen technical hitches can significantly affect the project completion date, and hence the payback period.

Figure 1, based on Pitts and Hill (10), shows the basic elements required of a typical pipeline system. From the mine, assumed for present purposes to be iron ore, the product is

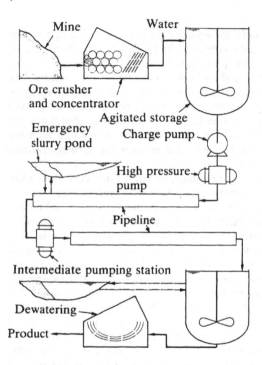

**Figure 1.** Schematic flow diagram of a long distance slurry pipeline (10)

crushed and concentrated. It is then fed into a mixing hopper where the water is added and the resulting slurry delivered to an agitated storage vessel. From here, it can be pumped into the pipeline by a high pressure pump, avoiding the need for frequently spaced pumping stations. High pressure slurry pumps, however, usually require a pressurised feed, supplied in this instance by a priming pump. Emergency slurry ponds must be provided at the end of each section of slurry line, if blockage at one portion of the line is not to be propagated along its entire length. In the case of iron ore concentrate, dewatering of the slurry is required before the fines can be pelletised in readiness for the blast furnace. From the above description, the following tasks might reasonably be assigned to the design team:

- Select, if the option is open, a size distribution to be used for slurry transportation. If the size is fixed by process requirements, this should be monitored and taken into account for the transportation system.
- Determine the design throughput of solids, and find a source of water capable of handling it at the desired concentration, also to be fixed by the designer.
- Select an agitated storage system to ensure a constant throughput for intermittent

production. Agitation must maintain a well-mixed suspension for minimal cost in power or agitator wear.

● Design the pipeline for its internal diameter, wall thickness and length between pumping stations. This requires the selection of a pumping system of suitable flow and head capability.

● Protect the system as much as possible from wear, corrosion, blockage, leakage, cavitation and pressure surges. Minimise the extent of any damage resulting from failure of such protection.

Clearly, design of the pipeline itself is one component of the total problem. Nevertheless, it is central and impinges upon the design of other parts of the system. Much of the technology relevant to slurry lines can be borrowed from that developed by the water supply and petroleum industries. Indeed, Sandhu and Weston (11) considered the feasibility of conversion of existing crude oil or natural gas lines into coal slurry lines. They concluded that in many cases it could be done at a cost fourfold less than that required for a new system. New pumps would have to be installed at the pumping stations, but only 2 per cent of the line length would have to be replaced to cope with slurry service.

The basic features of a long pipeline with a downward sloping ground profile between mine and outlet are now described. The line must be capable of handling the pump head plus a surge allowance. Its wall thickness is, therefore, required to be greatest just downstream of the pump. The hydraulic grade line falls linearly along the pipe to the next pumping station needing progressively less wall thickness to cope with the pressure. It makes sense, however, to vary the thickness in stages rather than attempt to follow the hydraulic grade too closely. It is worth pointing out that an upwards slope makes more severe demands upon the pipe section downstream (up slope) of a pumping station.

The above process depends upon reasonably accurate predictions of head gradient for the design flow and concentration of slurry. Some savings can be made if small-scale loop tests can be scaled up to prototype. Thomas (12) reviewed a number of scale-up methods, some involving a simple power correlation of pipe diameter and others of a more complicated polynomial form. For semi-heterogeneous non-Newtonian slurries, five constants were required to be determined, demanding numerous tests. It may be possible, however, to split the non-Newtonian and heterogeneous aspects in such a way that the analysis is simplified. Formulations attempting to deal with everything at once have rarely been proven successful.

At a Symposium of the Canadian Society of Chemical Engineers, two papers were presented highlighting another way in which pipeline design can be subdivided for clarity. Stevens (13) focused on long-distance pipelines, defining them as anything longer than ten miles. Shook (14) looked at short-distance pipelines, defining them as anything that could be designed without an extensive test programme. He concluded that there was no such thing as a 'short-distance pipeline' at the time of writing (1969)! Now that test loops are beginning to look like commercial projects, it is worth considering how money might be saved for the design team, ultimately for their clients and all those associated with the product being shipped.

## 1.3  NOMENCLATURE

In an attempt to save the tedium of continually needing to refer to a list of terms and subscripts, these are defined in the text when the occasion arises. Common usage is employed where possible, such as $u$ or $V$ for velocity, $\rho$ for density, $S$ for specific gravity, $\mu$ or $\eta$ for viscosity, and commonly accepted symbols for dimensionless numbers such as Reynolds number (Re) and Froude number (Fr). Variables may be prefixed with a 'delta' ($\Delta$) to denote difference or a 'del' ($\nabla$) to denote gradient, a feature applying principally to pressure ($p$). Although the use of $\nabla$ stems from the mathematics of vectors, this is the nearest the reader will be subjected to vector mathematics. Head gradient ($J$) may be taken to refer to the carrier fluid (water) rather than the mixture, unless otherwise specified. Departures from common usage include $m$ for pseudoplastic consistency coefficient, in preference to $K$, one of the more overworked constants in the English alphabet. Friction factor is always $f$, not $\lambda$, and usually interpreted as Darcy, rather than Fanning, friction factor. Section 2.1 explains this in more detail. In any situation where the symbols are not explicitly defined in the text, the reader can refer to the nomenclature given in an appendix. This is set out chapter by chapter to overcome the problem of multiple definitions with the range of subjects covered.

## 2.  SLURRY PIPELINE FLUID DYNAMICS

Fundamental to the problem of slurry transportation are the principles of fluid mechanics for single-phase Newtonian systems. Despitë he fact that new theories of turbulence and creeping flow, are continually being published, enough is understood about fluid flow in pipes for it to be considered a textbook subject. A brief presentation of it is given here for its value as a basis for looking at more complex systems. Settling and stable suspensions of solid particles each depart in a special way from single-phase Newtonian behaviour. The dynamics of a settling slurry can be most readily understood by treating it as a two-phase mixture of liquid and solids. In a manner similar to that of gas-liquid two-phase flow, there arise numerous flow regimes, each representing a distinct distribution of liquid and solids within the pipeline. Such systems also behave differently in vertical and horizontal flow, and are otherwise sensitive to pipeline slope. Without the two-phase complications, stable slurries often display a non-Newtonian response to shear, requiring two or more parameters to replace viscosity. The Reynolds number must be reformulated and the laminar-turbulent transition is frequently altered. Considering that some industrial slurries consist of coarse solids unstably suspended in a non-Newtonian medium, the need for a sound fluid mechanical basis becomes obvious.

The reader of this section is to look at single-phase, steady-state, Newtonian flow in pipes. Departures from this ideal are then taken in turn, beginning with settling mixtures and the flow regimes they generate. Non-Newtonian media follow, with attention given to basic definitions and practical examples of substances displaying the various properties. Departures from the steady state are represented by pressure surges (water-hammer), generated either by a pump trip, a sudden valve closure, or a burst pipeline. They are of engineering significance, in that measures must be provided to deal with them without damage to the pipeline or its ancilliary components. The fluid transients of interest are treated first in basic terms and then in the context of solids suspensions.

## 2.1 FLUID FRICTION

Fluids inside pipelines invariably require a pressure gradient to maintain them in steady flow. The pressure gradient ($\nabla p$) can be determined from the shear stress ($\tau$) developed by the fluid at the pipeline wall. For a Newtonian fluid, this is in turn governed by fluid viscosity ($\mu$) and wall shear rate ($\dot{\gamma}$). Pressure gradient is usually expressed non-dimensionally, known as the Darcy friction factor ($f$), for which the normalising scale is the product of velocity pressure ($\frac{1}{2}\rho u^2$) and the inverse of pipeline diameter ($D$), where $\rho$ and $u$ represent the density and mean velocity of the fluid concerned. For laminar flow through a pipe of circular section, the parabolic velocity profile form can be used to calculate wall shear rate and close the problem:

$$\dot{\gamma} = 8uD^{-1} \qquad \text{Application of parabolic velocity profile} \qquad (1)$$

$$\nabla p = 4\tau D^{-1} \qquad \text{Force equilibrium of pressure and shear stress} \qquad (2)$$

$$\tau = \mu\dot{\gamma} \qquad \text{Definition of Newtonian viscosity} \qquad (3)$$

From equations 1, 2 and 3, we can develop an expression for pressure gradient in terms of mean velocity, pipe diameter and fluid viscosity. In dimensionless form, this becomes an expression of friction factor in terms of a Reynolds number, the ratio of inertial to viscous fluid forces:

$$\nabla p = 32\mu uD^{-2} \qquad \text{Combination of above three equations} \qquad (4)$$

$$f = 2D\nabla p\rho^{-1}u^{-2} = 64\,\text{Re}^{-1} \qquad (5)$$

where $\text{Re} = \rho uD\mu^{-1}$

Equations of the form of equation 5 have great universal significance in nearly all fluid mechanical phenomena. Drag effects on bodies in flight, power requirements for fluid mixing equipment and the flow of non-Newtonian substances are all described by formulae like equation 5. The form of the Reynolds number may vary, as may that of the friction factor and the value of the numerical coefficient. The equation is only valid, however, for laminar flow. It is worth noting at this point that some researchers prefer the Fanning friction factor ($f'$), defined in terms of the wall shear stress instead of pressure gradient:

$$f' = 2\tau\rho^{-1}u^{-2} = 16\,\text{Re}^{-1} \qquad \text{for laminar flow} \qquad (6)$$

It has been argued that the Fanning friction factor is more fundamental in concept, as it relates shear stress directly to velocity and fluid properties. The Darcy definition, however, is more convenient in terms of pipe flow.

Most practical commercial pipelines operate in the turbulent flow regime, for which momentum transfer takes place by means of a spectrum of eddies. Much bulk mixing takes place across the diameter of the pipe and fluid inertia becomes significant. An approximation for the friction factor in smooth pipes is given by the Blasius formula:

$$f = 0.316 \, \text{Re}^{-0.25} \tag{7}$$

if $\text{Re} \geqslant 3000$

It is of interest to note that a Reynolds number of 3000, equation 5, gives a friction factor of 0.0213, half that of the prediction of equation 7 of 0.427. It is typical of the transition from laminar to turbulent flow that the friction factor increases. Turbulence in pipes is essentially born of instability; eddies formed at the wall tend to grow rather than decay. As the Reynolds number is increased beyond the transition value, the friction factor will continue to fall (equation 7), though at a slower rate than in laminar flow.

Equation 7 is not quite accurate, especially at Reynolds numbers above about 30 000. An improved formula for turbulent friction factor is the Karman–Nikuradse equation:

$$f^{-0.5} = 2 \log_{10} (\text{Re}.f^{0.5}) - 0.80 \tag{8}$$

Unfortunately, equation 8 gives an implicit expression for friction factor, difficult to evaluate on a pocket calculator. Furthermore, it is only valid for smooth pipes. As the Reynolds number is increased, the significance of pipe roughness increases as the thickness of the laminar boundary layer at the wall decreases. For a given scale of roughness $(k)$, an approximate formula may be given for friction factor, in terms of a new dimensionless parameter, the relative roughness $(k/D)$:

$$f = 0.25 \, (\log \, (0.27 \, k/D + 5.74 \, \text{Re}^{-0.9}))^{-2} \tag{9}$$

Equation 9, taken from Miller (15), is an improvement on equation 8 in that it takes account of roughness, gives friction factor directly and is more accurate over the full parameter range. A more rapid alternative may be to use the 'Moody Chart', a friction factor graph as plotted in Figure 2. It gives the friction factor over a wide range of Reynolds numbers and pipe roughnesses and has been developed by iteration of the Colebrook–White equation, to which equation 9 is a close approximation. Figure 2 also gives approximate $uD$ values for air and water at ambient temperature and pressure.

Most pipelines consist not only of straight lengths of pipe, but also bends, valves, tees and other assorted fittings. Miller (15) gives a comprehensive compilation of the pressure drop $(\Delta p)$ across such components, usually expressed in the following terms:

$$\Delta p = K \rho u^2 / 2 \tag{10}$$

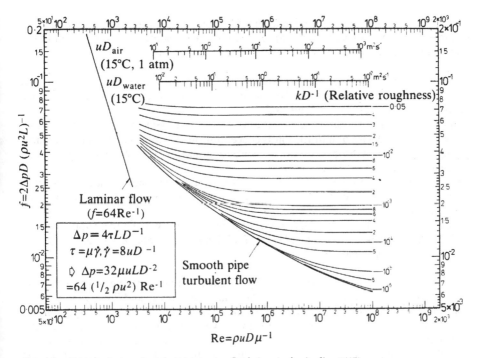

**Figure 2.** Friction factor chart for Newtonian fluids in rough pipelines (15)

The term $K$ is a dimensionless 'loss coefficient', which can also be expressed in terms of an equivalent length ($L$) of straight pipe:

$$L = KD/f \tag{11}$$

## 2.2 SETTLING SUSPENSIONS

This is covered more fully in section 3, so the treatment here will be brief and introductory, yet hopefully covering all the major issues. Much of the basic information is available in general reviews, so specific references to the literature will be few. As for most other pipelining applications (e.g. crude oil, natural gas), prediction of the head loss as a function of independent design variables is a primary objective. Amongst the important independent parameters for settling slurry systems are pipeline diameter ($D$), operating velocity ($V$), mean particle size ($d$), particle size distribution (PSD), and properties of the carrier fluid as set out in the previous subsection. Kazanskij (8) has provided a useful guide to the predominant flow regimes, expressed in terms of particle size alone but assuming that water is the carrier fluid. With some abbreviation and re-arrangement, these are set out below and are applicable to quartz sand:

- *Homogeneous* ($d \leqslant 40 \ \mu$m): The mixture behaves as a single-phase liquid, Newtonian at low solids concentrations and non-Newtonian at higher concentrations. Gravitational forces on the particles can be neglected.
- *Pseudohomogeneous* ($40 \ \mu$m $\leqslant d \leqslant 150 \ \mu$m): Under turbulent conditions, the mixture can be transported with a uniform solids concentration distribution across the pipeline. During such transport the mixture can be viewed as homogeneous. When brought to rest, however, gravity will settle the particles.
- *Heterogeneous* ($0.15$ mm $\leqslant d \leqslant 1.5$ mm): For acceptable transport velocities, a solids concentration gradient exists over the depth of the pipe, and some particles may be sliding along the bed.
- *Fully Stratified* ($d \geqslant 1.5$ mm): The pipeline contents can be divided into an upper layer of liquid and a sliding bed, or 'contact layer' of solid particles. Heterogeneous transport is only possible at high velocities.

The list above indicates the importance of liquid velocity in determining the flow regime. Figure 3 illustrates the effects in more detail. For a given mean size and PSD, 3a shows how the solids might be distributed inside the pipeline. At low velocities, no transport occurs and the solids lie in a stationary bed at the pipeline invert, apart from the very fine particles which might be suspended. As the fluid velocity is increased, solids are lifted up in 'dunes', in much the same manner as wind develops waves on the sea. They are then transported from dune to dune, in a fashion sometimes known as 'saltation'. Saltation can also occur at low solids concentration without dunes, by a mechanism more fully explained in section 3.1. It is not an especially efficient transport mode and some danger exists that a dune might block the entire pipeline. Further increasing the fluid velocity brings one into the heterogeneous regime, in which concentration and size distribution gradients exist over the pipeline depth. Although some of the coarsest particles may be transported along the bed, the regime is considered acceptable for many operations. Increasing the velocity to give homogeneous, more precisely pseudohomogeneous, flow can exact too costly a head loss penalty.

Figure 3b shows the influence of flow regime on head gradient. Of special significance is the velocity for minimum head loss, marked as the boundary between the frictional contributions of sliding particles and turbulent fluid; at higher velocities fluid dominates; at lower speeds the particles take over. Although the minimum point may represent a technical optimum, it is usually advisable to run the pipeline at somewhat higher velocities. This avoids any instabilities that may arise with a system characteristic curve parallel to the pump curve. Expressions for mixture head loss are often given in the form below, with subscript s for slurry and w for water:

$$\nabla p_{\mathrm{s}} - \nabla p_{\mathrm{w}} = f(\mathrm{Fr}) \ \nabla p_{\mathrm{w}} C_{\mathrm{v}} \tag{12}$$

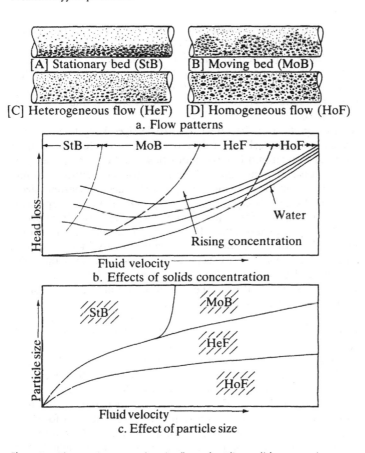

[A] Stationary bed (StB)   [B] Moving bed (MoB)

[C] Heterogeneous flow (HeF)   [D] Homogeneous flow (HoF)

a. Flow patterns

b. Effects of solids concentration

c. Effect of particle size

**Figure 3.** Flow regime maps for pipeflow of settling solids suspensions

where
$$Fr = V (gD)^{-0.5} (S-1)^{-0.5},$$  (13)

$C_v$ = Volumetric solids concentration, and

$S$ = Specific gravity (often referred to as SG) of solids. (The term relative density is now preferred.)

The philosophy behind equation 12 is to divide the total frictional gradient into a fluid and solids contribution. Fr represents a densimetric Froude number, based on the weight of the solids in the water. This is designed to reflect the competing effects of inertia and gravity. The particle friction contribution, on the right-hand side of equation 12, is a strongly falling function of velocity in the heterogeneous regime as shown by the merging curves of Figure 3b. It means that suspended transport is more efficient than sliding transport, and can be recommended, provided fluid friction can be kept within acceptable limits. Another restriction upon mixture velocity is pipeline wear, proportional to velocity cubed under some circumstances.

Figure 3c gives a qualitative picture of the effect of particle size. Very fine particles ($\leq 40$ $\mu$m) tend to be carried homogeneously, regardless of the mixture velocity. As size is increased the velocity required to maintain homogeneous flow can be seen to rise quite sharply with size. Much, of course, depends upon the criteria for distinguishing homogeneous from heterogeneous flow, but the principle remains unchanged. As particle size is further increased, there arises the possibility of a stationary bed below the deposition velocity. The particles are still too small to form stable dunes and a direct transition occurs between stationary deposit and heterogeneous flow. For yet larger particles, the transition occurs between stationary and moving bed at a velocity substantially independent of particle size. The transport mechanism in this case is particle saltation, whereby the solids undergo alternate suspension and deposition. Suspension can occur by means of the 'Magnus effect', whereby particles are lifted from the bed into the fluid boundary layer. Section 3.1 explains how this is also independent of particle size. Turbulent eddies may also sweep the particles into suspension. Coarse particle transport has strong economic advantages, especially for coal, and it is worth noting that in many cases the moving bed may be the most economic mode.

Duckworth (9) has analysed the data of Durand and Condolios (27), with the aim of providing more generalised relations, and given limits on the values of particle Reynolds number, $Re_d$, for which solids will be transported either as non-settling or settling slurries. The Reynolds number is based on particle diameter, terminal velocity under conditions of free settlement and fluid density and viscosity. The maximum Reynolds number for which non-settling slurries will be formed is approximately 2. The minimum Reynolds number for which particles travel by saltation without continuous suspension is approximately 525. These values are similar to the maximum values of Reynolds number for which a laminar boundary layer around the particles is maintained ($Re_d = 1$) and the minimum Reynolds number for a fully turbulent boundary layer ($Re_d = 1000$). These latter values are also quoted as the demarcations for non-settling and settling slurries. Such limits should be regarded as approximate since a detailed analysis of the interaction between particles and turbulent carrier fluid shows complex mechanisms at work. Slurries for which the particle-size distribution spans the range from non-settling to settling are termed mixed flow slurries. If fine particles are present they may change the properties of the carrier liquid to form a mixture, either Newtonian or non-Newtonian, which has an increased viscosity. Run-of-mine (ROM) coal might be carried this way, allowing lumps as large as 50 mm (2 inches) into the pipeline. The fines may themselves be transported for end use, or re-routed to the source. Their concentration requirements are a function of coarse particle concentration and can be expressed on a 'ternary diagram', as sketched in Figure 4. This consists

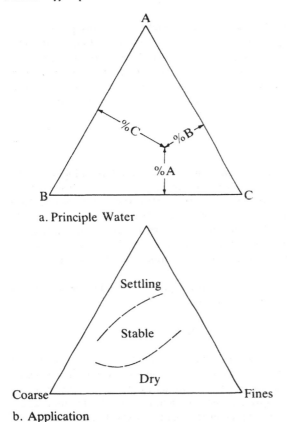

a. Principle Water

b. Application

**Figure 4.** Use of the ternary diagram to predict settling behaviour

simply of an equilateral triangle, with each vertex labelled with one of the three components. The sum of the altitudes from any point can represent the percentage of a given component as illustrated in Figure 4a. Figure 4b illustrates the general form of the lines separating settling, stable and dry suspensions. A 'dry' mixture, hardly a suspension, consists of insufficient water to fill the interstices between the particles. Of obvious interest is the stable slurry, enabling coarse material to be pumped in laminar flow. Its properties, however, are complex, and represent one 'edge' of the state-of-the-art of solids transport.

## 2.3  NON-NEWTONIAN SUBSTANCES

The purpose of this subsection is primarily definition, as the term 'non-Newtonian' serves simply to exclude Newtonian behaviour. The general term for the study of non-Newtonian fluids is 'rheology', derived from the Greek word for 'flow'. It concerns primarily the response of a substance to shear, applied at a constant rate of strain ($\dot{\gamma}$). Apart from superfluids, the general response is to develop a shear stress ($\tau$) in the opposite direction. For Newtonian fluids, most pure liquids and gases, there exists a direct proportion through the coefficient of viscosity ($\mu$). Non-Newtonian materials, suspensions of solids, polymer solutions and polymer melts, exhibit a different response, for which an 'apparent viscosity' ($\eta$) is often assigned:

$$\tau = f(\dot{\gamma}), \qquad \eta = f(\dot{\gamma}) \, \dot{\gamma}^{-1} \tag{14}$$

If the apparent viscosity falls with shear rate, the substance is termed 'pseudoplastic', if it increases it is called 'dilatant'. These are general terms, though they are used also to represent a particular form of the function $f$ (Ostwald-de-Waele power law). Pseudoplastic substances are also distinguished from the 'plastics', for which a stress limit must be overcome before motion can begin. Termed the 'yield stress', it can be difficult to determine without going to vanishingly small shear strain rates. The 'Bingham plastic' model is the simplest representation of this kind of behaviour, with a linear response to shear stresses above the yield.

Figure 5 illustrates non-Newtonian response on both logarithmic and linear scales. For pseudoplastic and dilatant fluids obeying the power law:

$$\tau = m\dot{\gamma}^{n}, \; \eta = m\dot{\gamma}^{n-1} \quad \text{pseudoplastic } (n < 1), \text{ dilatant } (n > 1) \tag{15}$$

The constants m and n are termed respectively the 'consistency coefficient' and 'flow behaviour index', sometimes abbreviated to 'consistency' and 'index'. Although consistency represents a kind of viscosity ($m = \mu$ for Newtonian fluids), equation 15 predicts zero apparent viscosity for pseudoplastics at infinite shear rate and for dilatants at limiting zero strain rate. This is plainly unrealistic and equation 15 might be expected to hold for real substances only between well-specified strain rate limits.

Amongst the models for plastic substances, Figure 5c, are the Bingham and Casson models, the latter developed by Casson (16) for certain types of suspensions such as blood and printer's ink:

$$\tau = \tau_y + \mu_p\dot{\gamma} \quad \text{(Bingham)}$$

$$\tau = (\sqrt{\tau_y} + \sqrt{\mu_p\dot{\gamma}})^2 \quad \text{(Casson)} \tag{16}$$

They bear strong similarities to each other; both are specified by two parameters, the yield stress ($\tau_y$) and the 'plastic viscosity' ($\mu_p$). Nevertheless, substitution of identical values of these parameters leads to very different curves for the two models. The Casson model is increasing in popularity and appears to have some physical basis (16). Nevertheless, it suffers the same disadvantages of all two-parameter models that it cannot be made to fit all sets of rheological data. For this reason, three-parameter models are available, such as the

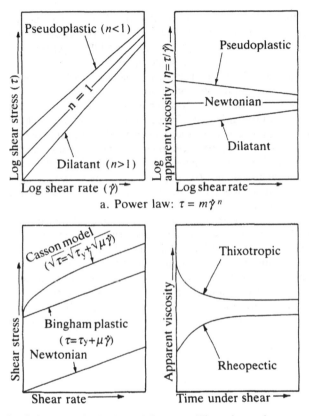

**Figure 5.** Definition sketch for non-Newtonian fluid properties

'generalised Bingham', or Herschel-Bulkley class, consisting simply of a superposed yield stress and power law:

$$\tau = \tau_y + m\dot\gamma^n \quad \text{(Herschel-Bulkley)} \tag{17}$$

As explained in more detail in section 4.1, these models achieve their curve fitting power at the cost of high sensitivity of the parameters to small changes in the data.

We have so far considered only time-invariant responses to the shear rate. Some materials are 'thixotropic', in that application of shear over a period of time causes a decrease in their apparent viscosity, often expressed in terms of a yield stress history.

Thixotropy is usually associated with plastic or pseudoplastic materials but this need not be the case. Thixotropic substances have also been known to revert to their original properties upon being stored at rest, an 'ageing' effect. This usually takes considerably longer than the original thixotropic breakdown and may not always reach completion. The opposite to thixotropy, 'rheopexy', corresponds to a build-up of structure (thickening) with prolonged shear application, understandably enough, a rare phenomenon.

In addition to shear stresses, some substances, notably polymer solutions and melts, are capable of developing normal stresses in the direction of applied shear strain rate. These are known as 'viscoelastic', as the normal stresses represent a stored elastic strain energy. Unlike solid elasticity, these elastic stresses cannot be developed in the absence of viscosity. The phenomenon is discussed in detail in section 4.3, in terms of its definition, mechanisms and modelling. Despite the fact that slurries do not frequently display viscoelasticity, the property underlies the drag reducing effects of polymer additives, which have found some application in the solids transport industry.

Perry's Chemical Engineer's Handbook (17), provides examples of some substances with various non-Newtonian properties. They list drilling muds and rock suspensions as Bingham plastic, polymer solutions and clay suspensions as pseudoplastic, de-agglomerated and highly concentrated suspensions (e.g. quicksand) as dilatant, and certain types of polymer solutions (e.g. polyacrylamide) as viscoelastic. Paints, inks and mayonnaise were also classed as thixotropic and suspensions of bentonite, gypsum or vanadium pentoxide as rheopectic. In the context of slurry transport, none of these categories should be viewed in rigid isolation, and each model must be taken on its merit, be this simplicity, accuracy or flexibility.

## 2.4  PRESSURE SURGES

Pressure surges can arise whenever a pipeline carrying liquid is suddenly shut-down for any reason. The shut-down may be due to pump trip, emergency valve closure or some other cause; if it is sudden, surges usually result. In the case of a pump trip, the liquid mass is likely to draw a vapour cavity somewhere near the pump, or at some elevated point in the line. Upon cavity collapse, the surge is born. On the other hand, the pump may be fitted with a non-return valve, which, if it slams, can give rise to surges in the event of a pump trip. There are means of mitigating the effects of a pump shut-down by installing air-admission valves at points of likely cavitation, and by selecting non-return valves resistant to slamming.

Emergency valve closure also requires careful design. Such valves are a necessary safeguard against excessive spillage in the event of a break in the line. They must be designed to close sufficiently quickly to perform their function, yet sufficiently slowly to avoid excessive surge pressures. For a sudden change ($\Delta u$) in fluid velocity, the following formula gives the surge pressure rise ($\Delta p$) in terms of the fluid density ($\rho$) and speed of sound ($a$).

$$\Delta p = \rho a \Delta u \qquad \text{Joukowski surge pressure} \qquad (18)$$

In the event of valve closure, some of the surge pressure is relieved by the return of a reflected negative wave from the other end of the pipeline (pumping station). Unfortunately, however, most valves develop little flow resistance until the last few degrees of closure, allowing little time for the relief wave to arrive. This can be cured by the use of a two-part closure sequence, in which the first eighty degrees of closure takes place in ten seconds, and the last ten degrees in forty seconds, for example. Figure 6 illustrates a sample prediction for a ball valve subjected to this sequence in terms of its flow and head history. The head rise was about half the value predicted for a simple fifty-second sequence. Further details on this particular example are available from the report by Silvester and Langley (18).

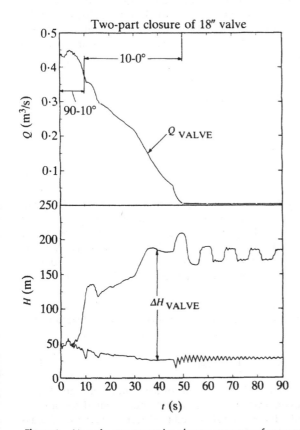

**Figure 6.** Use of a two-part valve closure sequence for surge protection (18)

In slurry lines, surge problems can be exacerbated by a number of factors. The lines are usually very long, so that the negative relief wave must be allowed considerable time. Furthermore, the presence of solid particles can increase both speed of sound in the medium and its effective density, both of which contribute to the Joukowski pressure rise. Thorley (19), noted also that solids can become lodged in the seats of non-return and air-admission valves. He also raised the question of effluent disposal, should relief valves or bursting disks be used to combat the surge problem. In experiments with suspensions of glass beads, sand and coal in concentrations up to 40 per cent by volume, Thorley and Hwang (20) demonstrated that surge pressures up to twice that of clear water are possible. The total surge usually took place in two stages, corresponding to deceleration of the liquid followed by that of the solids. Fortunately, the effect of hold-up, a lower mean velocity for the solids in the line than the liquid, was claimed by Thorley (21) to mitigate the surge.

Kao and Wood (22) also considered the surge problem, and developed a mathematical model for the secondary surge pressure in terms of hindered particle drag coefficients in the liquid. They used the following expression for speed of sound in the medium, where $s$ represents solids, $f$ fluid, $p$ pipewall, $C$ and '$C_f$' volumetric concentration of solids and fluid respectively, $\rho$ density and $E$ bulk modulus (Young's modulus for pipewall), with $t$ the pipewall thickness:

$$a^2 = [C/\rho_s + (1 - C)/\rho_f] \, [C/E_s + C_f/E_f + D/(tE_p)]^{-1} \tag{19}$$

From the above, they proposed that the initial pressure wave can be determined by assuming a homogeneous fluid at the mixture density:

$$\Delta p_i = V_o a \rho_s \rho_f [(1 - C)\rho_s + C\rho_f]^{-1} \tag{20}$$

This pressure rise is insufficient to bring the denser solids to rest, and more than sufficient to stop the liquid, now reversed in flow direction. The phases now interact to bring each other to rest, giving rise to the secondary surge, adding between five and ten per cent to the surge pressure over a time-scale between ten and twenty times that of the initial surge. Reasonable agreement between predictions for sand-water and limestone-water and corresponding test results was reported. The solids specific gravity, though predicted to be significant, was not varied experimentally.

Bechteler and Vogel (23) criticised the physical meaningfulness of the density term in equation 19. Instead, they proposed the speed of sound expression below, in line with the choice of Thorley and Hwang (20):

$$a^2 = \{[C/E_s + C_f/E_f + D/(tE_p)][C\rho_s + (1 - C)\rho_f]\}^{-1} \tag{21}$$

The German authors then applied the Kao-Wood concept of separated solid and liquid flows, comparing it with the homogeneous model (20) and their own development incorporating added mass. Their results have been sketched in Figure 7. The added mass model appears to work very well for sand near 50 per cent volumetric concentration, whilst

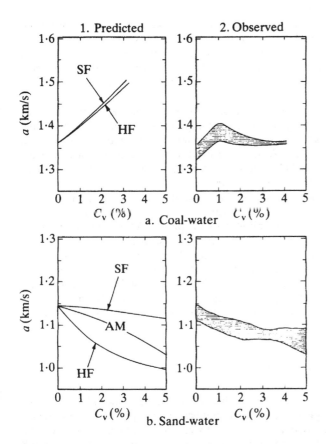

**Figure 7.** Velocities of sound in various solid-liquid suspensions (23)

the homogeneous model works well at the lower concentration. The discrepancy for coal-water slurries beyond 10 per cent was attributed to gas in the system. Fortunately, the model overpredicts the results, an error very much in the right direction for a surge analysis.

## 3. SETTLING SLURRY SYSTEMS

An understanding of fluid flow in pipelines, though helpful, must be coupled to that of particle dynamics if settling slurries are considered for solids transportation. Principles of solid mechanics, such as dry solid friction, are involved, along with 'aerodynamics', in which context the particle can be seen as a flying body, subject to drag and lift from the surrounding fluid. In the initial development of these concepts for slurry systems, existing correlations are presented for the insight they provide, rather than their practical design significance. The opposite is true for the subsequent subsection on horizontal pipelines, upon whose design for settling suspension flow many long distance hydrotransport projects depend.

Sloped and vertical lines, though treated under the one heading, are interesting for different reasons. The study of sloping lines is important if line blockage due to excessive slope is to be avoided. Vertical transport, however, is a viable concept and has been used in hoisting coal from the pit to the surface. Vertical transport works so well in some instances, that it has been proposed that ground elevation effects be handled by sections of vertical, rather than sloping, pipe in long distance slurry lines.

Transport in flumes is one mode of flow available to solid-liquid mixtures unshared by liquid-gas systems. The flow is driven by an elevation gradient rather than a pressure gradient and can avoid the need for pumps or high pressure piping. A full flowing pipe is not the most efficient from the wetted perimeter viewpoint, and use can be made of the extra density imparted by the solids to the slurry. Of all systems, flumes handle coarse solids with the greatest ease, a feature of increasing economic importance. The final subsection is devoted to coarse solids transport in pipes, usually limited to a sliding bed over short horizontal distances and severely held-up flow in the vertical direction. The incentives of low preparation and separation costs, however, provide the impetus for further investigation into coarse particle slurries.

## 3.1   PRINCIPLES OF PARTICLE TRANSPORT

Regardless of the total number, mean size, or size distribution of the particles in a slurry, its settling behaviour ultimately depends upon the dynamics of each particle in interaction with its neighbours, the surrounding fluid and the pipe wall. We may begin by looking at a particle sitting on the pipe invert, subject to fluid flow and solid contact forces. Figure 8 illustrates the force balance, the submerged weight of the body balanced by a vertical normal reaction and fluid drag equilibrated by wall friction. Depending upon its shape and the proximity of neighbouring grains, its motion may take two forms, rolling or sliding. Rounded bodies tend to roll, unless there are many of them in close contact, under which conditions they are forced to slide. Thus, there exists a 'sliding bed' flow regime but no 'rolling bed' regime. However, in the heterogeneous flow regime, the particle population at the bed would be low enough to promote rolling. Rolling promotes lift (Magnus effect) and suspension is further eased. From these concepts, we may expect spherical particles to exhibit a sharper transition between sliding and heterogeneous regimes than those of a more angular shape.

Figure 9 illustrates the forces on a suspended particle, to which a spherical shape may now be assigned. The submerged weight, though unchanged in magnitude and direction, has now been overcome by a lift force. The lift may be generated by a number of

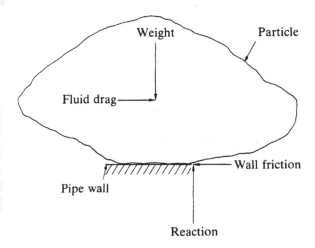

**Figure 8.** Force and moment equilibrium for a settled solid particle

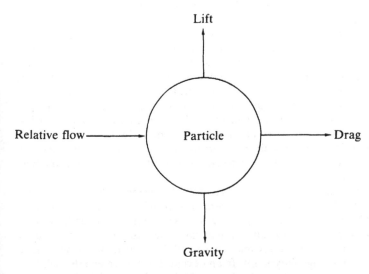

**Figure 9.** Unequilibrated forces on a suspended solid particle

mechanisms. A particle near the pipe invert may experience a rotation due to collision with the wall. The resulting combination of circulation around and relative flow past the object generates a lift force. For a rotational speed ($\Omega$) and relative velocity ($u$), the following proportion exists for lift ($F_L$) in terms of particle diameter ($d$):

$$F_L = K\rho\Omega ud^3 \tag{22}$$

where $K$ is a constant.

Known as the 'Magnus effect' it may be noted that lift is strongly dependent upon particle size. For vertical transport it means that large particles will migrate towards the axis of the pipe, a favourable effect from the wall friction viewpoint. For horizontal transport it means that the ratio of lift to weight is independent of particle size. The rotation may be considered proportional to boundary shear, itself proportional to $V/D$ for a mean velocity ($V$) in a pipeline ($D$), yielding a ratio (Fr) of lift to weight, for a given solids specific gravity ($S$):

$$\text{Fr} = K\,V^2\,(gD)^{-1}\,(S-1)^{-1} \tag{23}$$

where $K$ is another constant.

The force ratio now takes the form of a Froude number (squared) based on pipeline diameter, another reason why incipient motion of large particles is independent of particle size. The reader will realise that this is a simplified analysis; experiments have shown that the relation between lift, relative velocity and shear is more complicated. It was noted in section 2.2 that the transition from a stationary bed to a moving bed was independent of particle size, and explained in terms of moving dunes of stable form. That would apply to high particle concentrations but would not explain the same behaviour at low concentrations, where it is also observed. The Magnus effect is unable to keep particles suspended on its own account. As soon as one is lifted from the pipe invert, fluid drag then accelerates it to the speed of the fluid flow, at which point relative motion ceases, lift disappears, the particle settles and the entire process must begin again. This mechanism for particle transport is known as 'saltation' and may occur with or without dunes. Particles are in contact with the pipe wall for much of their transport life and contribute a solid frictional component to the total pressure gradient. A problem associated with saltation is the high rate of wear it tends to generate.

For particles to remain in stable suspension an alternative lifting mechanism is necessary. This can be provided by fluid turbulence. A turbulent eddy provides a fluctuating vertical velocity component, which can maintain a particle in suspension as long as it is much greater than the particle settling velocity. The mechanism in this case would be a vertically directed fluid drag. Alternatively, the eddy may 'surround' the particle, allowing relative horizontal flow to generate a Magnus-type lift. In any event, turbulence enables solids to remain suspended. Of special significance is the drag coefficient of the particles in question. The greater the drag coefficient ($C_D$), the greater the tendency for the particle to follow the fluid motion and the less its terminal settling velocity ($V_t$) given below in terms of particle diameter and specific gravity:

$$3V_t^2 = 4dg(S-1)C_D^{-1} \tag{24}$$

For spheres in slow viscous flow (very small particles), Stokes' law applies as follows for settling:

$$V_t = \rho g d^2 (S - 1) (18\mu)^{-1} \tag{25}$$

The drag coefficient for Stokes flow can be expressed in a fashion analogous to that for friction coefficient for laminar pipe flow:

$$C_D = 24 \text{ Re}^{-1} \quad \text{if Re} = \rho V d\mu^{-1} \tag{26}$$

Retaining the comparison with pipe flow, a chart for drag coefficient can be plotted for the entire range of Reynolds numbers, as shown in Figure 10, adapted from Streeter (24). Stokes flow applies up to particle Reynolds numbers of order unity. Beyond this the slope of the curve lessens and particle shape begins to show significance at Reynolds numbers over 100. The dip for spheres at Re = 300 000 corresponds to the transition from a laminar to a turbulent boundary layer, hardly likely to apply to slurry particles.

Interparticle effects can influence the effective drag coefficient. Some particles can experience enhanced drag if they become entrained in the wake of a large leading particle. On the other hand, many particles falling at different rates can interfere with each other's progress, leading to the phenomenon of hindered settling. The usual trend is for settling rate to increase with concentration up to a peak, followed by a decrease thereafter. It is of

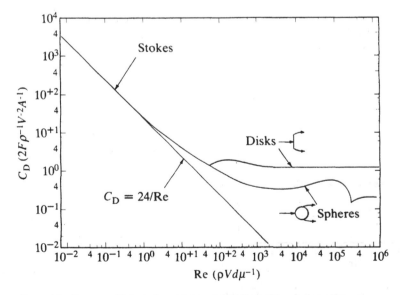

**Figure 10.** Drag coefficients for various particle Reynolds numbers (24)

interest in this context to consider how the presence of particles affects fluid turbulence. Rayan (25) observed both suppression and enhancement of turbulences by the presence of particles, and confirmed the presence of Magnus lift in laminar flow. Most of his data, however, was collected at lower concentrations than could be applied to slurry transport.

It is helpful to be able to relate the critical velocity for incipient suspension to the particle sedimentation velocity. With a river bed erosion application in mind, Yang (26) considered it in terms of a shear velocity particle Reynolds number. His approach has a fundamental basis in that shear velocity is related to shear stress at the wall as follows:

$$u_s^2 = \tau_w/\rho \tag{27}$$

The shear stress is in turn related to the strain rate, hence fluid rotation at the wall. The results are shown in Figure 11, and indicate clearly that the critical velocity must be many

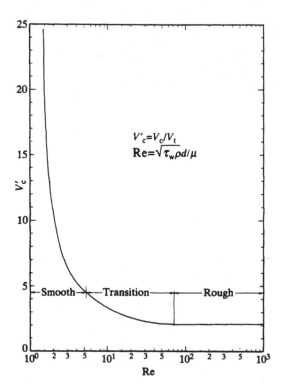

Figure 11. Critical velocity versus shear particle Reynolds number (26)

times the settling velocity for small particles in smooth pipes. It is worth noting that the critical to settling velocity ratio stands constant at 2.1 for large particle transport:

$$V_c = 2.42 \sqrt{dg(S-1)/C_D} \tag{28}$$

In contrast to the result for saltation, the stable suspension of large particles depends upon particle diameter rather than pipe diameter. Therefore, the ratio of the two must play some part in the transport process.

## 3.2 HORIZONTAL SLURRY PIPELINES

The design of a slurry pipeline entails predicting the power requirement per unit solids delivered over a unit distance. Whether or not this involves minimising the pressure gradient depends on additional economic factors. It is vital in this context to be able to relate head gradient to the independent design parameters. Durand and Condolios (27) pioneered the art with the following empirical correlation:

$$\phi = 81\psi^{-1.5} \tag{29}$$

where*

$$\phi = (J_m - J_w)/(C_v J_w) \tag{30}$$

and

$$\psi = V^2(S-1)^{-1} C_D^{-0.5}/(gD) \tag{31}$$

The relevant quantities are mixture head gradient ($J_m$) and clear water head gradient ($J_w$), both expressed in terms of metres water per metre of pipeline. In equation 30, the term $\phi$ represents a normalised difference between slurry and clear liquid head gradient, i.e. the particle contribution to the total. In equation 31 the term $\psi$ is the square of a weighted Froude number, with particle size accounted in terms of a drag coefficient. It is instructive to expand equation 29 in terms of a real pipeline with known Darcy friction factor ($f$) and flow velocity ($V$):

$$J_m = fV^2(2gD)^{-1} + 40fV^{-1}(gD)^{0.5}(S-1)^{1.5}C_D^{-0.75}C_V \tag{32}$$

The first term (in $V^2$) is the liquid contribution, whilst the second (in $V^{-1}$) is a solids frictional effect, the negative power of $V$ reflecting that suspension is more efficient than sliding. From equation 32, a velocity for minimum head gradient may be obtained by differentation:

$$V_c = 3.4 (gD)^{0.5} (S-1)^{0.5} C_D^{-0.25} C_V^{0.33} \tag{33}$$

---

* Different values for the multiplier of $\psi$ in equation 29 can be found in the literature; this value is considered to be correct (50). See also the work of Zandi and Govatos (34). $C_V$ represents delivered concentration.

For illustrative purposes, equation 32 has been evaluated for a 15 per cent by volume sand-water slurry ($S = 2.65$) flowing in a 300 mm (1 ft) pipeline with a friction factor of 0.02. Assuming a particle size of 100 $\mu$m gives us a settling Reynolds number of unity and a Stokes drag coefficient of 25. The resulting graph, compared with that for clear water is shown in Figure 12. The two curves approach each other at high velocities in what might be called the 'pseudohomogeneous' regime. At the critical velocity of 1.78 ms$^{-1}$ the head gradient reaches its minimum of 3.3 per cent. Below the critical velocity, the sliding bed regime might be expected to invalidate equation 32. Durand and Condolios (27) themselves provided a guide to the effect of particle size on deposition velocity. It is shown in graphical form in Figure 13 for solids concentrations of 2–15 per cent. It is notable that at high $C_v$ a decrease in $V_c$ with particle size is apparent between 0.5 and 2.0 mm. This could be due to the fact that as size increases the number of particles falls, allowing each to move more freely.

The approach of slurry pressure gradient to that of water at high velocities must be questioned in the limit of truly homogeneous flow. In homogeneous flow, the slurry can be

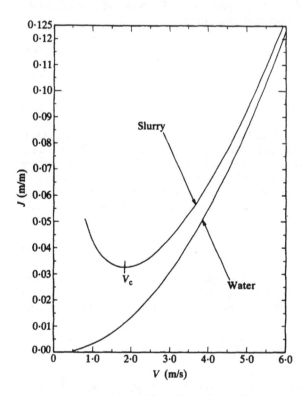

**Figure 12.** Sample applications of Durand's head loss correlation (27)

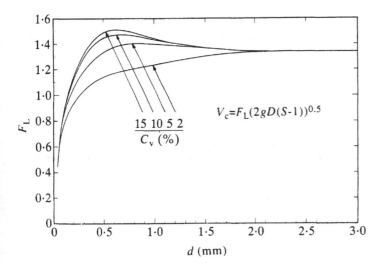

**Figure 13.** Critical velocity versus particle size and concentration (27)

considered as a single-phase liquid with a density and viscosity influenced by the presence of solids. In pseudohomogeneous flow, however, the slurry appears to behave as a fluid with the density of water. Weber (28) indicated limits for the pseudohomogeneous regime in terms of particle Reynolds number during settling. Beyond the upper boundary of $Re_p = 2.0$, heterogeneous flow prevailed with some particles travelling along the bed whilst below $Re_p = 0.1$ homogeneous flow could be assumed. Weber showed:

$$V = 0.0056\, V_t \quad \text{for Re} = 2.0 \tag{34}$$

and

$$V = 0.0015\, V_t \quad \text{for Re} = 0.1 \tag{35}$$

The addition of fines to water may result in a thicker carrier fluid which can support coarser particles. Care must be taken, however, to ensure that the use of fines has the desired effect. Kenchington (29) found that a 750 $\mu$m slurry of sand in water required a lower velocity for suspension than a similar mix with 40 per cent by weight of china clay. He concluded that the thicker china clay mixture suppressed turbulence, allowing the sand to settle more easily. Nevertheless, the increased lubrication effect of the china clay always allowed easy sliding of the sand bed, and, unlike water, never produced a stationary bed. Thomas (30), investigating the effect of thick Newtonian media (sucrose solutions), found that sand could be transported in laminar flow under special conditions. If the safety of full suspension provided by turbulence is desired, thickening the medium may be detrimental.

For large particles, such as crushed granite, the sliding bed regime may be the only economic mode of operation. Newitt *et al.* (31) pioneered the quantification of this regime with the following semi-empirical correlation, where $V_m$ is mean mixture velocity:

$$J_m - J_w = 66 \, C_v \, J_w \, V_m^{-2} \, (gD) \, (S - 1) \tag{36}$$

or

$$\phi = 66 \, \psi^{-1} \, C_D^{0.5} \tag{37}$$

The sliding bed is likely to be so rough that fully rough turbulent flow assumptions can be made, and Darcy friction factor assumed constant. Adjustment of the friction factor to apply to mixture, rather than fluid velocity yields:

$$J_w = f \, V_m^2 \, (2gD)^{-1} \tag{38}$$

therefore

$$J_s = 33 \, f \, C_v \, (S - 1) \tag{39}$$

if $J_s = J_m - J_w$

Equation 39 shows the solids contribution to frictional gradient to be independent of pipe diameter, particle drag coefficient and mixture velocity. In essence, the solids are simply hauled along the bottom of the pipe with a force proportional to the total weight of solids in the bed. The presence of a fluid friction factor in equation 39 is doubtful and might be profitably replaced with a coefficient of solid friction. Babcock (32), in an independent investigation of the sliding bed found a correlation identical in form to equation 36 to apply to sand slurries up to 40 per cent by volume. He replaced the coefficient of 66 by 61, but claimed this change to be of little significance. It may be noted that the use of low friction liners has been found to reduce the hydraulic gradient in a sliding bed. Wilson *et al.* (33) replaced the '33*f*' in equation 39 with '2$f_s$', where '$f_s$' is a particle-wall solid friction coefficient, and obtained a more generally useful result.

Other equations, usually modified versions of the Durand format, have been proposed. Zandi and Govatos (34) collected over 2500 points from various sources and compared three existing heterogeneous correlations with the data. They concluded that, of the three, Durand was by far the best, but required the following modification to improve the fit still further:

$$\phi = 81 \, \psi^{-1.5} \tag{40}$$

if $\psi \leqslant 40 \, C_v$

$$\phi = 280 \, \psi^{-1.93} \tag{41}$$

if $40 \, C_v < \psi \leqslant 10$

$$\phi = 6.3 \; \psi^{-0.354} \tag{42}$$

if $\psi \geqslant 10$

Thus Durand's correlation, (equation 40), fits best at the lower values of Froude number. Equation 42 also provides a better matching of the homogeneous and pseudohomogeneous regimes than Durand's formula.

In an attempt to improve the prediction of the flow-rate pressure-drop characteristics of settling slurries, with a wide particle size distribution, Wasp *et al.* (35,40) split the distribution into several parts. For each size fraction the ratio of the concentration at the top of the pipe ($C$) to that at the centre line ($C_A$) was calculated from the equation:

$$\log_{10} C/C_A = -1.8 \; (V_t/\beta k U^*)$$

where $V_t$ = terminal settling velocity, k = von Karman constant (0.4), $\beta$ = the ratio of mass transfer coefficient of the solids to that of momentum, and $U^*$ = friction velocity. The assumption was made that $\beta = 1$. This equation finds its source in the study of transport of sediment in rivers, etc., and is based on an equilibrium of upward transport, due to turbulance, and downward transport due to gravity.

The mean concentration of each particle size and the values of $C/C_A$ enable the values of C for each fraction to be obtained. It is assumed that these proportions of the slurry contribute to the 'vehicle' part and the remainder are carried as a heterogeneous suspension.

Initial conditions are provided by assuming that all the slurry forms the carrier, or vehicle, fluid and the pressure gradient is found by conventional methods for homogeneous fluids. An iterative calculation is then started whereby the $C/C_A$ values are used to split the fluid into carrier and suspended flow fractions. A new pressure gradient is determined by addition of the contributions from the new vehicle and suspended flow fractions. The contribution from the suspended portion is calculated using the Durand equation. A comparison of the two values of pressure gradient is made and, on the assumption that there is a significant difference, new values of $C/C_A$ can be calculated. The calculation procedure is repeated until a satisfactory convergence is achieved. The method lends itself to computerisation but hand calculation can be employed as the solution converges rapidly.

The technique was modified by Williamson (41) who used the Zandi correlation (equation 40) with the following modifications to $\phi$ and $\psi$:

$$\phi = (J_m - J_w) \; (C_v' J_w) \tag{43}$$

$$\psi = V^2 \sqrt{C_D'} \; (gD)^{-1} (S/S' - 1) \tag{44}$$

where $C_v'$ is the concentration of particles above a certain size limit, $C_D'$ is the weighted mean drag coefficient of such particles and $S'$ is the specific gravity of the slurry of solids finer than the size limit, in an assumed homogeneous suspension.

The cut-off point between homogeneous and heterogeneous suspension was found by iteration around $\psi = 40\ C_v$, although according to the foregoing discussion $\psi = 10$ might be more appropriate. The two criteria are identical at $C_v = 0.25$, a commonly used volumetric concentration. Williamson reported satisfaction with the method in comparison with data for magnetite, phosphate tailings and fluorspar tailings slurries.

The correlations discussed so far have been largely empirical. The developments begun by Wilson and Watt (36) have taken the alternative path based on fundamental principles. They used a Prandtl mixing length approach to define the scale of a turbulent eddy. Particles much smaller than this length scale would not affect the effective vertical velocity fluctuation ($V_e$) whereas larger particles would damp it to some degree. As the maximum mixing length is the pipe diameter, the particle to pipe diameter ratio becomes significant. Where k and $\alpha$ are constants, they proposed the following connection between effective turbulence and flow velocity:

$$V_e = k\ V f^{0.5} \exp\left(-\alpha d/D\right) \tag{45}$$

They developed this into a criterion for critical transport velocity ($V_c$) for a stable suspension in terms of the particle settling velocity ($V_t$):

$$V_c = 1.7\ V_t\ f^{-0.5} \exp\left(-45d/D\right) \tag{46}$$

This is the critical velocity for turbulent suspension and must not be confused with that for incipient motion or for minimum head loss in a Durand type of correlation. Developing the concept further, Wilson (37) presented a unified analysis for pipeline flow, able to cope with wide PSD by dividing the solids into specified size fractions. The model included a sliding bed, Figure 14, and the proportional split between the bed and suspended part was calculated with the aid of the above equation. The overall pressure drop is built up by contributions from both the sliding bed and the upper flow region. The value of these two components is influenced by the relative velocity between the bed and fluid above, which in turn is affected by the shear stress between these two regions of pipe flow. Iterative

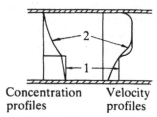

Concentration        Velocity
profiles               profiles

1. Two layer model
2. Distributed concentration
   model

**Figure 14.**   Elements of sliding bed model

techniques, best carried out on a computer, are required to provide solutions. Unfortunately, due to the complexity, the overall model could not be presented as a set of relatively straightforward design equations. In a later publication (38) the final results were presented as a set of nomographs. These are significant as design aids as the agreement between the theory and the test results was good. A number of assumptions are made concerning the friction between the moving bed and pipe wall and between the bed and flow in the upper region. In an attempt to apply the method to scale-up of pipeline data, Clift *et al.* (39) found the need to retain arbitrary exponents. The significance of this model is principally the increased degree of understanding it has promoted in the flow of slurries. Other research centres have constructed their own versions of this model; Jacobs and Tatsis (136), Tatsis and Jacobs (137) and Sieve and Lazarus (138).

Other work by Roco and Shook (42) has considered the case of flows where shear takes place within the bed. In the earlier work Wilson (33) analysed the case where the total submerged weight of the particles was carried by the 'contact load'. Bagnold (43) referred to a 'bed load' in which the immersed weight is transmitted to the pipe walls by dispersive stresses. Roco and Shook (42) considered that an additional shear stress was introduced by Coulombic friction between adjacent layers which in turn was proportional to the 'supported load'. Supported and bed loads are two separate components of contact load. Included in their model are terms which allow for turbulence alteration in the presence of solids and mixing effects due to particle interactions. A set of differential equations is formed and solved numerically. Various factors relating to diffusion coefficients and the influence of solids on the turbulence intensity must be determined by experiment. In general there is a good agreement between prediction and measurement of the velocity and concentration distributions for slurries having a relatively narrow particle size distribution. Previous work had indicated that the introduction of dispersive stresses made predictions acceptable for concentrations in excess of 5 per cent by volume but only for small particles. The introduction of the term due to friction enabled the predictions to be improved for larger size material. Since the prediction technique requires the estimation of certain factors by means of experiment, the analysis was limited to interpolation or extrapolation of results to other flow situations. The experiments were conducted in pipes ranging in diameter from 50–500 mm, at concentrations up to 40 per cent by volume and for particle sizes ranging from 0.165–13 mm nominal diameter. The model is complex and the good correlation with experiment represents a substantial achievement.

In a following paper, Roco and Shook (44) extended the analysis to cover a range of particles size distributions and applied the results to coal slurry flows in pipes up to 495 mm diameter.

It is worth noting the multiplicity of definitions of the critical velocity. If it is to be based on solids deposition, a visual criterion would seem to be fundamentally appropriate. However, as noted by Sinclair (45), the subjective nature of visualisation can lead to considerable scatter. Ercolani *et al.* (46) developed electrical and thermal probes to detect limiting deposition, extracting the information by frequency analysis. Sliding bed motion could be detected by a stick-slip response and the two probes agreed well together. Toda *et al.* (47) were interested in the transition between moving and stationary bed and used a saltation criterion in their numerical model. Weber and Gödde (48) noted the need for a critical velocity with some direct connection to pressure gradient. Gödde (49) correlated experimental data with some success using the Durand correlation and recommended a critical velocity based on the minimum hydraulic gradient. In an effort to optimise the transport of settling slurries, Lazarus and Nielson (50) criticised the concept of $V_c$ as optimum. They noted that in-situ concentration is generally greater than delivered concentration, and based their overall design equations with this in mind. In a manner similar to Wilson (38), Lazarus (51) converted them to a set of curves for design optimisation of any given transport project.

At the other end of the sophistication spectrum, a number of informal scaling schemes exist, based on such concepts as power law correlation. Whilst they might not be recommended for general use, they have been used in specific instances with some success. Kazanskij (52), for instance, found $V_c$ to be proportional to $D^{0.25}$ for slurries with fine particles whilst Middelstadt (53) suggested a $D^{0.33}$ relationship. Lokon *et al.* (54) limited his attention to fine particle (50 $\mu$m mean size) iron ore slurries and proposed the following scale rule:

$$J_m = 54\,900\ V^{1.63}\ D^{-1.42}\ C_v^{0.35} \tag{47}$$

It would appear that equation 47 corresponds to turbulent flow of a homogeneous slurry, as no critical velocity is apparent from the formula. Webster and Sauermann (55) note the conservatism of the Durand correlation under some conditions and recommend using the same exponent for diameter in both slurry and water flow. Once again, homogeneous assumptions would have to be made.

Vocadlo and Charles (56) divide the mixture head loss into the following components:

$$J_M = J_L + J_I + J_S \tag{48}$$

where the subscripts M, L, I and S represent the mixture total, liquid, interparticle effect and suspension energy contributions respectively.

They did not take solids hold-up into account and developed a final equation involving four free parameters. They obtained good agreement for coarse sand slurries in a 25 mm pipe but the general usefulness of their method must be offset by the number of free variables. Of all the correlations considered in this section, Durand's formulation (equation 29) has much to recommend it for initial feasibility studies. It seems to fit a wide range of data, is of a manageable simplicity and can be adjusted in ways that make physical sense.

## 3.3   SLOPED AND VERTICAL LINES

Vertical pipelines for solids transport are of direct interest in the mining and dredging industries, whereas sloping lines have some impact on long distance transport. If too steep a slope is used the line may block during shut-down and require considerable pressure to restart. Therefore, the sequence in this section will be to deal first with blockage, proceeding then to slope effects and finishing with hoisting applications.

Investigating the pressure gradient requirements for re-establishment of slurry flow following shutdown, Wood (57) found that gradients as high as twice those at the deposit velocity were required to restart the flow. He divided the total requirement into three components as follows:

- overcome boundary shear and turbulence,
- bulk acceleration of the total mixture, and
- acceleration of particles leaving the bed.

If the heterogeneous regime cannot be restarted sufficiently rapidly, saltation may develop, with dune formation and a more serious blockage resulting. Okada *et al.* (58) experimented with restart conditions in the event of plug formation, and found that a plug formed at the bottom of a downward sloping pipeline required a higher pressure difference to shift it than one at the foot of an upwards slope. They related their results to the mean porosity (inverse compaction) of the plug and developed the following expressions:

$$\Delta p = 10.5 \text{ bar} \times (0.43 - \epsilon) \, L/D \qquad \text{for ascending pipe} \qquad (49)$$

$$\Delta p = 22 \text{ bar} \times (0.42 - \epsilon) \, L/D \qquad \text{for descending pipe} \qquad (50)$$

where porosity ($\epsilon$) is less than 0.41 and $L$ is length of plug. These pressures are in addition to the hydrostatic effects. The authors also found the dynamic sliding angle to be less for a downslope than the static angle, due to downwards solids momentum effects on shutdown. A static angle of 26°, for example, corresponded to dynamic angles of 25° and 32° for upslope and downslope respectively. A line inclined more steeply than the sliding angle is liable to blockage. As noted by Takaoka *et al.* (59), such a situation would be disastrous for an underwater line.

For long distance pipelining, Gandhi and Aude (60) raised the point that slope limitations can add 10–16 per cent to the construction cost. They referred to tests on the Brazilian Samarco iron concentrate line under shutdown conditions. Samples were taken at a 1.8 km long section at a 10 per cent slope. They found that a semi-consolidated bed could form at the lowest point, filling up to 75 per cent of the pipe cross-section. Nevertheless, they found that the system could be restarted without unreasonable pressure requirements. It would appear, then, that 10 per cent is a safe slope for iron concentrate lines of the Samarco type. It is still less than the 14 per cent used for the Savage River iron ore line in Tasmania.

Kao and Hwang (61) looked at sand, glass beads, walnut shells and coal transport in sloping pipelines. They found that for upwards slopes, the steady pressure gradient first increased with slope, reached a peak, and then fell away as the vertical inclination was approached. This might be expected from the fact that initial slope increases tend to raise the degree of slip, hence the in-situ solids concentration. Near the vertical, however, the solids tend not to contact the pipe wall and losses are consequently low. Steady flow in a downsloping pipe invariably produced a sliding bed even at lowest velocities. Concerning shut-down effects, the authors found that plugging angles of 23° (glass beads) to 37° (coal) were appropriate. They claimed that they could be determined from static tests but the means of achieving this was not stated.

Investigating iron ore and limestone slurries, Shook *et al.* (62) pointed out that strongly settling systems may not give the worst case. The mechanism they considered for plug formation involved the following sequence. At the point of shut-down, particles would begin to settle out, creating a density gradient over the pipe section. If the pipe is set at a slope, a counter flow would be set up between the dilute mixture in the upper half of the pipe and the concentrated mixture in the lower half. Under the right conditions this current would continue until the concentration at the bottom of the slope were to rise sufficiently to form a plug. Coarse particles would not participate in the density current, having already settled, and very fine particles would tend to remain in homogeneous suspension. Therefore, there may be a 'worst case' degree of fineness for plugging in sloped pipelines after shutdown.

A vertical slurry line may well have been used in the first application of hydrotransport, back in the gold rush days of the mid 1850s. It has since been taken over by interest in horizontal transport with the advent of commercial long-distance pipelining. Nevertheless, the vertical line offers the following possibilities:

(a) It represents a means of negotiating rugged terrain, as a combination of horizontal and vertical legs may offer better steady state and restart characteristics than the equivalent sloping line. The entire contents of a sloping line may slump into a plug at the bottom, whereas the smaller volume within the equivalent vertical line would pose a less serious shut-down problem.
(b) It is the logical means of material transport from wet mines, from which water must be continuously drained to the surface, or from mines in which the material has been won by hydraulic jetting.
(c) The relatively new and untried application of deepsea mining requires some means of bringing the ore nodules to the surface. Air-lifting and jet-pumping are primary contenders in this area.

Relatively little has been published concerning application (a), and (c) is discussed more fully in chapter 7. Regarding (b), Kortenbusch (63) has reported some of the experience gained at the Hansa Hydromine in West Germany, during the three years of its operation. It was ideally suited to hydraulic hoisting as the 850 m lifting depth could be matched to the 100–120 bar required for the jet to loosen the coal. The system allowed solids to be transported in both directions so that worked out tunnels could be refilled with dirt and blocks of ice could be brought down for cooling purposes. As noted by Harzer and Kuhn (64), the pumps, pipelines and overflow basins were over-dimensioned, and the coal could have been pumped at higher concentrations and lower velocities without the risk of plugging. They estimated that excavation volume reductions from 24 000 to 10 000 $m^3$ (60 per cent) and installed power reductions from 8200 to 4500 kW (45 per cent) could have been made. They found the greatest number of problems to be associated with the various solids handling flumes around the mine. These, covering a total distance of 50 km, involved numerous joints and curves, at which points blockages occurred. The transport system handled both coal and tailings, and Maurer and Mez (65) reported some difficulties involved in lifting the latter (SG 2.3). They remcommended the use of fines to avoid the formation of 'clouds' of coarse particles. These were frequently larger than usually found in run-of-mine coal (topsize 50 mm) and 100 mm lumps were not uncommon.

Newitt *et al.* (66) pioneered the correlation of pressure drop for vertical solids transport and developed the following from tests with sand, zircon, manganese dioxide and perspex particles in 13 m × 25 mm and 8 m × 50 mm columns:

$$J_m - J_w = 0.0037 \, C_v J_w \, (gD)^{0.5} V^{-1} D d^{-1} S^2 \tag{51}$$

or

$$\phi = 0.0037 \psi^{-0.5} S^2 D/d \tag{52}$$

if $\psi = V^2/(gD)$

The above equations have been written in a form to enable comparison with correlations for horizontal lines. Features to be noted are the high power of specific gravity ($S^2$), the absence of particle drag coefficient ($C_D$) and the presence of the pipe to particle diameter ratio ($D/d$). In contrast to the horizontal case, vertical transport of coarse particles incurs less friction than that of fine slurries. To explain this, it must be realised that the Magnus effect in vertical flow would move large particles away from the pipe wall. Unopposed by gravity, this would concentrate the solids near the pipe centreline and reduce particle-wall collisions and hence the amount of solids friction contribution. Increasing the particle relative density would tend to raise both in-situ concentration and the solids contribution to slurry density. This could be why it appears to such a high power in equation 51. The particle size range for which the equation is valid was stated to be 0.1–3.0 mm. Finer particles tended to be homogeneously transported whilst coarser particles, at high velocity, returned a friction gradient little different from that of the carrier fluid alone. For the latter case, Newitt *et al.* (66) observed a clear water annulus surrounding a core of suspended solids.

Wilson *et al.* (67) experimented with 0.3–0.7 mm sand in a 25 mm test loop, with a view to studying friction mechanisms at high concentrations (50 per cent by volume). They noted that the Magnus effect would tend to concentrate large particles near the centreline, reducing particle friction at the wall yet possibly increasing interparticle collisions. At high concentrations, however, the violence of these collisions was markedly reduced and the solids tended to slow as a plug. The frictional implications were obviously very favourable and agree with the results of Newitt *et al.* (66) for coarse perspex particles, found to add negligibly to the clear water friction gradient and not to fit equation 52. In a development of the same concepts for 75 mm and 300 mm pipelines, Streat (68) recommended dense phase vertical transport of coarse particles, and suggested indirect pumping (e.g. jet pump), as a means of avoiding the presence of large lumps in contact with the pump impeller.

Other models of vertical flow have included that by Ayukawa *et al.* (69), who developed separate equations for the solid and liquid phase motion and solved them by a finite difference method. In their experiments with polycarbonate (SG 1.14) and 'Silbead' (SG 1.88) particles, they also noted a high central concentration of particles of order 3 mm and above. Sellgren (70) considered the problem of design velocity for transport, basing head gradient calculations on the following formula:

$$J = [1 - C_v(S - 1)][1 + fV^2(2gD)^{-1}] \qquad (53)$$

In essence, equation 53 means that total gradient is the sum of a gravitational component based on slurry density and a frictional component also based on this density. It is therefore only valid for slurries in homogeneous flow. From experiments with crushed granite, Sellgren suggested that mean mixture velocities of 4–5 times the settling velocity of the largest particle should be used, and that a drag coefficient as low as 0.1 should be used in calculating this figure. He quoted supporting evidence from industrial experience with slurries at 15–30 per cent by volume. It is worth speculating whether gains are to be made by increasing the concentration to the point where hindered settling occurs. Higher solids contents would seem to be the favoured direction for development of vertical transport technology.

## 3.4   SOLIDS TRANSPORT IN FLUMES

Flumes, essentially open channels, have already been mentioned in the context of coal transport in the Hansa Hydromine. In another sense, the flume represents a river bed and has been used in civil engineering research to model the natural watercourse. Blench *et al.* (71) note the contribution made by civil engineers to the science of solids transport, yet admit that the artistic requirement remains, with nearly as many transport correlations as there have been researchers. There is a basic distinction to be made between the civil engineer's interest and the slurry system engineer's interest in flumes. The civil engineer wants to know about incipient motion with a view to predicting erosion patterns, whereas the objective in slurry transport is to suspend the entire bed if possible.

Based on previous developments for transport in pipeflow (equations 40 and 41), Wilson (72) produced a numerical model for mixed suspended and sliding transport of solids in flumes. If a sliding bed motion is desired it can be of benefit to maximise the shear stress, given below:

$$\tau = \mu\sigma_n \qquad (54)$$

where $\sigma_n$ = normal stress

Bed shear stress can be raised by increasing either bed slope or hydraulic radius. If, on the other hand, the entire solids load can be suspended, it pays to maximise the total flow rate, achieved at a wetted half angle of 150°. By the use of his model, Wilson (72) developed an optimum slope prediction for any given particle size or pipe diameter. By optimum, he meant that which gave the greatest product of transport rate and distance for any given drop in elevation. For a fixed section diameter of 0.25 m, it is seen that particles below 100 $\mu$m in size are optimally transported at a slope of 0.25 per cent, whilst those above 1.0 mm are best carried in the sliding mode at the very steep grade of 40 per cent. The steep variation in optimum slope between these points represents the heterogeneous regime, with some particles in sliding and others suspended. Increasing pipe diameter widens the boundaries of the heterogeneous regime. The above predictions have been run for sand-weight solids, though the effect of particle S.G. has been incorporated into the model.

The flume geometry need not be circular. The Hansa Hydromine in Dortmund used flumes of over twenty different types. In an investigation of the alternatives, Kuhn (73) found trapezoidal sections to be the most effective, especially when fitted with a plastic liner. This reduced both friction and wear rate, the former to a value below one third that of steel. Wood (74) has indicated how the hydraulic radius of various basic section geometries might be optimised by an appropriate choice of liquid depth. This procedure has most relevance to sliding bed flow, which if Wilson (72) is correct, applies to most solids ($d \geqslant$ 1 mm) of interest for transport in flumes. In a review of the subject, Lau (75) concurs with Wilson that solids transport in flumes has received scant attention in the literature. Considering their virtues of simplicity, low cost, suitability for mines, and easy maintenance, it is to be hoped that this situation will change in the near future.

## 3.5 COARSE PARTICLE TRANSPORT

The term 'coarse particle' is subjective and can mean anything from a 3 mm gravel particle to a 200 mm lump of coal. Another way of viewing it is to consider coarse slurries to be those requiring special measures to cope with the particle size. Such measures may be to operate the pipeline or flume in a sliding bed mode, to install special wear or low-friction liners, to run the system at a velocity sufficient to keep all material suspended, or to stabilise the mixture.

The influence of particle size has been studied by Elsibai *et al.* (76), who compared coal in two size distributions with upper limits of 50 mm and 1.2 mm. They concluded that with inclusion of preparation and dewatering costs, the distance at which the finer grind became more economical was 40 km at a 2.27 Mtpa throughput and 80 km at a 4.55 Mtpa load. The respective velocities recommended were 3.2 ms$^{-1}$ for ROM and 1.7 ms$^{-1}$ for fine coal. They moderated their conclusions by the thoughts that particle attrition could raise the cost of coarse coal dewatering, and the availability of extraction steam could lower the drying costs of the fine coal system. Nevertheless, 40–80 km is a commercially significant distance range.

Boothroyde *et al.* (77) looked at the transport of colliery waste (mudstone, shale, siltstone and sandstone) in coarse slurry form, attractive from the short distances involved. They found a sliding bed to be the only feasible transport regime, and addition of fines was of benefit only when the fines themselves were required to be transported. For crushed granite transport, on the other hand, fines can have the additional benefit of reducing the impact of a serious wear problem, which could not be solved with liners (78). The separated fines suspension could be returned to the mixing (preparation) vessel in such a way as to help maintain the granite in suspension. For the dredging industry, Schaart and Verhoog (79) noted very different problems, such as prediction of the particulate properties of the dredged material. It could be dredged in consolidated lumps, unconsolidated grains or a mixture of the two extremes. They pointed to the need for a unified theory of coarse particle transport.

Coal and mine waste transport are the applications most likely to gain from widespread development of coarse slurry technology. As long ago as 1978, Lawler *et al.* (80) presented a scheme for stabilising coarse coal mixtures with fines addition. They found a trade-off between the stability and fluidity of the mixture and concluded that developments in pumping technology could make the most needed contributions. The thick mixture required a high pressure pump, even for laminar flow, and few high pressure pumps were available that could cope with 50 mm solids. The upper limit of 3 mm was more usually accepted. In a more recent study, Sakamoto *et al.* (81) found that coarse slurries stabilised with fines could be pumped in turbulent suspension with very much lower friction loss than predicted by the accepted theories for non-Newtonian mixtures. They suggested that a wall migration effect might be at work, and proposed the construction of a several kilometre pilot facility to look at the idea on a more commercial scale.

There would appear to be three main options for coarse particle transport in horizontal flow, i.e. sliding bed, stabilised mixture, or dilute phase. Dilute phase transport involves high velocities and large quantities of water in order to suspend the solids in heterogeneous or pseudohomogeneous flow. Due to the high wear rates and head losses incurred, it is rarely used outside the dredging industry, where the convenience outweighs its disadvantages. Outside this context, sliding bed offers the most economically promising regime, especially if the carrier fluid can be given lubricating properties by the addition of fines. Stabilised flow is considered in more detail in section 4.6.

## 4.  NON-NEWTONIAN RHEOLOGY

The term non-Newtonian is a negative descriptor; any departure from Newtonian behaviour qualifies as a non-Newtonian property. The basic definitions have been given in section 2.3, along with some examples from the real world. At the possible expense of repetition, this section begins with a treatment of the problems inherent in classifying non-Newtonian fluids.

There follows a presentation of fundamental aspects of mechanisms for development of a non-Newtonian viscous response, covering plastic, pseudoplastic and dilatant behaviour.

Viscoelasticity is treated next, not merely for the sake of completeness, but also as an introduction to drag-reducing polymers which have found some application in slurry transport systems. Simple slurries rarely display viscoelastic properties. Next in sequence is the pipeflow, both laminar and turbulent, of the various types of non-Newtonian fluids hitherto described. Transition Reynolds numbers, friction factors and drag reduction mechanisms are to be given attention. Finally, the application of these various principles to slurry transport systems is treated, with emphasis both on drag reduction techniques and the use of properties such as yield stress to support coarse solid particles in laminar flow.

## 4.1 FLOW BEHAVIOUR CLASSIFICATION

Figure 15 illustrates some of the models appropriate to shear-thinning substances. The two simplest are the Bingham plastic and power law, represented by straight lines on linear and double-logarithmic scales respectively. They derive most of their advantages from their simplicity and are usually the first choices for any hitherto unknown non-Newtonian substance. The power law enjoys the widest use as most rheological data is compiled on logarithmic paper to cope with the wide range of viscous behaviour displayed by suspensions of variable concentration. It has the disadvantage that consistency ($m$) and flow behaviour index ($n$) are difficult to visualise in physical terms. The yield stress ($\tau_y$) of the Bingham

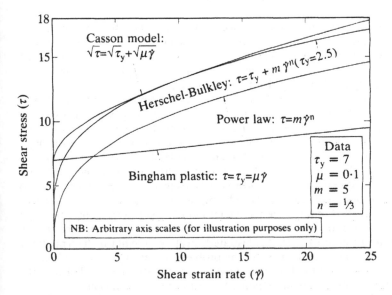

**Figure 15.** Choice of models for a shear-thinning stable solids suspension

plastic is, by contrast, easily related to the principles of solid mechanics, and the plastic viscosity ($\mu$, $\mu_p$) is the limiting viscosity at infinite shear rate.

Ostwald-de-Waele (power law) and Bingham models are useful only if they fit the data with the required precision. The more scattered the data, the more appropriate the use of a simple model. A systematic misfit, however, cannot be accepted and an alternative model must be used. Of the two parameter models, the Casson model is gaining popularity. Its similarity to the Bingham formula can be seen by the following comparison:

$$\tau^{1.0} = \tau_y^{1.0} + (\mu_p\dot{\gamma})^{1.0} \quad \text{(Bingham)} \tag{55}$$

$$\tau^{0.5} = \tau_y^{0.5} + (\mu_p\dot{\gamma})^{0.5} \quad \text{(Casson)} \tag{56}$$

They share a yield stress and a limiting (infinite shear) viscosity. However, as Figure 15 shows, the similarities end there. Use of identical parameters in Bingham and Casson models leads to very different curves and the Casson fluid approaches its limiting viscosity at much higher shear rates than does the Bingham plastic. Such are the dangers of interpreting fluid behaviour in terms of its limiting properties.

If a good fit is required, and the Casson model does not quite fit the data, it may be advisable to select a three-parameter model, such as the Herschel-Bulkley or, as it is often termed, the 'generalised Bingham' model:

$$\tau = \tau_y + m\dot{\gamma}^n \quad \text{(Ostwald-de-Waele : } \tau_y = 0, \text{ Bingham : } n = 1) \tag{57}$$

As Figure 15 shows, despite the fact that none of the Herschel-Bulkley parameters correspond to those of the Casson model, the curves can be made to fit very closely over a limited shear rate range. Outside this range, however, the curves diverge, illustrating one of the dangers of extrapolating beyond the measured range. As explained below, three-parameter models are especially susceptible to this problem.

The extra degree of freedom enables the Herschel-Bulkley model to be fitted to most data better than any of the two-parameter formulae. In essence, Bingham and power law fluids are special cases of the generalised model. As pointed out by Johnson (82), such data fitting power comes at some cost. Figure 16 shows various attempted fits through rheological data between strain rate limits $A$ and $B$. The Herschel-Bulkley curves represent two independent algorithms, both directed towards a least squares fit. As shown, they are coincident between the stated limits, but divergent outside them. This illustrates two points. Firstly, extrapolation beyond the measured range is dangerous, even for unscattered data. Secondly, in the context of the three-parameter model, yield stress, consistency and index are strongly sensitive to shear rate range, and may not correlate well against such variables as solids concentration or PSD. As illustrated, an attempt to fit Bingham plastic or power law curves to the data might not solve the problem, especially if extrapolation is attempted. It may be noted that the Bingham curve underestimates the data between the limits and overestimtes it outside them, opposite to the power law behaviour.

To show that the above problems can be overcome, Colombera and Want (83) took

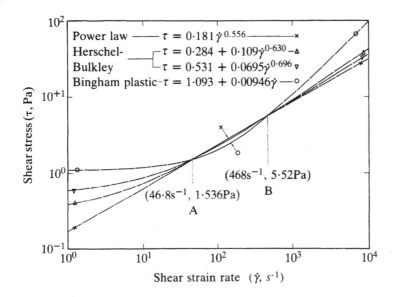

**Figure 16.** Effects of extrapolating beyond the measured shear rate range (82)

rheological data for a bauxite residue mud over three and a half decades of strain rate, and fitted Bingham, Casson and Herschel-Bulkley models to obtain a yield stress, comparing them with a direct measurement by stress relaxation. They obtained good agreement, and a good correlation against concentration between Casson, Herschel-Bulkley and relaxation values. As might be expected, the Bingham yield stress exceeded the others but showed an encouragingly similar trend. It was important for some confidence to be attached to the yield stress as it was used later in estimates of thixotropy and ageing. Between 30 and 70 per cent concentration by weight, it was found to be proportional to the 8.6th power of concentration; illustrating the need for confidence!

It is sometimes the practice to mix rheological models as alternative descriptions of the one substance. Solomon *et al.* (84), for instance, wanted to model a xanthan broth with a transparent water soluble polymer. To ensure accurate modelling, pseudoplastic power law parameters were fitted to both broth and polymer solution. To obtain a 'yield stress' a Casson model was also fitted and extrapolated to zero shear rate. Though lacking a certain rigour, such a process may be justified in the context of fluid-to-fluid modelling. The dangers of extrapolation can be lessened if two rheological models are used to offset each other's excesses.

None of the above models is of use if the non-Newtonian fluid in question exhibits a finite viscosity at zero shear. Boger *et al.*(85), in a general review of mineral slurries considered the following:

$$\tau = \dot{\gamma} \left[ \eta_\infty + (\eta_0 - \eta_\infty)/(1 + (\tau/\tau_m)^{\alpha - 1}) \right] \qquad \text{(Meter model)} \qquad (58)$$

$$\tau = \dot{\gamma} \, \eta_0 \left[ 1 - (\tau/\tau_m)^{\alpha - 1} \right] \qquad \text{(Ellis model)} \qquad (59)$$

where $\eta_0$ is zero shear viscosity, $\eta_\infty$ the infinite shear viscosity, $\tau_m$ a shear stress constant (not yield stress) and $\alpha$ an exponent of the implicit shear stress.

Besides their implicit character, the four parameters of the Meter model and the three of the Ellis model introduce the usual problems associated with multiple parameters, the Meter model requiring as many as six decades of shear rate. Whilst an awareness should be kept that these models exist, they are rarely used for design purposes.

## 4.2   DATA REDUCTION METHODS FOR VISCOMETERS

In general, a shear-stress shear-rate relationship for the fluid under test is required and must be derived from pairs of values of pressure and flow rate or torque and speed in tube or rotational instruments respectively. Experience suggests that people new to the field use incorrect methods of data reduction, frequently based upon Newtonian fluid flow assumptions which are clearly erroneous for non-Newtonian fluids.

The data reduction methods will be considered here under headings of first and second order effects. First order effects include derivation of shear stresses and shear rates assuming ideal flow conditions in the viscometer. Second order effects include calculations to allow for non-perfect flows caused by end effects, wall slippage and thixotropy, centrifugal separation of phases, etc.

### 4.2.1   Derivation of stress–strain relationships

The calculation of shear stress in tube and rotational instruments is simple since it can be derived from a balance of forces and is related to pressure or torque and the viscometer dimensions.

The derivation of shear rate is more complex since it depends on flow rate or rotational speed, the viscometer dimensions and the characteristics of the fluid under test. Figure 21 illustrates how the shear rate at the wall of a pipe varies with pseudoplastic index. The shear rate at the wall is given by the gradient of the velocity profile and it will be seen that non-Newtonian properties, indicated by values of $n$ less than 1.0, give increased shear rates at the wall. It will be appreciated that the shear rate at the tube wall is thus a function of the fluid rheology. An analogous situation exists within all other types of viscometer. For this reason, 'calibration' values of shear rates based on Newtonian fluid assumptions, frequently supplied by manufacturers of rotational viscometers, give erroneous results on non-Newtonian fluids.

Two different approaches to the problem of calculating the shear rate within a viscometer may be identified. The older methods assume a fluid model (e.g. Bingham Plastic) and a series of simultaneous equations may then be derived relating yield stress and viscosity (for example) to the viscometer geometry and recorded data points. These methods which can be used for both tube and rotational instruments are not completely satisfactory since the fluid under test may not be represented truly by the chosen model and hence, erroneous results will be produced. Subsequently, data analysis techniques have been produced which make no prior assumption regarding the fluid model. However, after obtaining the shear rate values it is still necessary to fit them to a suitable correlating equation.

Van Wazer (86) gives a good summary of the techniques for both tube and rotational instruments, while Schummer and Worthoff (87), Huang (88) and Harris (89) give further methods. Huang (88) also deals with cone and plate geometries. In general, these techniques require the first and second derivatives of the torque and speed results (or pressure and flow rate in tubes). Some sets of data, particularly data of poor quality or for unusual fluids, cannot be analysed using these methods. Johnson (82) recommends that shear-stress shear-rate relationships, derived by any method, should be checked against original results by a numerical integration of the viscometer flow field.

## 4.2.2 End effects

Practical viscometers of finite dimensions generally exhibit effects due to the ends of the tube or bob. These effects can be deduced by conducting tests with a range of tube lengths, or bob lengths respectively (90). It has been shown that these effects are a function of fluid rheology and hence will need investigation for each fluid.

## 4.2.3 Phase separation effects

Several workers have observed the creation of a more dilute layer in the area of greatest shear in both tube and rotational viscometers. Several calculation methods for correcting for this have been proposed (e.g. 90 and 91) and tested with varying success. Johnson (82) describes a method for distinguishing wall slip or phase separation from thixotropy in rotational instruments.

## 4.2.4 Thixotropy and time dependency

This is included here as a 'secondary effect' but can be a very important aspect in such fluids as clays, very fine coal slurries and kaolin. The reduction in apparent viscosity with shear (thixotropy), or increase with shear (rheopexy) and the subsequent build up or decrease of viscosity on standing are frequently characterised by time constants describing the change in the fluid yield stress or consistency index. This means that to describe a thixotropic

Bingham Plastic model, for example, six or more constants must be experimentally determined. The extensive experimental work required for this means that thixotropic properties are frequently ignored.

Cheng (92) describes an investigation of thixotropic properties using tube viscometers on sewage sludges. Bhattacharya (93) also describes studies on the time dependent recovery of coal–oil suspensions.


## 4.3     PHYSICS OF PSEUDOPLASTICITY AND DILATANCY

The emphasis in this subsection is upon the physical mechanisms whereby substances develop non-Newtonian viscous properties. 'Pseudoplasticity' is here taken to include any kind of shear-thinning effect, whether or not a yield stress is involved. It will be seen that fine particle agglomeration is responsible for the shear-thinning effects of fine slurries, and de-agglomerating the suspension can turn the system Newtonian or even dilatant. Non-agglomerating suspensions, such as cornflour in water, are known for their dilatancy at high concentrations.

Pure liquids, dilute suspensions, and solutions of relatively low molecular weight solutes generally exhibit Newtonian behaviour. This is due to the fact that shearing imparts a momentum gradient, and momentum, like any other property, is diffused by molecular means at a rate proportional to the gradient. Newtonian viscosity can thus be viewed as a molecular diffusion coefficient for momentum. It can be assumed only when the structure of the medium remains unchanged by the shearing, as would be expected for the fluids listed above.

Non-Newtonian behaviour, most commonly pseudoplastic, is found for polymer solutions, polymer melts and suspensions of agglomerating solid particles. Polymer melts cannot be considered pure liquids as they generally contain a wide distribution of molecular weights, especially at part polymerisation. Each of the media in question possesses a structure which is changed as a result of shearing action. The phenomenon is best understood in the context of suspensions since these encompass the mechanisms at work in polymer liquids.

Figure 17a illustrates a suspension, e.g. fine clay in water, either at rest or subject to a low shear rate. Electrostatic effects can cause the particles to agglomerate in such a way that, instead of settling, they form a kind of solid matrix throughout the fluid. If this is sufficiently rigid, a finite yield stress may be required to break it, and the system would be appropriately described by a Bingham, Herschel–Bulkley or Casson model. Most fine suspensions contain a broad particle size distribution (PSD), resulting in the more gradual breakdown of structure typical of psuedoplastic media. Figure 17b shows the disintegrated structure. Particle chains have been broken down by the shear, and those that have survived are aligned to minimise their interaction with other units. The alignment is due to the presence of rotation (vorticity) within most shear fields, e.g. those in a mixing vessel. Polymer liquids develop pseudoplasticity by the alignment mechanism alone. Disintegration of a polymer molecule is usually irreversible and requires violent or prolonged shear. The process is

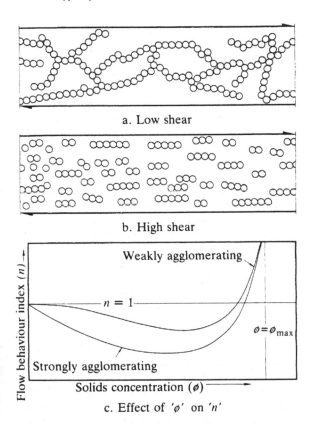

a. Low shear

b. High shear

c. Effect of '$\phi$' on '$n$'

**Figure 17.**  Basic features leading to non-Newtonian behaviour of suspensions

usually expressed as thixotropic behaviour. Reversal of thixotropy, if it occurs, generally takes much longer than the original breakdown.

The above discussion holds only for agglomerating suspensions. If these are treated with a deflocculating agent or concentrated to a sufficient strength, dilatancy may result. In support of this observation, Ackermann and Hung (94) have analysed suspensions of solid particles in Newtonian liquids in the absence of agglomerating phenomena. They found that up to about 50 per cent volume fraction of solids the suspensions should behave like Newtonian liquids and reported excellent agreement with other workers' experimental data. Beyond this concentration the particles are forced into mutual contact; 52 per cent is sufficient for a cubic lattice of monodisperse spheres and 75 per cent requires one of the close-packed arrangements, higher concentrations needing fines to fill the interstices. The fluid dynamic result is inertial interaction between particles, following a velocity squared law. However, the phenomenon is not dependent upon the bulk velocity, but rather upon the shear rate. A shear stress proportional to shear rate squared corresponds to dilatancy. Ackermann and Hung (94) recommend the use of a parameter, representing the ratio of inertial to viscous shear stress, to determine degree of dilatancy. It is thus conceivable for a medium to be pseudoplastic at low shear and dilatant at high shear; indeed some pigment pastes have been found to exhibit such a 'viscosity' minimum. Although the shear stress ratio has the form of a Reynolds number, particle concentration is the most important variable and influences the flow behaviour index in the manner indicated in Figure 17c. Although the above is by no means the only theory for dilatancy, the phenomenon is virtually unknown outside the context of concentrated suspensions.

A complementary explanation for dilatancy, set out below, has been based on the early fundamental work of Osborne Reynolds. In a concentrated suspension subject to low shear rates, the particles can retain their close packing arrangement without coming into contact. At higher strain rates, the configuration is disrupted and the suspension expands or 'dilates' as a result of the less efficient packing. There being no longer sufficient liquid to fill the voids results in rheological behaviour now being dominated by solid contact friction. Even if liquid could be supplied to the voids, the suspension structure would still exhibit direct interparticle contact. In Figure 17a and 17b, shearing is seen to increase the packing efficiency from that allowed by simple agglomeration. In a dilatant suspension it has the reverse effect. Both inertial and frictional interparticle effects might be at work in the one mixture, giving it marked dilatant properties.

The chemistry of agglomerating suspensions has considerable influence upon the degree of agglomeration. Fine ground silica, for example, is a surface acid, absorbing hydroxyl ions from the water in which it is suspended. The suspension behaves like an acid until separated into solids and supernatant. In suspension, each particle develops a net negative charge, contributing to an electrostatic repulsion between itself and its neighbours. This counteracts the attractive Van-der-Waals forces tending to agglomeration. The attractive forces, however, are also electrostatic in origin, arising from local charge migration. Adding acid to the suspension neutralises particulate charge, allowing attractive forces to work and agglomerate the solids. Adding alkali, on the other hand, intensifies the charge, dispersing the particles, and destroying the system plasticity. Viscosity reducing chemical additives can be quite specific to the system concerned, as discussed in more detail in section 4.7.

## 4.4   FUNDAMENTALS OF VISCOELASTIC BEHAVIOUR

Viscoelastic fluids can be defined as capable of developing normal stresses in the direction of an applied shear strain. Although they achieve this by means of elastic effects, the phenomenon must be distinguished from solid elasticity. Indeed, some materials exhibit solid elasticity if stressed below their yield point, followed by viscoelasticity if the stress is increased sufficiently. They are known as elastic plastic viscoelastic fluids, as distinct from rigid plastic viscoelastic media. Their theory, developed by White and Hwang (95), finds its application in the processing of filled polymer melts. The filler is usually a particulate, such as clay, added to give bulk to the product without seriously impairing its mechanical properties. It may be responsible for imparting a yield stress to the polymer in the molten state.

This section is confined to a discussion of viscoelastic liquids in stirred tanks. They range from dilute solutions ($\sim$ 20 ppm) of 'drag-reducing' agents, through the more concentrated ($\sim$ 0.2 per cent) polymer solutions used for experimental modelling, to the highly concentrated ($\sim$ 50 per cent) polymer-in-monomer solutions found in industrial reactors. As with the previous one, this section begins with an explanation of viscoelastic stresses; how they are measured, where they act and how they arise. Viscoelastic effects are of primary interest in the mixing industry and it is in this context that most of the research has been done. Drag-reducing polymer additives in the pipeflow field are treated in section 4.5.

The measurement of viscoelasticity provides a useful starting point for an understanding of its nature. The most popular device for the purpose is called a rheogoniometer, illustrated schematically in Figure 18. It is identical to the cone-and-plate viscometer, able to impart a uniform shear rate to the fluid sample, with an additional transducer for measuring the thrust developed by viscoelastic fluids. For a given angle ($\beta$) between cone

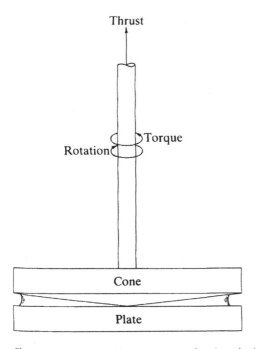

**Figure 18.** Schematic rheogoniometer for viscoelastic property measurement

and plate, rotational speed ($N$), sample radius ($r$), shear stress ($\tau$) and first normal stress difference ($F_1$), the shear rate ($\dot{\gamma}$), reaction torque ($T$) and thrust ($F$) are given below:

$$\dot{\gamma} = 2\pi N/\beta, \qquad T = \tfrac{2}{3}\pi r^3 \tau, \qquad F = \pi r^2 F_1 \tag{60}$$

The first normal stress difference acts like a tensile hoop stress, pressurising the fluid sample, thus forcing the cone and plate apart. Appreciation of this feature is essential to an understanding of the fluid mechanics of viscoelasticity in mixing vessels.

Figure 19 defines the planes on which the normal stresses are developed. The planes are numbered according to the way the shear is applied, planes '1' undergoing rotation, planes '2' parallel motion and planes '3' distortion, as indicated by the dotted lines. In the context of a shaft rotating in the medium, plane 1 is radial, plane 2 cylindrical and plane 3 normal to the shaft. Shear stresses ($S_{12}$, $\tau$) are developed on planes 1 and 2 as one might expect for a viscous fluid. A viscoelastic fluid will also develop a tensile stress normal to plane 1. If the hydrostatic pressure is superimposed, the tension is expressed as a normal stress difference ($F_1 = S_{11} - S_{22} \geq 0$). Some fluids are able to develop a second normal stress difference

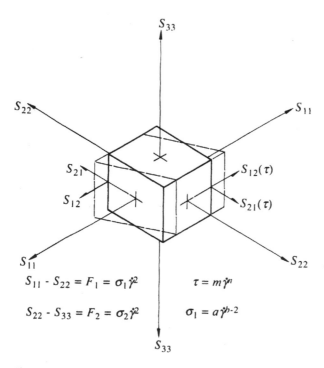

Figure 19. Definition sketch for normal stresses in a viscoelastic fluid

($F_2 = S_{22} - S_{33} \leq 0$) with tension normal to plane 3. They are known as third order viscoelastic fluids, referring to their one viscous and two elastic responses to shear.

Elastic fluids are characterised by their 'first and second normal stress coefficients' ($\sigma_1$, $\sigma_2$), multiplied by the square of the shear rate to yield their respective normal stress differences. The use of $\dot\gamma^2$ is motivated by the fact that reversal of the strain rate makes no difference to the elastic response, and by the constancy of $\sigma_1$ at low strain rates. At higher strain rates the stress difference tends to follow a lower power of shear, like the shear stress in pseudoplastic media:

$$F_1 = \sigma_1 \, \dot\gamma^2 \quad \text{at low } \dot\gamma, \qquad F_1 = a\dot\gamma^b \quad \text{at high } \dot\gamma \, (\sigma_1 = a\dot\gamma^{b-2}) \tag{61}$$

Most viscoelastic fluids are also pseudoplastic, excepting dilute polymer solutions (drag-reducing) and some silicone oils. Indeed, the mechanism for normal stress development has much in common with pseudoplastic phenomena. It is illustrated schematically in Figure 20 and takes place in four stages. The first requirement is for the presence of long, compliant, 'rubbery' molecules. If these are initially aligned as in Figure 20a, the fluid rotation carries them into a 45° alignment where they can be stretched by the tensile component of the stress field (Figure 20b). The stretched units are then further rotated into the horizontal alignment (Figure 20c), where intermolecular linkages create a net tension in the shear direction. The relaxation of these linkages (Figure 20d) limits the magnitude of the stress difference. Equations 1–3 represent the major governing factors: the driving stress is proportional to the viscous stress (1), the rate of alignment of stretched units must equal the relaxation rate (2) and the net normal stress difference is proportional to the product of driving stress and the population of aligned stretched molecules.

It is noteworthy that the viscoelastic stress is inversely proportional to molecular stiffness ($E$), by virtue of the time constant for relaxation (spring and damper in series). The relaxation time constant can also be expressed in the following way, which, when multiplied by the impeller rotational speed, yields the 'Weissenberg number' (Wi), the ratio of elastic to viscous effects, where the impeller speed ($N$) is characteristic process frequency:

$$t_c = \sigma_1 \, \eta^{-1} \tag{62}$$

at high shear rates

$$\text{Wi} = Nt_c = N\sigma_1 \, \eta^{-1} = Nam^{-1}\dot\gamma^{b-n-1} \tag{63}$$

Ulbrecht (96,97) measured the values of $\eta$ and $\sigma$ for polymer solutions and melts. It is noted that silicone oil is a viscoelastic fluid with a high Newtonian viscosity (40 000 cP). Newtonian viscosity is usually displayed by substances of moderate molecular weight, longer chains conferring some pseudoplasticity. The viscoelasticity of the silicone oil must thus be the result of a small concentration of ultra-high molecular weight components, too few in number to confer appreciable shear-thinning effects. Therefore, the silicone oil is appropriately called a polymer solution. The polyethylene melts (data taken at 180° C)

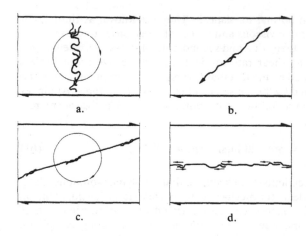

Driving stress (45°):        $F_0 = \tau = \eta \dot{\gamma}$                    (1)

Population equilibrium:   $C_1 \phi_0 \dot{\gamma} = \phi'$                    (2)

Normal stress at 0°:        $F_1 = C_2 F_0 \phi$                    (3)

$F_1 = C_2 F_0 \phi / k$   where   $k = E / \eta$                    (4)

Hence $F_1 = C_3 \phi_0 \eta^2 \dot{\gamma}^2 / E = \sigma^2 \dot{\gamma}^2$                    (5)

**Figure 20.**   Mechanism for developing tension in the direction of shear

show time constants of order ten seconds, indicating the presence of strong viscoelastic effects in the most gently stirred mixing vessels.

The shape of the $\eta$, $\sigma$, $v$ shear rate functions precludes the use of a two-parameter model for either elastic or viscous coefficients over the whole shear rate range. This problem was solved by Carreau *et al.* (98) who proposed three-parameter models of the following form ($\sigma_1$ – unsubscripted):

$$\sigma = \sigma_0 (1 + t_1^2 \dot{\gamma}^2)^{-S}, \qquad \eta = \eta_0 (1 + t_2^2 \dot{\gamma}^2)^{-R} \tag{64}$$

Taking viscosity as our example, the three parameters are the 'zero shear' viscosity $(\eta_0)$, a 'time constant' $(t_2)$ and a kind of flow behaviour index $(R)$. Although $t_2$ has the units of time it is not to be confused with $t_c$, the relaxation time constant.

## 4.5 PIPEFLOW OF NON-NEWTONIAN FLUIDS

In the design of a slurry pipeline the pipeflow characteristics may be of interest from a number of standpoints. For instance, the transport of coarse particles in a thin suspension of fines would require one to know the laminar to turbulent transition point if settling is to be avoided. Were more fines to be added, sufficient to support the coarser fractions stably, the laminar head loss characteristic itself would be needed. In this section, the laminar flow theory is developed first, as this gives rise to a definition of Reynolds number, in terms of which turbulent friction factors are also expressed. The discussion closes with a brief, qualitative, treatment of the drag reducing action of polymer additives, indicating why they only work in turbulent flow.

Like the Newtonian case, the velocity profile of a pseudoplastic in laminar flow through a pipe of circular section can be obtained by direct integration. In terms of mean velocity $(V)$, the local velocity $(u)$ is given as follows as a function of distance $(r)$ from the axis of the pipe (radius $- r_0$):

$$u = V(3n + 1)(n + 1)^{-1}[1 - (r/r_0)^{(n+1)n^{-1}}] \tag{65}$$

$$u = 2V(1 - r^2/r_0^2) \tag{66}$$

if $n = 1$      (Newtonian)

$$u = V \tag{67}$$

if $n = 0$      (plastic)

Equation 65 has been plotted on Figure 21 for various $n$ between zero and one. Increasing pseudoplasticity is seen to increase the shear rate at the wall, so that a wall shear rate of $8V/D$ can no longer be applied. However, a revised definition of Reynolds number can allow us to express friction factor in a familiar way:

$$f = 64/\mathrm{Re} \quad \text{if } \mathrm{Re} = \rho D^n V^{2-n} m^{-1} 8[n/(6n + 2]^n \tag{68}$$

This is the generalised Reynolds number for a power-law fluid. Equation 68, though strictly accurate, is somewhat inconvenient if a tube viscometer is to be used to obtain the basic data. There is a way around this problem, as discussed in detail later in this subsection. Bingham plastics in laminar flow display similarly shaped profiles, yet with a core of pure

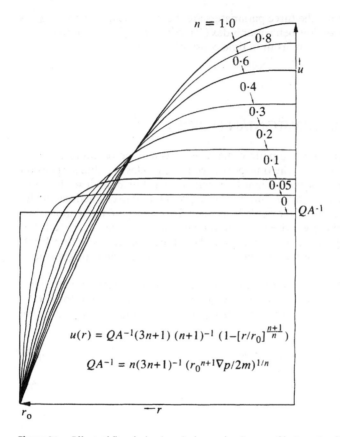

**Figure 21.** Effect of flow behaviour index on laminar profile in a circular pipe

plug flow around the axis of the pipe. Figure 22 shows that the radius of this core ($r_c$) can be related as follows to the yield stress:

$$r_c = 2\tau_y/\nabla p \qquad \text{shear stress pressure balance} \qquad (69)$$

Outside the core, where the shear stress exceeds the yield, there exists a shear flow from which the wall shear rate can be determined. It is possible for the vorticity in this region to propel large particles towards the axis (Magnus effect), thinning the layer near the wall and reducing friction. If this can be encouraged, plug flow could become a useful transport

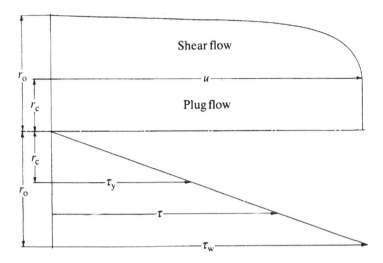

**Figure 22.** Development of plug flow for substances with a yield stress

mode. In the absence of wall migration, the following friction factor relationship can be written:

$$f = 64/\text{Re} \qquad \text{if Re} = [\mu_p(\rho VD)^{-1} + \tau_y (8\rho V^2)^{-1}]^{-1} \qquad (70)$$

In this case, an assumption has been made that the shear rate at the wall is equal to $8V/D$. Though lacking rigour, it simplifies the algebra as will later become apparent. It is not a universally favoured approximation, some workers showing a preference for a $6\rho V^2$ in place of $8\rho V^2$. This can lead to a more accurate prediction for substances with a relatively low yield stress. Applying the same approximation to Casson and Ostwald-de-Waele (power law) media yields the following Reynolds numbers for laminar friction factor:

$$\text{Re} = [\mu_p(\rho VD)^{-1} + \tau_y(8\rho V^2)^{-1} + (8\tau_y\mu_p V^{-3}D^{-1}\rho^{-2})^{-0.5}]^{-1} \qquad (71)$$

$$\text{Re} = \rho V^{2-n}D^n \, m^{-1} \, 8^{1-n} \qquad (72)$$

(NB The Generalised Reynolds number is given by

$$\text{Re}_g = \rho V^{2-n'}D^{n'}(m')^{-1}8^{1-n'}, \quad \text{see below})$$

In examining any formula for the flow of a non-Newtonian fluid, it is worth simplifying it to the two limiting cases, one of which is the Newtonian limit ($n = 1$ or $\tau_y = 0$), and the other the simple plastic ($n = 1$ or $\tau_y = 0$), and the other the simple plastic ($n = 0$ or $\mu_p = 0$). For the latter situation all the above expressions for Reynolds number become:

$$\text{Re} = 8\rho V^2/\tau_y \quad \text{and} \quad m = \tau_y = \text{wall shear stress} \tag{73}$$

There remains a difference between equations 68 and 72 despite their agreement at the limits. It is represented by the following function of $n$:

$$f(n) = [4n/(3n + 1)]^n = 8^n[n/(6n + 2)]^n \tag{74}$$

Evaluation of this function shows equation 72 to under predict the friction factor by 13 per cent at the worst case of $n = 0.24$.

One way around the problem has been proposed by Dodge and Metzner (99), who used a tube viscometer to obtain their basic data, relating pressure gradient and mean velocity as follows:

$$\nabla p = (4/D) \, m' \, (8V/D)^{n'} \tag{75}$$

This equation is applicable to any fluid whose properties are independent of time. In general $m'$ and $n'$ are functions of $8V/D$. Equation 72 applies with the constants $m$ and $n$ replaced by the variables $m'$ and $n'$. The dimensionless quantity so formed is the Generalised Reynolds number, with no limitations relating to specific mathematical models, other than that the properties should be independent of time.

Equation 75 involves no errors in the transfer of data from one set of laminar tube-flow parameters to another. If a rotational viscometer such as the cone-and-plate type is used, the true stress versus strain rate curve may be obtained, at the expense of a more complex transfer of the information to pipe friction prediction. For a three-parameter (Herschel–Bulkley) model, the resulting complexity can be extreme, even for laminar flow calculations. Cheng (100) put forward a design equation for the friction factor for such a generalised condition ($\tau = \tau_y + m\dot{\gamma}^n$):

$$f = 64/(\text{Re } A(1 - B\,(1 + C\,(1 + D)))) \tag{76}$$

where

$$\text{Re} = \rho D^n V^{2-n}\, m^{-1}\, [4n/(3n + 1)]^n\, 8^{1-n}, \tag{68}$$

$$A = 1 - 8\text{He}(f\,\text{Re}^2)^{-1}, \tag{77}$$

$$B = 8\text{He}\,(2n - 1)^{-1}\,(F\text{Re}^2)^{-1}, \tag{78}$$

$$C = 16n\text{He}\,(n + 1)^{-1}\,(f\text{Re}^2)^{-1}, \tag{79}$$

$$D = 1 + 8n\text{He}\,(f\text{Re}^2)^{-1} \tag{80}$$

and

$$He = \tau_y \, Re^2 \, \rho^{-1} \, V^{-2} \quad \text{(Hedstrom number)} \quad (81)$$

In view of the above implicit and complicated expression for $f$, and the fundamental difficulties involved in identifying the 'correct' Herschel–Bulkley parameters, the use of a tube viscometer for the prediction of tube flow friction has much to recommend it. For the limiting case of $n = 1$, equation 76 represents the exact solution for Bingham plastics. The exact Casson solution, however, cannot be obtained by any such simplification. It is worthy of note that equation 76 can be transformed into an explicit equation for flow rate in terms of pressure gradient, but not the other way around.

Much of the tedium associated with solving the above equations can be avoided by using a computer. It is also possible to simplify the programming such that the flow rates can be obtained by numerical intergration of simpler functions. With the aid of a computer it is now no longer necessary to use simplified analytical formulae since numerical solutions can be obtained quickly. The above is included however to illustrate the background and enable the reader to understand some of the papers written on the subject.

The turbulent flow theory for Bingham plastic and power law fluids was pioneered by the respective contributions of Hedstrom (101) and Dodge and Metzner (93). Their results appear in Figure 23, showing both laminar and turbulent regimes. Instead of using equation 70 for the Bingham plastic Reynolds number, it is preferable to formulate it from the plastic viscosity alone. This was supported by both Hedstrom (101) and Bowen (102) who claimed that the turbulent flow of Bingham plastics differed little from the Newtonian case. A more exact treatment by Hanks and Dadia (103) found this to be strictly true for Hedstrom numbers around $10^6$. Below this value, the Bingham plastic friction factors were less than those for Newtonian fluids, and above if they were greater. However, since the total range of $f$ at $Re = 10^6$ was less than a factor of two between $He = 10^4$ and $He = 10^9$, the Newtonian curve could be used with confidence. The effect of pipe roughness would be expected to improve the agreement.

Power law media were correlated by Dodge and Metzner (99) using equation 75 to model the rheology, and equation 72 to provide the Generalised Reynolds number. With good agreement over a limited Reynolds number range (approximately that indicated by Figure 23), they proposed the following semi-theoretical equation in terms of Fanning friction factor ($f' = f/4$):

$$f'^{-0.5} = [4n'^{-0.75} \log_{10}(Re_g \, f'^{(1-n'/2)})] - 0.4/n'^{1.2} \quad (82)$$

Equation 82 reduces to the Karman–Nikuradse limit for smooth pipe Newtonian flow (equation 8, section 2.1) if $n' = 1$. The non-Newtonian fluids used to check the above equation included two types of polymer solution (CMC and Carbopol from the Hercules and Goodrich Chemical Companies), and a fine clay suspension (Attasol–Goodrich). The CMC produced some anomalously low friction factors attributed to its viscoelastic properties. Carbopol is also slightly viscoelastic so that some of its results may have underestimated the friction factor despite agreement with equation 82.

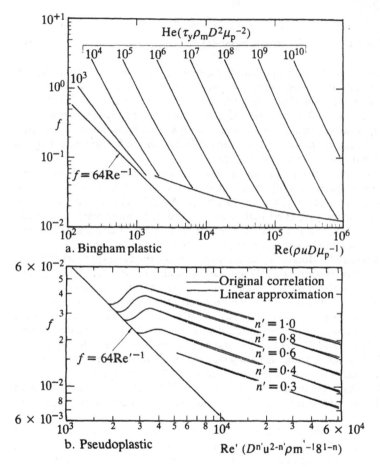

**Figure 23.** Non-Newtonian friction factor for laminar and turbulent flow (101, 99)

In an assessment of the various methods available for predicting turbulent non-Newtonian friction, Kenchington (104) favoured the Kemblowski and Kolodziejski (105) formula, developed from experiments with kaolin slurries. They contended that above a Reynolds number of $3 - 4 \times 10^4$, pseudoplastic fluids should behave like Newtonian fluids and developed their correlations around this idea. Use of Newtonian correlations requires the identification of a representative shear rate, presumably appropriate in fully turbulent flow. The underprediction of the Dodge-Metzner correlation could have been due to pipe roughness, not included in the Kemblowski–Kolodziejski formula. Furthermore, the latter appears to be based on the Blasius, rather than the Karman–Nikuradse equation for Newtonian flow.

For Casson model fluids, Hanks (106) developed a theory for laminar friction factor and transition from laminar to turbulent flow, based on a generalised theory for the flow stability of fluids with a yield stress (107,108). In terms of Darcy friction factor, his formula for laminar flow was as follows:

$$f = 64\,Re^{-1}\,(1 - 6.465\,Ca^{0.5}\,f^{-0.5}\,Re^{-1} + 2.667\,Ca\,Re^{-2} - 195\,Ca^4\,f^{-4}\,Re^{-4})^{-1} \quad (83)$$

where $Re = \rho V D \mu_p^{-1}$ and $Ca = D^2 T_y \rho \mu_p^{-1}$   (Casson number)

There is some similarity in the basic features of equations 83 and 76, indicating how the assumption of a yield stress complicates laminar flow calculations, unless tube viscometers are used for pipeflow predictions. Hanks (106) commented that there is a need for viscometers with much wider strain rate ranges. Small errors in the basic rheological model could compound seriously in a calculation of the complexity represented by equation 83.

The conclusion of this section concerns the drag-reducing features of viscoelastic polymer solutions, beginning with the mechanics whereby they achieve these effects, illustrated in Figure 24. Imagine an eddy, the central unit of turbulences, newly formed at the solid boundary of a flowing stream. Viscosity responds to the local increase in shear by suppressing the eddy, whilst inertial effects, tend to cause the eddy to grow and to be lifted from the wall (Figure 24a). This is why turbulence occurs at a critical Reynolds number. The elastic influence (Figure 24b) tends, like viscosity, to suppress eddy formation and growth by virtue of the curvature of streamlines along which normal stresses have developed. Unlike viscosity, the normal stress makes no contribution to the drag and the

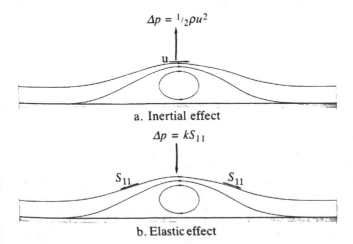

$$\Delta p = {}^1/_2 \rho u^2$$

u

a. Inertial effect

$$\Delta p = k S_{11}$$

$S_{11}$    $S_{11}$

b. Elastic effect

**Figure 24.**  Mechanism for turbulence suppression by drag-reducing polymers

net result is an extension of the Reynolds number range of laminar flow and reduced drag in turbulent flow. These features have proven of interest as a means of reducing mixing impeller power consumption as well as friction factor reduction in pipe flow. The phenomenon must be distingished from turbulence suppression by a pseudoplastic medium whose high viscosity away from the boundary layer decreases the effective Reynolds number.

Drag reduction by viscoelastic fluids occurs only in transition or turbulent flow. Viscoelasticity will not necessarily suppress the wake behind a bluff body, such as a mixing impeller blade; nor will it reduce the friction due to laminar shear. In a study of such effects, Sylvester and Rosen (109) demonstrated that viscoelastic fluids in fully developed laminar flow produced the same drag as an inelastic fluid of the same apparent viscosity. In the entrance regions, however, the friction loss imposed by a viscoelastic fluid was of order five times its Newtonian equivalent. This is understandable considering that normal stresses represent the storage of elastic energy. Initially, the source of such energy must be the driving pressure difference. Due to the fact that the stored energy cannot be returned to the flow, viscoelastic entrance effects must be written off as a loss.

As Maskelkar *et al.* (110) note from their experiments with smooth turbulent flow at 5–100 ppm polyethylene oxide solutions, drag reduction factors of order four can be achieved. Unfortunately, however, viscoelastic polymers lose their properties under prolonged or vigorous shear, as might occur in a pump. Therefore, they must be injected at some point downstream. As noted in a study by El Riedy and Katto (111), the polymer can tend to become more concentrated at the pipe axis than the walls, with a concentration profile not entirely matched to that of the turbulence. In parallel with the polymer effect, Zagustin and Power (112) noted drag reduction factors of order two in experiments with 15 ppm suspensions of finely ground sand (7 $\mu$m). In common with many fluid mechanical phenomena, its mechanism for achieving this was not entirely clear. If, however, the effect can be repeated in commercial slurry transport, there will be a good reason for exploring it further.

## 4.6   COARSE PARTICLE TRANSPORT IN A NON-NEWTONIAN CARRIER FLUID

Bingham plastic flow has found application in the prediction of flow-rate pressure-loss functions for coarse particle coal and colliery waste. Duckworth (113) has described how the flow of slurries composed of coal, top size approximately 20 mm, and supported in a mixture of coal fines and water (maximum size 90 $\mu$m) could be correlated by the use of the Buckingham equation; the pipe flow solution for a Bingham plastic. It was found that the plastic viscosity and yield stress of the mixture appeared to increase as the proportion of coarse coal was increased. It was considered that these effects were apparent rather than absolute, nevertheless they were sufficiently systematic for simple correlations to be effective. Gandhi (114) suggested that the coarse material was being transported within a central plug and it was this that resulted in apparent changes to the carrier fluid viscosity. Brown (115) has shown that the Herschel–Bulkley equation can be used to describe both a coarse and a fine coal mixture, as well as the fine component by itself. This led to three flow prediction methods being proposed. Two of the three methods involved making rheological measurements on the mixtures of coarse and fine material. The first prediction method regarded the mixture as a homogeneous slurry in a manner similar to that of Duckworth. The second supposed that the material migrated towards the pipe axis to form a central plug which was lubricated by a coarse-particle depleted layer at the wall. It was suggested that the migration effect also takes place in the viscometer, which is operated in the infinite-sea mode, so that the shear stresses recorded can be attributed to a similar layer. The third supposed that each coarse particle was surrounded by a layer of fines which modified the size of the plug which could be formed. The lubricating annulus was considered to be composed of the fine component of the slurry. None of the models was entirely satisfactory.

In separate work (116) it was found that a pseudoplastic equation could be applied to a colliery waste mixture. However, this paper also describes a method in which the shear stresses between the mixture and a moving surface can be measured and these data employed in a sliding bed model which supposed that the flow in the upper region was composed of carrier fluid alone. The results of this work were not entirely conclusive, partly due to the slurry tending to be of the homogeneous type. Further analysis based on the flow regime being described by plug flow (117) suggested that for this slurry it was not the case that the pressure loss could be described by shear in an annulus of carrier fluid alone. This indicated that the technique of simulating the shear stresses at the pipe wall experimentally should be pursued as a method of predicting pressure loss characteristics of these slurries.

## 4.7 APPLICATION TO SLURRY ENGINEERING

If the rheological parameters of a particular suspension are known, they can be applied to the prediction of head gradient in pipeflow. The concerns of this subsection are rather broader, with the emphasis on using rheological ideas in the engineering of head loss reduction. The following possibilities emerge:

- Produce a carrying medium with sufficient yield stress to stabilise an otherwise settling suspension, now pumpable in laminar flow.
- If the suspension can be made thixotropic, select a pump with strong shearing action to break down its structure prior to transport.
- Employ a deflocculating agent to reduce the laminar flow friction factor, unless coarse particles need to be supported.
- Inject air into the line to reduce the effective friction length and increase the velocity, reducing the friction for shear thinning slurries.
- Reduce pipe friction in turbulent flow with soluble polymer additives, ensuring, however, that the suspension remains stable.

An alternative to fines as a means of thickening the carrier is a novel proposal by Bradley (118) to use a concentrated emulsion of water-in-oil. Known as HIPR (high-internal-phase-ratio) emulsions, they acquire a yield strength by a development of structure in the external phase. At a water concentration of order 90 per cent, the internal phase droplets take up so much volume that the 10 per cent of oil phase must form a structured network in order to remain external. The system requires specialised emulsifying agents to stabilise it at such high concentrations and the risk of inversion is great. Nevertheless, at the downstream end, inversion could be used to ease coal separation, provided the carrier could be reformed without too high a cost in chemicals.

Investigating the effect of fines on sand-water flows, Bruhl and Kazanskij (119) found that sodium bentonite in suspension could reduce the drag in dredge spoils by up to 50 per cent. They also noted a 4–6 per cent reduction in turbulence intensity. Perhaps the reason that the pressure loss reduction did not match the turbulence reduction could be due to settling effects as noted by Kenchington (29) for sand in kaolin suspensions. In later experiments with the same medium, Kenchington (120) found that the kaolin greatly reduced the frictional effects of sand in the sliding bed regime. The total solids content in his system was very high, with 40 per cent by weight kaolin and 43 per cent sand. He attributed the reduction in excess pressure gradient to be due to the lubricating effect of the kaolin suspension in the bed. As Wilson *et al.* (33) found contact friction to be the dominant feature of a sliding bed, the observed improvement is not surprising.

When settling is important, the solids contribution to head loss can often be assumed to be proportional to volumetric concentration. In stable non-Newtonian suspensions the relationship is rather more complex. Sims (121) noted that iron ore suspension could be maintained quite fluid up to 70 per cent by weight, beyond which both Bingham yield stress and plastic viscosity increased sharply with concentration. Looking at the optimisation of phosphate ore transport, Lazarus (122) recommended high concentrations (40 per cent by volume) and low velocities. It should be stressed that porous solids may raise the effective volumetric concentration, which, considered in terms of interparticle contact, will increase viscosity and perhaps yield stress.

Want *et al.* (123) sought maximum pumpable concentrations of red mud, a waste product from the process of alumina recovery from bauxite. This entailed environmental, as well as commercial, benefits as the size of the settling lagoons could be reduced, along with the risk of seepage of caustic soda into the water table. At concentrations of order 70 per cent they found the mud to be strongly thixotropic. They utilised this property in their preparation/pumping system, ensuring a well sheared mixture for transport. They could also utilise the converse process, increase of yield stress under undisturbed ageing, by discharging the waste as one layer of a self-supporting tailings dam for which structural strength was an advantage. The ageing effect is not always desirable, as reported by Bhattacharya (124) for certain types of coal-oil mixtures, for which agitated storage can be required for maintenance of pumpability.

Deflocculation represents the chemical means whereby suspension structure can be destroyed. Briefly recapitulating the mechanism explained at the end of section 4.3, the reader is referred to Figure 25, representing two particles in suspension. Each is surrounded by a negative surface charge, the electrostatic repulsive effect of which is termed the 'zeta potential' measured in millivolts. Increased magnitude of zeta potential leads to decreased

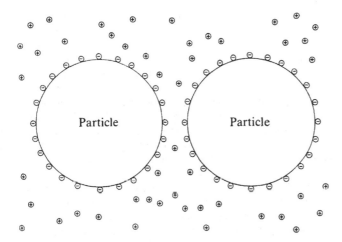

**Figure 25.** Surface charge on particles of a deflocculated suspension

agglomeration, hence a less viscous suspension for a given solids concentration. Horsley and Reizes (125) studied the effect of the suspension chemistry for sand (ground to below 100 $\mu$m) in aqueous media. The effect of increased pH on transition velocity (viscosity) can be quite dramatic. For a 33 per cent suspension by volume, a twofold reduction in pressure loss could be obtained, with a fivefold reduction for a 43 per cent mixture.

The chemistry of each system is quite specific and the right additive must be found for the suspended material. For gold slime slurries, Sauermann (126) recommends sodium polyphosphate ($Na_5P_3O_{10}$) at 0.2 per cent by weight for a 74 per cent by weight slurry. For the same system, Horsley (127) found that calcium hydroxide or calcium nitrate could be used to neutralise the additive at the downstream end, should the structural strength of the mixture need to be restored for dam wall support. Looking at a number of different slurries, Sikorski *et al.* (128) showed that the viscosity of limestone cement feed could be reduced from 7000 Cp at 70 per cent concentration to 2500 Cp at 80 per cent at additive (sodium phosphate and carbonate) levels of order 0.1 per cent. They found the same additive combination to work well for coal slurries.

The chemical cost of deflocculation can be avoided by making special use of the shear-thinning features of flocculated slurries. The head loss associated with the laminar flow of a Newtonian fluid is proportional to both pipeline length and flow velocity. A pseudoplastic fluid on the other hand, develops a pressure gradient proportional to a lower power of velocity. This means that if air is injected into a line carrying a pseudoplastic medium, the reduction in friction length more than offsets the increase in velocity needed to maintain the mixture flow rate. Carlton *et al.* (129) demonstrated this for the pumping of thick pastes. They developed a theory for the reduced friction by assuming perfect plug flow. Their results, indicating a departure from their theory at a total flow rate of about 0.006 $m^3$/s.

This was the point at which the authors suspected a departure from perfect piston flow, it also represented the optimum operating condition. A similar study by Heywood and Richardson (130) indicated drag reduction of as much as 85 per cent for pseudoplastic kaolin slurries and 80 per cent for Bingham plastic anthracite suspensions. They recommended that the slurry Reynolds number not exceed 1000 if turbulent flow problems are to be avoided.

It is to drag reduction in turbulent flow that viscoelastic polymer additives are directed. Kuzuhara and Shakouchi (131) found that addition of 500 ppm of polyethylene oxide could significantly reduce the friction loss for slurries of coarse sand, fine sand, or fly ash. The reduction in turbulence levels raised the critical velocity for settling but at all velocities above this the polymer reduced drag. At lower velocities, however, the reduced critical condition for the pure water carrier led to lower pumping losses for this system. Further research should lead to some quantification of the trade-off between increased critical velocity and reduced losses due to turbulence for polymer systems. Herod and Tiederman (132) investigating drag reduction for dredge spoils with polyacrylamide additive, noted an 80 per cent reduction to be attainable at the higher velocities. They recommended polymer injection downstream of the pump to avoid destroying its viscoelasticity.

Various applications of non-Newtonian physics have been outlined, including pumping at high solids concentrations, stabilisation of coarse particle systems, use of thixotropy, chemical viscosity reduction, drag reduction by air injection, and drag reduction by polymer injection. Each has its special advantages and careful thought must be given to the basic system before applying a drag reduction technique. Though few of the techniques would appear to work in combination with each other, they encompass a wide total range of slurry engineering applications.

## 5.   COMPUTATIONAL SOLUTION OF TWO-PHASE FLOW

Much of the work concerning slurries has stemmed from the fields of chemical, civil and mechanical engineering. In the field of fluid mechanics, which spans these engineering disciplines, there has been considerable emphasis on the computational solution of single-phase flow problems. These techniques are now being extended to two-phase slurry flow and two examples are given below. So far the physical representation of the slurries has only covered the simpler types. No doubt this will be rectified with the passage of time. When developed these methods will enable three-dimensional geometries to be analysed which will be most valuable for coal-conversion and other technologies.

An interesting computational approach has been undertaken by Abbas *et al.* (133). They considered the relative merits of Eulerian and Lagrangian techniques for two-phase flow analysis. Whereas the Eulerian system employs time averaged transport equations, the Lagrangian method involves tracking of the dispersed phase. The tracking may be of individual or representative particles. The authors prefer the Lagrangian system since they show that for multi-specie systems computer requirements are considerably reduced. The solution of the equations depends on a knowledge of the particle diffusion coefficients for which estimates were made. A comparison of predictions with experimental results from free jet flow was encouraging although the measured particle spread in the jet was greater than predicted.

Another example concerns where both the fluid and solid phases are regarded as continua, Chen and Wood (134). The mean velocities of both phases are taken as equal although a local velocity difference is assumed in order to calculate particle forces by velocity difference and response time. The time averaged equations are formed and hence further assumptions of the flow field must be introduced in order to generate sufficient equations for a solution to be obtained. This is the so called 'closure' requirement and is brought about by the loss of information caused by the time averaging of the differential equations. Time averaging is introduced in order to reduce the computational work to practical proportions. The flow field is characterised by the $k$–$\epsilon$ model, where $k$ represents the kinetic energy of the turbulent velocity fluctuations and $\epsilon$ is a measure of its dissipation rate. In this model the diffusivity is allowed to vary and is computed from a product of the turbulence length and velocity scales. The details of the model are those proposed by Jones and Launder (135).

In comparisons with experimental results for a spreading jet good agreement was obtained for velocity profiles. The turbulence distribution across the jet was not so well predicted, especially beyond the end of the potential core.

## 6. CONCLUSIONS AND RECOMMENDATIONS

The economic pressure towards increased solids concentration and particle size amounts to burning the candle at both ends. High concentrations can only be pumped if the viscosity and yield stress can be controlled. If excessive settling or flow velocities are to be avoided, coarse particle transport requires a relatively heavy or thick carrying medium. An understanding of the limits of concentration and size is one edge of the state-of-the-art in slurry transport. It is a multidisciplinary problem and might be tackled by a number of different routes. The items below indicate some of the areas in need of attention:

- For settling suspensions in heterogeneous flow, the Durand formula and its equivalents remain the most widely accepted and used. However, they do not match the limiting cases of sliding bed and homogeneous transport regimes. A sound, physically based model which could be simplified to a design formula would be a great contribution. Some solutions now divide the solids into contact load and suspended load, with further division of the latter into homogeneous and heterogeneous components. Provided all the forces act between the particles and between them and the fluid, and the resulting mixture and the pipe wall, then it should be possible to provide accurate predictions for settling slurries over a wide range of conditions.
- The turbulent flow of non-Newtonian media is inadequately understood. Similar to the situation for settling media, the correlations of greatest practical use appear to have a weak physical foundation. Very little is known about the flow of settling suspensions in a

non-Newtonian carrier. The critical velocity can be increased over that due to a pure water but friction under subcritical conditions (sliding bed) seems to be reduced.

• Many possibilities for pressure loss reduction exist. These include use of the wall migration effect for concentrated suspensions in plug flow, chemical deflocculation of agglomerating mixtures, air injection for shear-thinning media, pre-pump shearing for thixotropic systems, and the addition of drag-reducing soluble polymers for turbulence suppression.

## 7.  REFERENCES

1.   Stern, D. J. (June 1971) 'Pipeline transportation of solids', *The Certificated Engineer*, pp. 119–40.
2.   Parkes, D. M. and Lindsay, J. (March 1977) 'Hydraulic transportation and dewatering of coal', *Transactions of the AIME Society of Mining Engineers*, vol. 262, pp. 6–10.
3.   Rigby, G. R. (1982) 'Slurry pipelines for the transportation of solids', *Mechanical Engineering Transactions*, Paper M1173, pp. 181–9, I.E. Aust.
4.   Lee, H. M. (March 1982) 'An overview of proposed coal slurry technologies and their cost saving applications', *Proceedings Seventh International Technical Conference on Slurry Transportation, Lake Tahoe, Nevada*, pp. 217–23 Slurry Transport Association (USA).
5.   Shook, C. A. (February 1976) 'Developments in hydrotransport', *The Canadian Journal of Chemical Engineering*, vol. 54, pp. 13–25.
6.   Lazarus, J. H. (August 1978) 'Comparison of suspended sediment flow and capsule transportation in pipelines', *The South African Mechanical Engineer*, vol. 28, pp. 319–26.
7.   Thomas, A. D. and Flint, L. R. (December 1974) 'Pressure drop prediction for flow of solid-liquid mixtures in horizontal pipes', *Fifth Australasian Conference on Hydraulics and Fluid Mechanics*, Christchurch, N.Z., pp. 111–18.
8.   Kazanskij, I. (May 1978) 'Scale-up effects in hydraulic transport – theory and practice', *Hydrotransport 5*, Hanover, F.R.G., Paper B3, pp. 47–80, BHRA.
9.   Duckworth, R. A. (August 1978) 'The hydraulic transport of materials by pipeline', *The South African Mechanical Engineer*, vol. 28, pp. 291–306.
10.  Pitts, J. D. and Hill, R. A. (March 1978) 'Slurry pipeline technology – An update and prospectus', *Sixth International Pipeline Technology Convention (Interpipe 78)*, Houston, Texas, pp. 166–207.
11.  Sandhu, A. S. and Weston, M. D. (March 1980) 'Feasibility of conversion of existing oil and gas pipelines to coal slurry transport systems', *Proceedings Fifth International Technical Conference on Slurry Transportation*, Lake Tahoe, Nevada, pp. 85–92, Slurry Transport Association, USA.
12.  Thomas, A. D. (1970) 'Scale-up methods for pipeline transport of slurries', *International Journal of Mineral Processing*, vol. 3, pp. 51–69, (Elsevier).
13.  Stevens, G. S. (November 1969) 'Pipelining solids: The design of long-distance pipelines', *Symposium on Pipeline Transport of Solids*, Toronto, Canadian Society of Chemical Engineers.
14.  Shook, C. A. (November 1969) 'Pipelining solids: The design of short-distance pipelines', *Symposium on Pipeline Transport of Solids*, Toronto, Canadian Society of Chemical Engineers.
15.  Miller, D. S. (1978) *Internal Flow Systems*, BHRA Fluid Engineering Series, vol. 5, BHRA.
16.  Casson, N. (1959) 'A flow equation for pigment-oil suspensions of the printing ink type', *Rheology of Disperse Systems*, p. 59, Pergamon Press, London.
17.  Perry, R. H. (1973) *Chemical Engineers' Handbook (Fifth Edition)*, McGraw Hill Book Company.

18. Silvester, R. S. and Langley, P. (September 1981) *Pressure surge analysis of Balikpapan Refinery loading lines*, BHRA Research Report, Project Number RP 11103.
19. Thorley, A. R. D. (April 1980) 'Surge suppression and control in pipelines', *The Chemical Engineer*, pp. 222–4.
20. Thorley, A. R. D. and Hwang, L. Y. (September 1979) 'Effects of rapid change in flow rate of solid-liquid mixtures', *Hydrotransport 6*, Canterbury, UK, Paper D5, pp. 229–42, BHRA.
21. Thorley, A. R. D. (August 1980) 'Transient propagation in slurries with hold-up', *Proceedings ASCE, Journal of the Hydraulics Division*, vol. 106, HY8, pp. 1353–65.
22. Kao, D. T. and Wood, D. J. (May 1978) 'Pressure surge generation due to rapid shut-down of pipelines conveying slurries', *Hydrotransport 5*, Hanover, F.R.G., Paper E1, pp. 1–14, BHRA.
23. Bechteler, W. and Vogel, G. (August 1982) 'Pressure wave velocity in slurry pipelines', *Hydrotransport 8*, Johannesburg, S.A., Paper H2, pp. 383–98, BHRA.
24. Streeter, V. L. (1971) *Fluid Mechanics*, *(Fifth Edition)*, McGraw Hill-Kogakusha Limited, Tokyo.
25. Rayan, M. A. (1980) 'Influence of solid particles on some turbulent characteristics', *Multiphase Transport Fundamentals, Reactor Safety Applications* (T.N. Verizoglu ed.), vol. 4, pp. 1969–91.
26. Yang, C. T. (October 1973) 'Incipient motion and sediment transport', *Journal of the Hydraulics Division (ASCE)*, HY10, pp. 1679–1704.
27. Durand, R. and Condolios, G. (1952) 'The hydraulic transport of coal and solids in pipes', *Colloquium on Hydraulic Transport*, National Coal Board, London.
28. Weber, M. (February 1981) 'Principles of hydraulic and pneumatic conveying in pipes', *Bulk Solids Handling*, vol. 1, no. 1, pp. 57–63.
29. Kenchington, J. M. (May 1976) 'Prediction of critical conditions for pipeline flow of settling particles in a heavy medium', *Hydrotransport 4*, Alberta, Canada, Paper D3, pp. 31–48, BHRA.
30. Thomas, A. D. (September 1979) 'The role of laminar/turbulent transition in determining the critical deposit velocity and the operating pressure gradient for long distance slurry pipelines', *Hydrotransport 6*, Canterbury, UK, Paper A3, pp. 13–26, BHRA.
31. Newitt, D. M., Richardson, J. F., Abbot, M. and Turtle, R. B. (1955) 'Hydraulic conveying of solids in horizontal pipes', *Trans. I. Chem. E.*, vol. 33, pp. 93–110.
32. Babcock, H. A. (September 1970) 'The sliding bed flow regime', *Hydrotransport 1*, Warwick, UK, Paper H1, pp. 1–16, BHRA.
33. Wilson, K. C., Streat, M. and Bantin, R. A. (September 1972) 'Slip-model correlation of dense two-phase Flow', *Hydrotransport 2*, Warwick, UK, Paper C4, pp. 1–10, BHRA.
34. Zandi, I. and Govatos, G. (May 1967) 'Heterogeneous flow of solids in pipelines', *Proceedings ASCE, Journal of the Hydraulics Division*, vol. 93, HY3, pp. 145–59.
35. Wasp, E. J., Kenny, J. P. and Gandhi, R. L. (1977) *Solid-Liquid flow slurry pipeline transportation*, Clausthal, Germany, Trans Tech Publications.
36. Wilson, K. C. and Watt, W. E. (May 1974) 'Influence of particle diameter on the turbulent support of solids in pipeline flow', *Hydrotransport 3*, Colorado, USA, Paper D1, pp. 1–9, BHRA.
37. Wilson, K. C. (May 1976) 'A unified physically based analysis of solid-liquid pipeline flow', *Hydrotransport 4*, Alberta, Canada, Paper A1, pp. 1–16, BHRA.
38. Wilson, K. C. (September 1979) 'Deposition-limit nomogram for particles of various densities in pipeline flow', *Hydrotransport 6*, Canterbury, UK, Paper A1, pp. 1–12, BHRA.
39. Clift, R., Wilson, K. C., Addie, G. R. and Carstens, M. R. (August 1982) 'A mechanistically based method for scaling pipe-line tests for settling slurries', *Hydrotransport 8*, Johannesburg, S.A., Paper B1, pp. 91–101, BHRA.
40. Wasp, E. J., Regan, T. J., Withers, J., Cook, D. A. C., and Clancey, J. T. (July 1963) 'Cross-country coal pipeline hydraulics', *Pipeline News*, vol. 35, no. 7, pp. 20–8.

41. Williamson, J. R. (August 1978) 'Application of theory to practice in the heterogeneous conveying of solids in pipelines', *The South African Mechanical Engineer*, vol. 28, pp. 327–34.
42. Roco, M. C. and Shook, C. A. (August 1983) 'Modelling of slurry flow: the effect of particle size', *The Canadian Journal of Chemical Engineering*, vol. 61, no. 4, pp. 494–503.
43. Bagnold, R. A. (1956) 'The flow of cohensionless grains in fluids', *Proceedings of the Royal Society A*, 249, 235–97.
44. Roco, M. C. and Shook, C. A. (1984) 'Computational method for coal slurry pipelines with heterogeneous slurry distribution', *Powder Technology*, vol. 39, no. 2, 156–76.
45. Sinclair, C. G. (November 1961) 'The limit deposit-velocity of heterogeneous suspensions', *London I. Chem. E. Meeting on Interaction between Fluids and Particles*, pp. 78–86.
46. Eroclaņi, D., Ferrini, F. and Arrigoni, V. (September 1979) 'Electric and thermic probes for measuring the limit deposit velocity', *Hydrotransport 6*, Canterbury, UK, Paper A3, pp. 27–42, BHRA.
47. Toda, M., Konno, H. and Saito, S. (November 1980) 'Simulation of limit-deposit velocity in horizontal liquid-solid flow', *Hydrotransport 7*, Sendai, Japan, Paper J2, pp. 347–58, BHRA.
48. Weber, M. and Gödde, E. (May 1976) 'Critical velocity as optimum operating velocity in solids pipelining', *Hydrotransport 4*, Alberta, Canada, Paper D2, pp. 17–30, BHRA.
49. Gödde, E. (May 1978) 'To the critical velocity of heterogeneous hydraulic transport', *Hydrotransport 5*, Hanover, F.R.G., Paper B4, pp. 81–100, BHRA.
50. Lazarus, J. H. and Nielson, I. D. (May 1978) 'A generalised correlation for friction head losses of settling mixtures in horizontal smooth pipes', *Hydrotransport 5*, Hanover, F.R.G., Paper B1, pp. 1–32, BHRA.
51. Lazarus, J. H. (August 1982) 'Optimum specific power consumption for transporting settling slurries in pipelines', *Hydrotransport 8*, Johannesburg, S.A., Paper B4, pp. 123–32, BHRA.
52. Kazanskij, I. B. (September 1979) 'Critical velocity of depositions for fine slurries – new results', *Hydrotransport 6*, Canterbury, UK, Paper A4, BHRA.
53. Middelstadt, M. (March 1981) 'Hydraulic transport of coal and iron ore slurries', *Seatec*, vol. 3, Asian ports Development and Dredging Seminar, Singapore, Oxted, UK, Paper 7.4.
54. Lokon, H. B., Johnson, P. W. and Horsley, R. R. (August 1982) 'A "scale-up" model for predicting head loss gradients in iron ore slurry pipelines', *Hydrotransport 8*, Johannesburg, S.A., paper B2, pp. 103–10, BHRA.
55. Webster, I. W. and Sauermann, H. B. (August 1978) 'Pressure gradient scale-up methods for slurry pipelines', *The South African Mechanical Engineer*, vol. 28, pp. 312–17.
56. Vocadlo, J. J. and Charles, M. E. (September 1972) 'Prediction of pressure gradient for the horizontal turbulent flow of slurries', *Hydrotransport 2*, Warwick, UK, Paper C1, pp. 1–12, BHRA.
57. Wood, D. J. (September 1979) 'Pressure gradient requirements for re-establishment of slurry flow', *Hydrotransport 6*, Canterbury, UK, Paper D4, pp. 217–28, BHRA.
58. Okada, T., Hisamitsu, N., Ise, T. and Takeishi, Y. (August 1982) 'Experiments on restart of reservoir sediment slurry pipeline', *Hydrotransport 8*, Johannesburg, S.A., Paper H3, pp. 399–414, BHRA.
59. Takaoka, T., Hisamitsu, N., Ise, T. and Takeishi, Y. (November 1980) 'Blockage of slurry pipeline'; *Hydrotransport 7*, Sendai, Japan, Paper D4, pp. 71–88, BHRA.
60. Gandhi, R. L. and Aude, T. C. (May 1978) 'Slurry pipeline design – special considerations' *Hydrotransport 5*, Hanover, F.R.G., Paper J1, pp. 1–12, BHRA.
61. Kao, D. T. Y. and Hwang, L.Y. (September 1979) 'Critical slope for slurry pipelines transporting coal and other solid particles', *Hydrotransport 6*, Canterbury, UK, Paper A5, pp. 57–74, BHRA.
62. Shook, C. A., Rollins, J. and Vassie, G. S. (June 1974) 'Sliding in inclined slurry pipelines at shut-down', *The Canadian Journal of Chemical Engineering*, vol. 52, pp. 300–5.
63. Kortenbusch, W. (August 1982) 'Latest experience with hydraulic shaft transportation at the

Hansa hydromine', *Hydrotransport 8*, Johannesburg, S.A., Paper J5, pp. 571–483, BHRA.

64. Harzer, J. and Kuhn, M. (August 1982) 'Hydraulic transportation of coarse solids as a continuous system from underground production face to the end product in the preparation plant', *Hydrotransport 8*, Johannesburg, S.A., Paper J4, pp. 461–70, BHRA.

65. Maurer, H. and Mez, W. (August 1982) 'Hydraulic transport in coarse-grain material in the hard coal mining industry and experimental test results', *Hydrotransport 8*, Johannesburg, S.A., Paper J3, pp. 445–60, BHRA.

66. Newitt, D. M., Richardson, J. F. and Gliddon, B. J. (1961) 'Hydraulic conveying of solids in vertical pipes', *Trans. I. Chem. E.*, vol. 39, pp. 93–100.

67. Wilson, K. C., Brown, N. P. and Streat, M. (September 1979) 'Hydraulic hoisting at high concentration: A new study of friction mechanisms', *Hydrotransport 6*, Canterbury, UK, Paper F2, pp. 269–82, BHRA.

68. Streat, M. (August 1982) 'A comparison of specific energy consumption in dilute and dense phase conveying of solids-water mixtures', *Hydrotransport 8*, Johannesburg, S.A., Paper B3, pp. 111–22, BHRA.

69. Ayukawa, K., Kataoka, H. and Hirano, M. (November 1980) 'Concentration profile, velocity profile and pressure drop in upward solid-liquid flow through a vertical pipe', *Hydrotransport 7*, Sendai, Japan, Paper E2, pp. 195–202, BHRA.

70. Sellgren, A. (August 1982) 'The choice of operating velocity in vertical solid-water pipeline systems', *Hydrotransport 8*, Johannesburg, S.A., Paper D3, pp. 211–26, BHRA.

71. Blench, T., Galay, V. J. and Peterson, A. W. (November 1980) 'Steady fluid-solid flow in flumes', *Hydrotransport 7*, Sendai, Japan, Paper C1, pp. 89–100, BHRA.

72. Wilson, K. C. (November 1980) 'Analysis of slurry flows with a free surface', *Hydrotransport 7*, Sendai, Japan, Paper C4, pp. 123–32, BHRA.

73. Kuhn, M. (November 1980) 'Hydraulic transport of solids in flumes in the mining industry', *Hydrotransport 7*, Sendai, Japan, Paper C3, pp. 111–22, BHRA.

74. Wood, P. A. (November 1980) 'Optimisation of flume geometry for open channel transport', *Hydrotransport 7*, Sendai, Japan, Paper C2, pp. 101–10, BHRA.

75. Lau, H. H. (August 1979) 'A literature survey on slurry transport by flumes', Technical Note TN 2564.

76. Elsibai, N. G., Snoek, P. E. and Pitts, J. D. (July 1982) 'How particle size affects the cost of short distance coal-slurry pipelines', *Oil and Gas Journal*, vol. 80, no. 30, pp. 253–8.

77. Boothroyde, J., Jacobs, B. E. A. and Jenkins, P. (September 1979) 'Coarse particle hydraulic transport', *Hydrotransport 6*, Canterbury, UK, Paper E1, pp. 405–28, BHRA.

78. Jacobs, B. E. A. (December 1980) 'A recent study of developments in the hydraulic transport of coarse particles', *I. Mech. E. Conference and Workshop on Hydraulic Transport of Solids*, London Headquarters, Paper C319/80, pp. 31–6.

79. Schaart, J. and Verhoog, C. (December 1980) 'Operating experience of hydraulic transportation of solids in dredging', *I. Mech. E. Conference and Workshop on Hydraulic Transport of Solids*, London Headquarters, Paper C317/80, pp. 7–12.

80. Lawler, H. L., Pertuit, P., Tennant, J. D. and Cowper, N. T. (March 1978) 'Application of stabilised slurry concepts of pipeline transport of large particle coal', *Third International Technical Conference on Slurry Transportation* (Slurry Transport Association), pp. 164–78.

81. Sakamoto, M., Uchida, K. and Kamino, K. (August 1982) 'Transportation of coarse coal with fine particle-water slurry', *Hydrotransport 8*, Johannesburg, S.A., Paper J2, pp. 433–44, BHRA.

82. Johnson, M. (August 1982) 'Non-Newtonian fluid system design – some problems and their solutions', *Hydrotransport 8*, Johannesburg, S.A., Paper F3, pp. 291–306, BHRA.

83. Colombera, P. and Want, M. (December 1982) 'Bauxite residue disposal and rheology', *Chemical Engineering in Australia*, vol. ch. E7, no. 4, pp. 36–40.

84. Solomon, J., Nienow, A. W. and Pace, G. W. (March 1981) 'Flow patterns in agitated plastic

and pseudoplastic viscoelastic fluids', *Fluid Mixing Symposium (I. Chem. E.)*, Bradford, UK, Paper A1.

85.   Boger, D. V., Sarmiento, G., Tiu, C. and Uhlherr, P. H. T. (September 1978) 'Flow of mineral slurries', *The Australian Institution of Mining and Metallurgy Conference*, North Queensland, pp. 291–8.

86.   Van Wazer, J. R. *et al.* (1963) *Viscosity and flow measurement – A laboratory handbook of rheology*, J. Wiley & Sons.

87.   Schummer, P. and Worthoff, R. H. (1978) 'An elementary method for the evaluation of a flow curve', *Chem. Eng. Sci.*, vol. 33, no. 6, pp. 759–63.

88.   Huang, C.-R. (1971) 'Determination of the shear rates of non-Newtonian fluids from rotational viscometric data. Parts 1 and 2,' *Trans. Soc. Rheol.*, vol. 15, no. 1, pp. 25–30.

89.   Harris, J. (1971) 'Viscometers with curvilinear flow patterns. Part 1: Concentric cylinder viscometer', *Bull. Brit. Soc. Rheol.*, vol. 17, pp. 70–7.

90.   Jastrebski, Z. D. (August 1967) 'Entrance effects and wall effects in an extrusion rheometer during the flow of concentrated suspensions. *I & EC Fundamentals*, vol. 6, no. 3, pp. 445–54.

91.   Vinogradov, G. V., Froishteter, G. B. and Trilisky, K. K. (1978) 'The generalised theory of flow of plastic disperse systems with account of the wall effect', *Rheol. Act.*, *17*, pp. 156–65.

92.   Cheng, D.C.H. (September 1971) 'Appendices to report on sludge pipeline scheme: Appendix J', Binnie and Partners.

93.   Bhattacharya, S. N. (August 1980) 'Some observations on the rheology of time dependent suspensions on long-term storage', *Chemeca 80. Process Industries in the 80's, 8th Australian Chem. Eng. Conf.*, Melbourne, Australia, Section 5, pp. 95–7.

94.   Ackermann, N. L. and Hung, T. S. (March 1979) 'Rheological characteristics of solid-liquid mixtures', *A.I.Ch.E. Journal*, vol, 25, no. 2, pp. 327–32.

95.   White, J. L. and Huang, D. C. (1981) 'Dimensional analysis and a theory of elastic recovery following flow of elastic plastic viscoelastic fluids with application to filled polymer melts', *Journal of Non-Newtonian Fluid Mechanics*, vol. 9, pp. 223–33.

96.   Ulbrecht, J. (1975) 'Influence of rheological properties on the process of mixing', *Verfahrenstechnik*, vol. 9, no. 9, pp. 457–563.

97.   Ulbrecht, J. (June 1974) 'Mixing of viscoelastic fluids by mechanical agitation', *The Chemical Engineer*, pp. 347–67.

98.   Carreau, P. J., Patterson, I. and Yap, C. Y. (June 1976) 'Mixing of viscoelastic fluids with helical ribbon agitators I: Mixing time and flow patterns', *The Canadian Journal of Chemical Engineering*, vol. 54, pp. 135–42.

99.   Dodge, D. W. and Metzner, A. B. (June 1959) 'Turbulent flow of non-Newtonian systems', *A.I.Ch.E. Journal*, vol. 5, no. 2, pp. 189–204.

100.  Cheng, D. C. H. (September 1970) 'A design procedure for pipeline flow of non-Newtonian dispersed systems', *Hydrotransport 1*, Warwick, UK, Paper J5, pp. 77–97, BHRA.

101.  Hedstrom, B. O. A. (March 1952) 'Flow of plastics materials in pipes', *Industrial and Engineering Chemistry*, vol. 44, no. 3, pp. 651–6.

102.  Bowen, R. L. B. (July 1961) 'Scale-up for non-Newtonian fluid flow', *Chemical Engineering*, vol. 68, no. 14, pp. 147–50.

103.  Hanks, R. W. and Dadia, B. H. (May 1971) 'Theoretical analysis of the turbulent flow of non-Newtonian slurries in pipes', *A.I.Ch.E. Journal*, vol. 17, no. 3, pp. 554–7.

104.  Kenchington, J. M. (May 1974) 'An assessment of methods of pressure drop prediction for slurry transport', *Hydrotransport 3*, Colorado, USA, Paper F1, pp. 1–20, BHRA.

105.  Kemblowski, Z. and Kolodziejski, J. (April 1973) 'Flow resistances of non-Newtonian fluids in transitional and turbulent flow', *International Chemical Engineering*, vol. 13, no. 2, pp. 265–71.

106.  Hanks, R. W. (December 1981) 'Laminar-turbulent transition in pipeflow of Casson model fluids', *Trans. ASME*, vol. 103, pp. 318–21.

107. Hanks, R. W. (May 1963) 'The laminar-turbulent transition for fluids with a yield stress', *A.I.Ch.E. Journal*, vol. 9, no. 3, pp. 306–9.
108. Hanks, R. W. (January 1969) 'A theory of laminar flow stability', *A.I.Ch.E. Journal*, vol. 15, no. 1, pp. 25–8.
109. Sylvester, N. D. and Rosen, S. L. (November 1970) 'Laminar flow in the entrance region of a cylindrical tube', *A.I.Ch.E. Journal*, vol. 16, no. 6, pp. 964–72.
110. Mashelkar, R. A., Kale, D. D. and Ulbrecht, J. (July 1975) 'Rotational flows of non-Newtonian fluids: Part I – Turbulent flow of inelastic and viscoelastic fluids around discs', *Transactions of the Institution of Chemical Engineers*, vol. 53, no. 3, pp. 143–9.
111. El Riedy, O. K. and Latto, B. (December 1981) 'Dispersion of aqueous polymer solutions injected from a point source into turbulent water flows', *The Canadian Journal of Chemical Engineering*, vol. 59, pp. 662–7.
112. Zagustin, K. and Power, H. (May 1976) 'Drag reduction in turbulent rigid particle suspensions', *Hydrotransport 4*, Alberta, Canada, Paper B3, pp. 29–38, BHRA.
113. Duckworth, R. A., Pullum, L. Lockyear, C. F., and Addie, G. R. (October 1986) 'The pipeline transport of coarse materials in a non-Newtonian carrier fluid', *Hydrotransport 10*, Innsbruck, Austria, pp. 69–88, BHRA.
114. Gandhi, R. L. (March 1987) 'Effect of the rheological properties of fines fraction on slurry hydraulics', *Twelfth International Conference on Slurry Technology*, Washington DC, USA, Slurry Technology Association.
115. Brown, N. P. (October 1988) 'Three scale-up techniques for coal-water slurries', *Hydrotransport 11*, Stratford-upon-Avon, UK, BHRA.
116. Tatsis, A., Jacobs, B. E. A., Osborne, B. and Astle, R. D. (April 1988) 'A comparative study of pipe flow prediction for high concentration slurries containing coarse particles', *Thirteenth International Conference on Coal and Slurry Technology*, Washington DC, USA, Coal and Slurry Technology Association.
117. Jacobs, B. E. A. (May 1989) 'Interpretation of laboratory test data for prediction of pipeline pressure gradients of coarse particle slurries in laminar flow', *Sixth International Conference on Freight Pipelines*, Columbia-Missouri, USA Paper 6B/3, Hemisphere Publishing Co.
118. Bradley, G. M. (March 1982) 'Novel slurry transport medium', *Proceedings Seventh International Technical Conference on Slurry Transportation*, Lake Tahoe, Nevada, pp. 411–18, Slurry Transport Association (US).
119. Bruhl, H. and Kazanskij, I. (May 1976) 'New results concerning the influence of fine particles on sand-water flows in pipes', *Hydrotransport 4*, Alberta, Canada, Paper B2, pp. 19–28, BHRA.
120. Kenchington, J. M. (May 1978) 'Prediction of pressure gradient in dense phase conveying', *Hydrotransport 5*, Hanover, F.R.G., Paper D7, pp. 97–102, BHRA.
121. Sims, W. N. (December 1980) 'Evaluation of basic slurry properties as design criteria', *I. Mech. E. Conference and Workshop on Hydraulic Transport of Solids*, London Headquarters, Paper C324–80, pp. 73–82.
122. Lazarus, J. H. (November 1980) 'Rheological characterisation for optimising specific power consumption of a phosphate ore pipeline', *Hydrotransport 7*, Sendai, Japan, Paper D1, pp. 133–48, BHRA.
123. Want, F. M., Colombera, P. M., Nguyen, Q. D. and Boger, D. V. (August 1982) 'Pipeline design for the transport of high density bauxite residue slurries", *Hydrotransport 8*, Johannesburg, S.A., Paper E2, pp. 249–62, BHRA.
124. Bhattacharya, S. N. (August 1980) 'Some observations on the rheology of time dependent suspensions on long term storage', *Eighth Australian Chemical Engineering Conference (I.Chem.E., I.E.Aust.)*, Melbourne, Victoria, pp. 95–7.
125. Horsley, R. R. and Reizes, J. A. (August 1978) 'Variation in head loss gradient in laminar

slurry pipe flow due to changes in zeta potential', *The South African Mechanical Engineer*, vol. 28, pp. 307–11.

126.  Sauermann, H. B. (August 1982) 'The influence of particle diameter on the pressure gradients of gold slimes pumping', *Hydrotransport 8*, Johannesburg, S.A., Paper E1, pp. 241–8, BHRA.

127.  Horsley, R. R. (August 1982) 'Viscometer and pipe loop tests on gold slime slurry at very high concentrations by weight, with and without additives', *Hydrotransport 8*, Johannesburg, S.A., Paper H1, pp. 367–82, BHRA.

128.  Sikorski, C. F., Lehman, R. L. and Shepherd, J. A. (March 1982) 'The effects of viscosity reducing chemical additives on slurry rheology and pipeline transport performance for various mineral slurries', *Seventh International Technical Conference on Slurry Transportation* (Slurry Transport Association), Lake Tahoe, Nevada, pp. 163–73.

129.  Carleton, A. J., Cheng, D. C. H. and French, R. J. (September 1973) 'Pneumatic transport of thick pastes', *Pneumotransport 2*, Guildford, UK, Paper F2, pp. 11–19, BHRA.

130.  Heywood, N. I. and Richardson, J. F. (May 1978) 'Head loss reduction by gas injection for highly shear-thinning suspensions in horizontal pipe flow', *Hydrotransport 5*, Hanover, F.R.G., Paper C1, pp. 1–22, BHRA.

131.  Kuzuhara, S. and Shakouchi, T. (May 1978) 'Hydraulic transport of solids in special pipes', *Hydrotransport 5*, Hanover, F.R.G., Paper H3, pp. 37–50, BHRA.

132.  Herod, J. E. and Tiederman, W. G. (December 1974) 'Drag reduction in dredge-spoil pipe flows', *Proceedings ASCE* (Hydraulics Division), vol. 100, HY12, pp. 1863–6.

133.  Abbas, A. S., Koussa, S. S. and Lockwood, F. C. (August 1980) 'Prediction of the particle laden gas flows', *Eighteenth Symposium (International) on Combustion, Waterloo, Canada*, The Combustion Institute,pp. 1427–38.

134.  Chen, C. P. and Wood, P. E. (June 1985) 'A turbulence closure model for dilute gas-particle flows', *Canadian Journal of Chemical Engineering*, vol. 63, no. 3, pp. 349–60.

135.  Jones, W. P. and Launder, B. E. (1972) 'The prediction of laminarization with a two-equation model of turbulence', *International Journal of Heat and Mass Transfer*, vol. 15, no. 301.

136.  Jacobs, B. E. A. and Tatsis, A. (October 1986) 'Measurement of wall shear stresses for high concentration slurries', *10th International Conference on the Hydraulic Transport of Solids in Pipes*, Innsbruck, Austria, pp. 267–73, BHRA.

137.  Tatsis, A. and Jacobs, B. E. A. (March 1987) 'Slip model correlation for semi-stabilised slurries', *12th International Conference on Slurry Technology*, New Orleans, USA. Coal and Slurry Technology Association.

138.  Sieve, A. W. and Lazarus, J. H. 'Computer based mechanistic approach for modelling mixed regime slurry flows in pipes', *Hydrotransport 11*, Stratford-upon-Avon, UK. pp. 309–32, BHRA.

# 2. PUMPS AND PUMPING SYSTEMS

## 1. INTRODUCTION

The choice of pumps or pumping systems for slurry transport will depend not only on the flow and head required, suction conditions, type of installation and location, as for any other pump application, but also on the slurry flow regime and properties, i.e. mixture concentration, apparent viscosity, particle size and abrasivity. Hence, the range may be quite wide or relatively limited, depending on the particular application involved; as pointed out in Odrowaz-Pieniazek (1), each application must be considered individually, as no pump type is universally suitable.

This chapter deals with pumping equipment for handling mineral slurries only, e.g. sand, gravel, coal, china clay, ores and concentrates, furnace ash, chalk/clay, etc.; pumps for sewage duties and food products are not included. It covers pump types and construction, including the main conventional designs and some special types of both rotodynamic and positive displacement pumps, and also various feeder systems, with a brief discussion of the basic performance aspects and application to pipeline systems.

## 2. GENERAL ASPECTS OF PUMPING EQUIPMENT SELECTION

### 2.1 TYPES OF PUMPING SYSTEMS

There are two main groups from which pumps, or pumping systems, may be selected:

1. direct transport of the slurry by solids-handling pumps, and
2. indirect transport by a displacement or feeder system, using high pressure clean water pumps, or air compressors (e.g. air-lift pumps).

These groups may be sub-divided into various types of pumps and systems:

1. solids-handling pumps
   - rotodynamic
   - positive displacement
2. feeder systems
   - lock-hopper
   - rotary
   - pipe-chamber
   - jet pump
   - air-operated

Each type will be considered briefly in turn in sections 3, 4 and 5.

## 2.2  DISCUSSION OF EQUIPMENT SELECTION

In general, solids-handling pumps have the advantage of relatively low capital cost for short pipelines, but are subject to wear and hence maintenance costs may be high; they will also cause some degradation of the solids passing through.

Rotodynamic pumps, of which the centrifugal or radial-flow type is the most common in slurry service, are usually considered for the higher flow, lower head duties, whereas conversely, positive-displacement reciprocating types tend to be used for the lower flow, high pressure applications, e.g. long-distance pipelines (1,2,3). However, relatively high pressures may also be achieved with centrifugal pumps, depending on casing pressure limitations, by arranging them in series (see section 6).

For a given duty, centrifugal pumps are usually cheaper, occupy less space and have lower maintenance costs than positive displacement types, and can handle much larger solids (1,2,4). However, in general, they have lower efficiencies, particularly in the smaller sizes, and their performance tends to be reduced with increasing solids concentration, notably where non-Newtonian slurries are involved. Wear is another important factor in the correct selection of pumps for abrasive duties, which is considered in more detail in chapter 3.

Feeder systems avoid the use of a large number of pumps, either in series or parallel, and the high-pressure pumps supplying the driving source are required to handle only clean water. However, the initial cost of most of these systems is relatively high, and they tend to be somewhat complex, particularly the pipe-chamber type. A brief discussion of the choice between solids-handling pumps and feeders for British Coal's Horden Colliery mine waste pipeline is given in Paterson and Watson (4); a centrifugal pump was chosen as a result of considerations of pressure, particle size and capital cost.

The following sections consider the various types and construction, limits of speed, head, mixture concentration and particle size, gland sealing, performance and problem areas for both rotodynamic and positive displacement pumps; a brief outline of the main types of pipe feeder system is also included.

## 3.  ROTODYNAMIC PUMPS

### 3.1  OPERATING PRINCIPLE

The term 'rotodynamic' implies the presence of a bladed impeller which imparts a tangential acceleration, and hence an increase of energy, to the liquid flowing through it. Most of this increase is in the form of pressure energy, the remainder being the kinetic energy corresponding to the tangential (or 'whirl') velocity with which the liquid leaves the impeller before entering the casing.

## 3.2 SPECIFIC SPEED

The specific speed of a rotodynamic pump is a numerical quantity derived from the flow rate, the total head and the rotational speed. It is used to classify and compare various pump types, and for analysis and graphical representation of design parameters.

It is given by the expression:

$$\text{Specific speed } N_s = \frac{N.Q^{\frac{1}{2}}}{H^{\frac{3}{4}}}$$

where $N$ is the rotational speed, e.g. in rev/min.
$Q$ is the flow rate, e.g. in gal/min, or m³/hr.
$H$ is the total head, e.g. in feet, or metres.

Using 'mixed' British units, the specific speed of the great majority of pumps lies within the approximate range 700–12 500; the equivalent metric range with $Q$ in m³/hr, $H$ in m, and $N$ in r.p.m. is 900–16 000.

The following points should be noted:

1. The specific speed is normally evaluated only for the design duty, or best efficiency point (b.e.p.).
2. In the form given, the specific speed is not dimensionless, and will have different values depending on the units used.

Conversion from one set of units to another involves only a numerical coefficient. However, the dimensionless form of specific speed in SI units, known as 'Type Number' $K$, (Referred to in International Standard ISO 2548), is defined as:

$$K = \frac{2 \pi n Q^{\frac{1}{2}}}{(gH)^{\frac{3}{4}}}$$

where $n$ = rotational speed, rev/s.
$Q$ = flow rate, m³/s.
$H$ = total head/stage, m.

3. Remember that a US gallon is *not* identical with a British Imperial one, and hence $N_s$ will require to be adjusted accordingly.
4. For multi-stage pumps, $N_s$ is evaluated for one stage only.

As pointed out in Odrowaz-Pieniazek (1) and Wilson (6), the effect of specific speed should also be considered when designing a slurry pump for minimum wear; for a given duty, a design with a lower value of $N_s$ should have a reduced wear rate, but will tend to have a lower efficiency and be more expensive than that with a higher value of $N_s$. Hence, the common practice of limiting the impeller tip speed to minimise wear (see section 3.3.3) can be misleading.

## 3.3   CURRENT PRACTICE – ROTODYNAMIC PUMPS

### 3.3.1   Conventional types

Nearly all solids-handling centrifugal pumps are single stage, single-suction, have volute or concentric type casings, and are of generally robust construction with thick material sections (3,6–8). This type has the particular advantage of ease of maintenance, which is an important consideration for solids-handling duties.

Impellers, Figure 1, are either fully-shrouded, semi-open or fully open, designed to give some degree of unchokeability (3,6–8) i.e. few blades, mainly 2–5. Figure 2 shows an

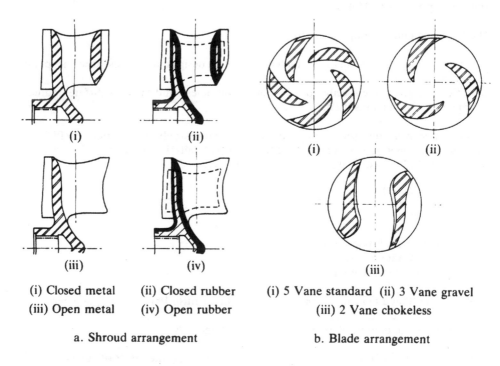

(i) Closed metal     (ii) Closed rubber          (i) 5 Vane standard  (ii) 3 Vane gravel
(iii) Open metal    (iv) Open rubber                        (iii) 2 Vane chokeless

a. Shroud arrangement                           b. Blade arrangement

**Figure 1.**   *Types of slurry and gravel pump impellers (Warman Intl.)*

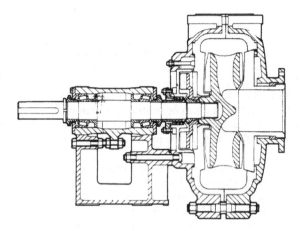

**Figure 2.** Heavy-duty slurry pump (Warman Intl.)

example of a Warman design (7). The 'Hazleton' design is an example of an all-metal slurry pump without a volute liner, although internal front and back wearing plates are fitted; the impeller has parallel shrouds, and appears to reflect conventional clean water pump practice except for fewer blades, which should also allow a reasonably high efficiency to be achieved, although limiting the size of solids which can be handled (see section 3.3.3).

The overall typical performance range varies widely according to the application; flow capacities range from 4.5 l/s (60 Imperial gallons/minute or Igpm) to 1140 l/s (15 000 Igpm) with heads for a single stage ranging from 7.6 m (25 ft) to 106.7 m (350 ft) (3). The corresponding branch size range of pumps in most frequent use varies from 19 to 600 mm (¾ to 24 inches).

Most slurry pumps handling mainly abrasive solids are electric-motor-driven (except in dredging applications), horizontal-shaft units, often with V-belt drive (3,6). Dredge pumps, which are among the largest in slurry service – up to about 2650 l/s (35 000 Igpm) for a 900 mm (36 inch) pump – are usually diesel-engine driven for marine applications; De Bree's series of articles (9) gives comprehensive data on the design, performance and application of such pumps. Examples of other types of sand dredging installation are described in White and Seal (10) and Read (11).

Portable pumps, also used for site drainage duties, may be either I.C. engine-driven horizontal-spindle, self-priming truck-mounted units or motor-driven (either electric or hydraulic) submersible types; typical sizes range from 38 to 300 mm (1½ to 12 inches) (3).

Certain applications, such as pumping slurry direct from a sump, may favour a vertical shaft arrangement (1). The pumps may be either of the shaft driven suspended type, both single- and double-suction units being available, or fully submersible. Both types have either volute or concentric casings; suspended pumps often have two discharge branches.

## 3.3.2   Special types

Several manufacturers are now offering recessed-impeller pumps in which the impeller is mounted on one side of the volute to reduce impact of solids (1,3,8). These are normally used where maximum unchokeability is required, but such pumps may also be offered for use in applications with relatively small particle size slurries. They are also claimed to be capable of handling liquids with a higher gas content than conventional pumps with closed impellers (12). However, their main disadvantage is a rather low efficiency (40–55 per cent) compared with more conventional types, and they may also be prone to abrasive wear (12), in spite of their design (1).

Single-bladed mixed-flow 'screw' pumps are claimed to be able to handle non-Newtonian slurries and sludges containing a higher proportion of solids than more conventional pumps, with less fall-off in performance for a given solids content (3).

Foster-Miller and Associates (USA) have reported the development of a boost pump with a single-bladed helical inducer type impeller (Figure 3) specifically for coarse or fine coal hydrotransport (13). It is claimed to be 'self-controlling', the performance being independent of slurry conditions, by introducing an air core from atmosphere at the impeller hub. The published performance of the present design shows a maximum pressure of approximately 7 bar (105 p.s.i.) at 3000 r.p.m.

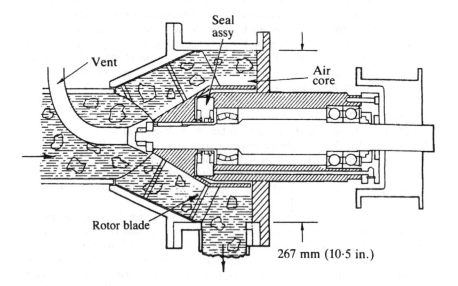

**Figure 3.**   Helical inducer slurry pump (Foster-Miller Assocs. Inc.) (13)

### 3.3.3   Limits of speed, head, mixture concentration and particle size

Few manufacturers quote specific limits on impeller speed or head for abrasive slurry duties (3); hence it must be assumed that, in general, the limit is effectively the maximum head stated for any particular range of pumps. However, most companies rely on their own experience for specific applications. Pumps are offered for duties up to 106 m (350 ft)/stage with both all-metal and rubber-lined wetted parts by certain firms, although some authors (e.g. 1,2,14,15) recommend lower limits of head or impeller tip speed for rubber-lined pumps. Usually, running speeds are below 1500 r.p.m., except in the smallest sizes. Limits on impeller eye speed may be necessary if suction conditions are arduous, particularly with non-Newtonian mixtures; Crisswell (16), suggests a general limit of 1.5 m/s (4.9 ft/s) if above the critical deposit velocity. The need to minimise wear is usually the main factor determining the maximum acceptable impeller tip velocity, which in turn governs the maximum head for a given rotational speed; limits of 25–28 m/s (82–92 ft/s) for slurry pumps generally (16), 30 m/s (98.5 ft/s) for hard metal impellers (1) and 20–23 m/s (65.5–75.5 ft/s) for rubber impellers (1,2,15) have been suggested in the literature.

Limits of mixture concentration vary considerably according to the application and may also be affected by pump design. For most settling slurries, the maximum concentration by volume $(C_v)$ is about 30–40 per cent (8,17). However, maximum $C_v$ for non-Newtonian mixtures is very dependent on the type of solids involved; for suspensions of fine granular particles, research on coal and kaolin slurries (18) has shown that it is possible to handle up to about 20–30 per cent $C_v$ with conventional centrifugal pumps with only moderate loss of performance (see section 3.4).

Particle size limits for most rubber-lined pumps vary between approximately 3 to 6 mm ($\frac{1}{8}$ to $\frac{1}{4}$ inch), depending on the manufacturer; however, Galigher A–S–H. in the USA offer a range of thick rubber-lined pumps which are claimed to withstand coarse particles and 'tramp' material. For metal pumps, the particle size limit merely depends on the degree of unchokeability of the impeller passages; if only relatively small particles are to be handled, then the passage shape may follow conventional water-pump practice more closely. The recessed impeller designs offer the maximum unchokeability of any type.

### 3.3.4   Shrouded vs. 'open' impellers

The selection of a fully-shrouded, semi-open, or fully-open impeller for a given duty is not always clear-cut, and often depends on a particular manufacturer's or user's experience; many makers offer more than one type to suit an individual pump range (3).

Slurry pumps tend to have either fully-shrouded or semi-open impellers (Figure 1a); some manufacturers suggest that the semi-open type is less prone to blockage and easier to manufacture in the smaller sizes (say <50 mm). However, certain manufacturers and users claim that for abrasive duties (where the risk of blockage is small), the fully-shrouded impeller is preferable (3,6,12); less wear occurs on both the impeller and suction liner bush, and higher efficiency can be obtained with this type. It is usual to incorporate a simple means of axial adjustment, enabling correct clearances between impellers and casing or wear plates to be maintained. Open and semi-open impellers are also said to be more capable of handling mixtures with high gas contents than closed impellers (12).

## 3.3.5  Casing and impeller materials

'Ni-hard' and soft natural rubber are still generally regarded by both manufacturers and users as the best materials currently available for abrasive slurry-duties, from an overall economic stand-point (1,3,6,14,15). Rubber-lining is recommended for slurries with small solids (<3 to 6 mm ($\frac{1}{8}$ to $\frac{1}{4}$ inch) particle size) in suspension; however, manufacturers now offer polyurethane impellers and linings for certain applications. Manganese steel may be used for large-particle slurries (e.g. gravel and dredging duties), where work-hardening can occur. High-chrome cast iron may also be supplied as an alternative to 'Ni-hard', but the choice often seems to rest on either the current delivery situation or customer preference; it is suggested that the wear resistance is similar (3). Because of the problems inherent in machining 'Ni-hard', manufacturers tend to reduce machining operations to a minimum. This can result in components with large clearances and, in consequence, reduced pump efficiency initially. However, any increase in pump efficiency where smaller clearances can be obtained with the softer, more readily available machineable materials, will be rapidly lost through abrasive wear, unless the material can be hardened after machining (3).

18/8 stainless steel in various grades is mainly used where corrosion is a greater problem than erosion, e.g. in chemical processing. 'Ni-resist' cast iron and several types of cast alloy steel are also offered by some manufacturers as 'standard' materials for corrosive slurry duties (3); alloys such as 'Ferralium' (25/5 Cr Ni alloy steel) and the Pettibone chromium carbide material, and the very hard materials such as metallic carbides and weld overlays, are generally both corrosion- and abrasion-resistant.

Cast iron (Grades 14 and 17) is the most common material for all applications where neither abrasion nor corrosion is usually a major problem. It is also used, together with spheroidal graphite iron, for duties where solids-handling is not continuous, such as site drainage; more frequent replacement of parts in a cheaper material, rather than an expensive, wear-resistant type, e.g. 'Ni-hard', becomes a preferable economic proposition (3).

A more detailed discussion of the erosion- and corrosion-resistance of various pump materials is given in chapter 3.

## 3.3.6  Shaft sealing

Soft packing ('greasy cotton' or similar) with clean water flushing is generally preferred for abrasive slurry duties, often with hard-surfaced shafts or sleeves (1–3, 6–8, 11, 14–16); good results were reported with Grade 316 stainless steel shaft sleeves and Teflon or PTFE packing (11). Most impellers have some form of 'scraper' or 'back' vanes to reduce the pressure at the gland. In some designs, these vanes are sufficiently powerful to create a vacuum; other types incorporate a separate dynamic seal or 'expeller' (Figure 2) to keep abrasive away from the gland region (1–3,6–8,16). In both these cases, the need for flushing water is removed, unless the suction head exceeds a certain limit (usually about 3.0 to 3.7 m (10 to 12 ft) above atmospheric pressure. Operating experience, including modification to gland flushing systems, with various types of slurries is reported in Refs. 11, 14 and 15.

Some manufacturers are prepared to offer mechanical seals, of either their own design or proprietary make, for abrasive applications, often the double type with clean water flush. However, user experience suggests that these seals are not always satisfactory. Multiple lip seals with grease injection have been used successfully in large dredge pumps either as a modification to an original packed gland (3), or fitted as standard modern practice by at least one manufacturer (9).

Soft packing also tends to predominate in the various process industries. However, where leakage cannot be tolerated, such as in some chemical processes, mechanical seals have to be used; again these are usually of the double type with tungsten carbide, 'Stellite' or ceramic faces and clean liquid flushing (3). Mechanical seals are more often offered as 'standard' on the smaller sizes of pump for site drainage, both self-priming and submersible, and general waste disposal duties (3).

## 3.4   PUMP PERFORMANCE

Most of the data on the effect of granular solids on centrifugal pump performance (e.g. 8,17–22) have been obtained from laboratory tests and subsequent derivation of empirical expressions for performance prediction. A BHRA review (17) of published literature on this topic attempts to correlate the data as far as possible, but notes that these tended to be limited to solids of similar specific gravity (S.G.) (e.g. sand, gravel, clay, etc.) and relatively small particle size, mostly<1 mm; also, little information is given on the design of the pumps tested. The review considers the effects of solids concentration, particle size and S.G., flow regime, pump speed, hydraulic design, wear and cavitation.

In very general terms, the presence of solids will reduce performance, in terms of both head and efficiency. Hence, it is usual to express this reduction compared to the clean water performance as a head ratio, $H_R$, (in terms of head of fluid pumped) and efficiency ratio, $\eta_R$ (or $E_R$) (17–20). The effect of flow regime, as determined by solids concentration and particle type, is particularly important, and two main groups of slurries have been considered in the literature (8,17):

1. Heterogeneous (or settling) slurries of medium-sized particles (approximately 0.2–2.0 mm), and
2. Homogeneous suspensions of fine particles (say <0.05 mm).

Figures 4 and 5 show Cave's results (19) on a 100 mm (4 inches) Warman pump handling various settling slurries, giving the effect of concentration by weight, $C_W$, on $H_R$ (Figure 4) and the head/flow characteristics (Figure 5). A collection of data from various sources (17)

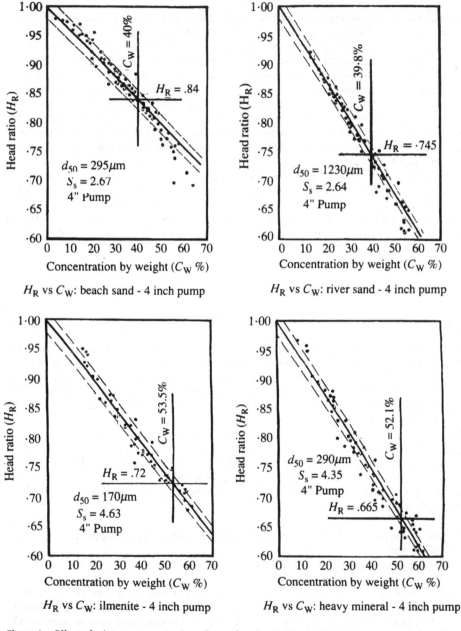

**Figure 4.** Effect of mixture concentration, $C_w$, on head ratio $H_R$, for 4 inch Warman pump handling various slurries (19)

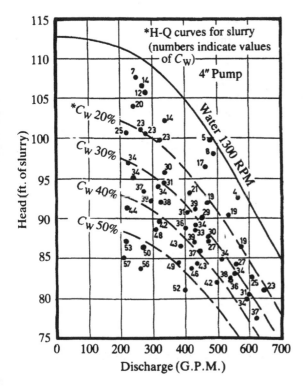

**Figure 5.** Effect of mixture concentration on head/flow characteristic for 4 inch Warman pump with beach sand slurry (19)

on the effect of volume concentration, $C_v$, on head reduction with solids of similar S.G. is given in Figure 6, showing the general reduction in head with increasing $C_v$. An analysis of limited data on the effect of particle size on head reduction showed no particular trend.

Figure 7 shows the effect of concentration (expressed as mixture S.G.) on the complete performance characteristics of a 100/75 mm pump handling a homogeneous 'china clay' slurry (18). These results demonstrate the effect of pumping a non-Newtonian slurry, where viscosity effects predominate; it was found that both $H_R$ and $E_R$ correlated reasonably well with the pump Reynolds No. (see Figure 8). It should also be noted that the $H$–$Q$ characteristics become unstable at the lower flow rates with the higher concentrations, i.e. higher apparent viscosities (see Figure 7).

Although there may be appreciable differences between measured and predicted values of $H_R$ and $\eta_R$ from the various sources, as shown by the correlation of data in the performance review (17), depending on the type and size of solids, concentration range and probably pump design also, some general points of agreement may be noted:

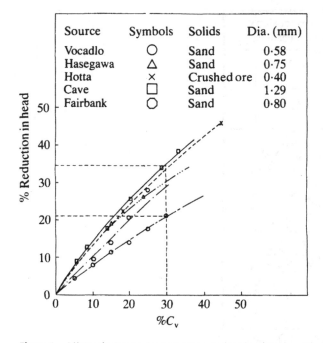

**Figure 6.**   Effect of mixture concentration on head reduction with solids of comparable S.G. (17)

1. For settling slurries of medium-sized particles, both head (of mixture pumped) and efficiency decrease as concentration increases (8,17,19–21). Most reports (17,19) suggest that the head and efficiency ratios ($H_R, \eta_R$) are approximately equal, although one investigation (20) found that the fall in efficiency became greater than the fall in head at concentrations > 20–25 per cent by volume, and with particle sizes up to about 8 mm (0.3 inch).

2. For homogeneous suspensions of fine particles, both head and efficiency will be unaffected at the lower concentrations, provided that the mixture remains Newtonian and there are no viscosity effects (8,17). However, at higher concentrations when viscosity effects become significant, both head and efficiency will fall as concentration increases, the fall-off in efficiency being more rapid than that for head, i.e. $\eta_R < H_R$ (17,18).

3. The b.e.p. flow rate is generally about the same for clean water and slurry (17,18).

4. The normal pump affinity laws apply for slurries, provided there are no viscosity or cavitation effects (17).

5. Power will increase in direct proportion to mixture S.G., provided there are no viscosity effects (17).

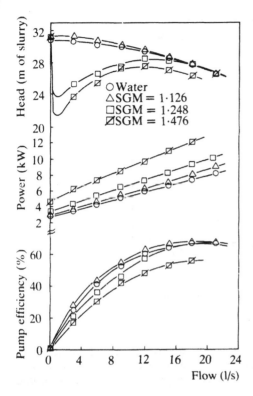

**Figure 7.** Effect of mixture concentration on performance characteristics of a centrifugal pump at 1200 r.p.m. with kaolin ('china clay') slurry (18)

6. Limited data suggest that cavitation characteristics may be unaffected by homogeneous slurries, even when non-Newtonian (17). Another report concerning dredging pumps (9), presumably involved mainly with settling slurries, states that suction performance will be impaired by the presence of solids, particularly large particles, though no quantitative data are given.

## 4. POSITIVE DISPLACEMENT PUMPS

### 4.1 OPERATING PRINCIPLES

In a positive displacement pump, discrete volumes of fluid are isolated between the moving and stationary parts, and moved from the suction to the discharge branch by direct

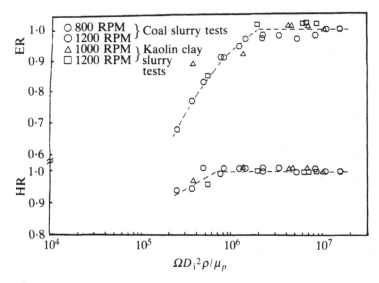

**Figure 8.** Effect of viscosity (expressed as pump Reynolds number) on head and efficiency ratios for a centrifugal pump at b.e.p. (18)

mechanical action. Hence the flow path is not continuous, as it is with a rotodynamic pump. The maximum pressure which can be generated depends on the amount of internal leakage, and hence a pressure relief valve is often fitted on the discharge side to prevent damage due to over-pressure.

In a reciprocating pump (e.g. piston, plunger, diaphragm), the pumping chamber alternately becomes part of the suction and discharge system volumes; hence inlet and outlet valves are required to isolate the suction and discharge cycles. Such pumps necessarily give a pulsating flow, although pulsations can be reduced with multi-cylinder pumps, or by using damping devices. Rotary pumps (e.g. helical rotor, gear, lobe rotor, vane, screw, etc), on the other hand, usually rely on the geometric shape of the rotor and casing, with relatively fine clearances between them, to create the displacement volume between suction and discharge. Thus, the resulting flow is substantially continuous. However, not all types of rotary pump are suitable for handling slurries, particularly if they are abrasive.

## 4.2 CURRENT PRACTICE – POSITIVE DISPLACEMENT PUMPS

### 4.2.1 Conventional types

Multi-cylinder reciprocating plunger and piston pumps are used for high-pressure applications, e.g. long pipelines, involving mineral slurries (1–3,8). Typical sizes of piston pumps (Figure 9) range from 140 to 200 mm ($5\frac{1}{2}$ to 8 inches) cylinder bore and 200 to 450 mm (8 to 18 inches) stroke (3), although much larger sizes are available (up to 300 to 325 mm (12 to 13 inches) bore), e.g. the Wilson-Snyder (USA) pumps used on the 'Black Mesa' pipeline (2). Typical flow rates vary from about 6.1 to 53 l/s (80 to 700 Igpm), and pressures from 26.5 to 176 bar (370 to 2550 p.s.i.) (2,3); however, the largest sizes of pump can give flows up to 170 l/s (2250 Igpm), though not at maximum pressure (2). These units are mainly of the horizontal cylinder 'duplex' or 'triplex' type, with geared motor drive. Some makes of piston pump use a piston flushing arrangement to reduce wear with abrasive duties.

Plunger pumps, which generally give lower maximum flow rates and somewhat higher maximum pressures than piston pumps, are available with up to six ('sextuplex') or more cylinders, and are recommended for the more abrasive duties. Odrowaz-Pieniazek (1) suggests that capacities of about 55.5 l/s (730 Igpm) are common, with new developments giving maximum flows up to 222 l/s (2930 Igpm), i.e. of the same order as piston pumps.

Small diaphragm pumps, mostly in the 25 to 50 mm (1 to 2 inches) branch size and both motor-driven and air operated, tend to be used for lower heads (up to 67 m (220 ft)) and flows (up to 8.7 l/s (115 Igpm)) (1,3). For less abrasive duties, such as general waste and effluent disposal, and some chemical processes, larger diaphragm pumps are used for flows up to about 38 l/s (500 Igpm) and relatively low pressures. This type of pump may also be used for pumping slurries of relatively high viscosity (1,3).

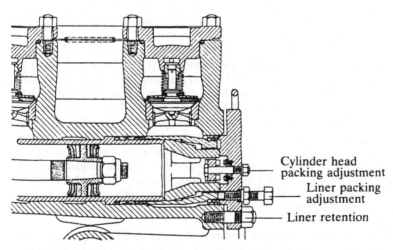

Cylinder head
packing adjustment
Liner packing
adjustment
Liner retention

Sectional arrangement of 'wet-end' parts

**Figure 9.** High pressure reciprocating slurry pump (National Supply Co.)

## 4.2.2 Rotary pumps

Both helical rotor, e.g. Mono (Figure 10), and lobe rotor pumps may also be used for the less abrasive, high viscosity applications with fine particle slurries (1,3,8). Typical branch sizes are from 25 to 150 mm (1 to 6 inches) for both types. Helical rotor pumps can deliver flow rates up to about 42 l/s (550 Igpm) and discharge pressures up to 5 bar (73 p.s.i.) at maximum speed for a single stage unit, with pressures up to 20 bar (290 p.s.i.) for 4 stages; special designs can give higher pressures/stage. Lobe rotor pumps can produce flows up to about 100 l/s (1320 Igpm) and pressures up to 10 bar (145 p.s.i.); rubber-lined rotors have recently been introduced for the more arduous duties. The main advantages are a reasonably steady output and no valves.

## 4.2.3 Special pump types

A number of designs are available, or under development, which isolate the abrasive slurry from the moving parts. One such is the Japanese designed 'Mars' reciprocating pump with its oil barrier 'piston'. Various types are offered for flow rates up to 57 l/s (750 Igpm) and pressures up to 157 bar (2275 p.s.i.). Operating experience, including problem areas and modifications to valve materials, at English China Clays Limited is discussed in Ref. 23.

An alternative approach is the piston-diaphragm pump (24) where the slurry is mechanically separated from the 'propelling fluid' (hydraulic oil) in which the pistons work by a rubber diaphragm. Standard sizes cover flows up to 55.5 l/s (730 Igpm) and pressures up to 100–150 bar (1450 p.s.i.). The main advantages claimed, apart from the separation of the slurry from the reciprocating parts, are longer valve life due to lower velocities through the valves and lower pressure fluctuations.

Another alternative is the hydraulically-operated cylindrical diaphragm pump, e.g. the USA 'Zimpro' design (25) shown in Figure 11, where hydraulic pumps transmit pressure to the slurry surrounding the diaphragm 'bags' in the main pumping chambers via a transfer valve and stroke control cylinders. Typical capacities are in the range of about 3–14 l/s (40–185 Igpm) and pressures are of the order of 35.5–83 bar (500–1200 p.s.i.), though units have been developed for flows up to 63 l/s (830 Igpm) and pressures up to 346 bar (5000 p.s.i.). The main advantages claimed are the avoidance of gland sealing problems, variable delivery, protection against over-pressure and, again, longer valve life due to the slow cycle rate. However, a separate low-pressure slurry charge pump is required to feed the unit. Somewhat similar pumps have been reported at the prototype stage in Sweden (26) and South Africa (27), but with the slurry inside the flexible tubes which are operated by external hydraulic pressure; in the South African design, hydraulically-operated pinch valves are also used.

Another version of the cylindrical diaphragm pump is the peristaltic type. Although these pumps have been made for many years in very small sizes (1,8), the Dutch 'Bredel' design (Figure 12) is available with branch sizes up to 100 mm (4 inches), for flow rates up to 17 l/s (220 Igpm) and pressures up to 15 bar (220 p.s.i.). It is suitable for abrasive, corrosive and

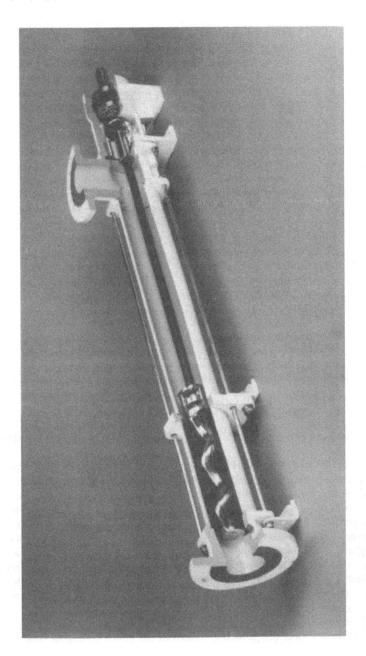

**Figure 10.** Helical rotor pump (Mono Pumps)

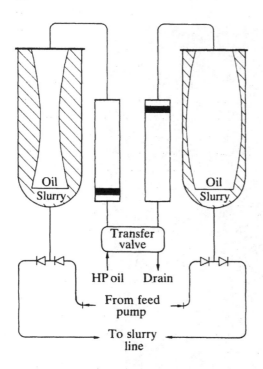

**Figure 11.** Operating principle of high-pressure cylindrical diaphragm pump (Zimpro/Hayward Tyler) (25)

viscous slurries, and can also run dry for long periods. Another advantage is the absence of valves.

In some other reciprocating pump designs, where the slurry is in contact with the piston, the conventional mechanical drive with crankshaft and connecting-rod is replaced by some form of oil hydraulic drive. One example is the concrete-handling pump, which may also be used for other types of high concentration slurries (28). A number of makes are available, most of the larger units being of West German origin. All are two-cylinder piston pumps, with the pumping and hydraulic cylinders arranged horizontally in tandem, and a common piston rod (Figures 13 and 14). Most types have hydraulic oil pressure supplied by a variable output pump, either electric motor or diesel driven; hence variable speed and slurry output may be obtained. The open ends of the pumping cylinders are connected alternately to the pump hopper or delivery pipe by a special valve mechanism; the design of this mechanism is the main distinguishing feature between one make and another. The larger types are capable of delivery pressures in the region of 100 bar (1450 p.s.i.), and have the additional advantage over conventional reciprocating pumps of the ability to handle relatively large

**Figure 12.** 'Bredel' peristaltic pump (Bredel B.V./Alpha Technical Services)

solids, due to the 'full way' valve design, as well as very viscous mixtures. Capacities are relatively low, usually of the order of 14–28 l/s (185–370 Igpm) maximum. Childs (29) describes a new type of vertical multi-cylinder plunger pump of American design with hydraulic drive; two versions are mentioned, one being a low pressure, high flow unit designed for 10 bar (145 p.s.i.) at 350 l/s (4620 Igpm), and the other for high pressure, low flow duties designed for 200 bar (2900 p.s.i.) at 17.7 l/s (234 Igpm). Among the advantages claimed are reduced pressure fluctuations due to the long, slow stroke, and variable output. It is not clear whether this pump is in production.

Boyle and Boyle (45) describe a rotary piston pump which will pump stabilised mixtures of coarse coal. The cylinders are contained in a rotating barrel. The rotation sequentially aligns the cylinders with a port plate. The relative motion between barrel and plate is capable of shearing coarse material present at the time of valve closure. The pistons are driven by high pressure water. Recently development has been undertaken by the ASEA company.

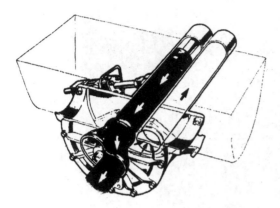

**Figure 13.** Concrete-handling pump showing valve mechanism (Schwing/Burlington Engrs.)

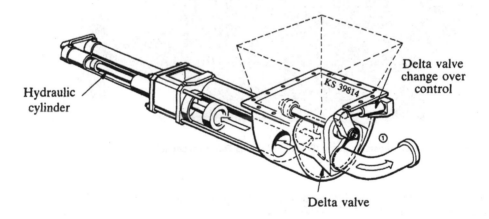

**Figure 14.** Concrete handling pump (Putzmeister GmbH)

## 4.2.4   Limits of speed, head, mixture concentration and particle size

In general, little data has been quoted on specific limits for most of these parameters. As for rotodynamic pumps, it must be assumed that, in the absence of any other information, the effective limit on head/pressure is the maximum stated for any given range of pumps, as determined by the design of the fluid end and the drive power rating (2,3). One manufacturer quotes limits of 276 bar (4000 p.s.i.) on pressure and 0.9 m/s (3 ft/s) on piston speed for high pressure reciprocating slurry pumps (3).

Higher mixture concentrations resulting in higher apparent viscosities can normally be handled by positive displacement rather than rotodynamic pumps, and the limit is more likely to be a function of the system or process rather than the pump. However, the types without valves, e.g. rotary, peristaltic, are generally more suited to handle the very viscous slurries.

The use of positive displacement pumps is mainly restricted to the smaller particle sizes <3 mm ($\frac{1}{8}$ inch), the maximum size governed by the valve design for reciprocating types. One firm claims to have successfully used piston pumps with special valve materials for sizes in the 3 to 6 mm ($\frac{1}{8}$ to $\frac{1}{4}$ inch) range, but only for pressures up to 10.3 bar (150 p.s.i.); however, solids up to 10 mm were reported to have been handled successfully by the same type of pump (National) with hardened stainless steel valve seats (10). As for centrifugal pumps with rubber wetted parts, both helical rotor and peristaltic pumps are unsuitable for handling large hard solids, as also applies to lobe rotor pumps. The only type of p.d. pump likely to be suitable for such solids is the concrete-handling variety.

## 4.2.5 Materials for stationary and moving wetted parts

High-pressure piston pumps may either have hardened steel or chromium liners, and rubber or polyurethane pistons; one type of plunger pump has a forged steel casing/liner and hard-faced steel plunger. Valves have either rubber, polyurethane or P.T.F.E. inserts or coatings, with some form of steel (including stainless) seat, depending on the type of valve and duty (3). Significant improvement in valve life has been reported with hardened 17/4 stainless steel seats in 'National' pumps (10) and with polyurethane valve rubbers and seats, and hard-faced valve bodies, in 'Mars' pumps (23). Most manufacturers offer a range of materials for all wetted parts to cover different applications.

Piston and plunger pumps for lower pressure duties may have either some grade of cast iron, mild steel or bronze for cylinders/liners, and cast iron, hardened steel, stainless steel or bronze for reciprocating parts. Valves tend to be rubber (or rubber-coated) ball type or chrome leather-faced flap type, with metal seats. A wider variety of materials is used for diaphragm pumps on similar duties, including various grades of cast iron (probably the most common), stainless steel, aluminium, and rubber and epoxy resin coatings. The diaphragms are usually either natural or synthetic rubber. Valves are mostly rubber (-coated) ball type with metal (e.g. cast iron, stainless steel) seats (3).

Of the various rotary types encountered, helical-rotor pumps usually have natural or synthetic rubber stators and hard-surfaced (e.g. chrome or 'Stellite') rotors (3). The Dutch-designed peristaltic pump has a special thick-wall rubber tube reinforced with braided steelcord.

## 4.2.6 Gland sealing

Packed glands, with either clean water flushing or grease lubrication, are normally used on the reciprocating types of pump; plunger pumps are fitted with a flushing system for the more abrasive duties (1–3,5). Some of the high-pressure piston and plunger pumps use composite rubber and fabric packings, whilst others have synthetic rubber (e.g. 'Viton') rings (3). Good results were reported with impregnated 'Teflon' packing and hard-chromed piston rods in National piston pumps handling a beach sand slurry (10).

Most rotary types also have packed glands, except on certain chemical duties where mechanical seals, often of the double type, are necessary to prevent excessive leakage. A few manufacturers of small pumps fit mechanical seals as standard (3).

## 4.3  PERFORMANCE

One of the main advantages of positive displacement (p.d.) pumps over rotodynamic types is that the hydraulic performance is basically unaffected by the presence of solids in suspension, provided that the particle size does not interfere mechanically with the satisfactory operation of the moving parts, including valves. Thus, a p.d. pump will deliver an approximate constant flow, depending on the amount of internal leakage, at any pressure up to the maximum rating at constant speed (8). Volumetric efficiency reduces gradually with increasing pressure, due to increased leakage, and hence the actual flow rate delivered will reduce correspondingly. Increasing viscosity will tend to reduce internal leakage, but could possibly affect valve operation adversely in reciprocating pumps, depending on the valve type, and cause cavitation on the inlet side if the net positive suction head (N.P.S.H.) is too low.

Since the maximum pressure delivered by a reciprocating pump is determined by the mechanical design of the fluid end and the power rating of the drive, the higher pressure pumps generally have the smaller piston diameters and give the lower flow rates, and vice versa.

## 5.  FEEDER SYSTEMS

For the purpose of this section, a solids 'feeder' will be considered as any system which uses an external source of clean high-pressure fluid, either hydraulic or pneumatic, to which the solids are added by various means before being pumped through a pipeline. Thus, this section will include jet pumps and air-operated systems, as well as 'positive displacement' feeders, such as lock-hopper, rotary and pipe-chamber types.

Various feeder systems and their selection are discussed in Refs. 8, 30 and 31. The main application area would appear to be for the transport of coarse solids at relatively high pressures, e.g. mine hoisting and long-distance pipelines (1,30,31). It has been suggested that such systems can be economical to operate, in spite of high capital cost (30,31), although experience with individual plant may not always confirm this view. The main development effort on pipe-chamber feeders has been in West Germany, Eastern Europe and Japan. The main types of feeder are outlined briefly in sections 5.1 to 5.5 below.

## 5.1 LOCK-HOPPER FEEDERS

The lock-hopper type consists basically of either one or two pressure vessels and a valve system which isolates the vessel(s) first from the pipeline while being filled with solids and water at low pressure, and then from the feed system while the slurry is transferred into the pipeline (8). The large pressure vessels involved tend to limit the maximum pressure rating to about 30 bar (435 p.s.i.), mainly on economic grounds. Although two-vessel designs (Figure 15) give improved feeding, some discontinuity still occurs, however, rotary valves can overcome this. Tianyu *et al.* (32) describe various feeders, including inclined designs, used in China for coal transport. Lock-hopper systems are tending to be superseded by pipe-chamber feeders.

## 5.2 ROTARY 'MOVING-POCKET' FEEDERS

These are devices having a series of 'pockets' which are alternately filled with solids and then moved to a position where high-pressure water transfers the slurry into the pipeline.

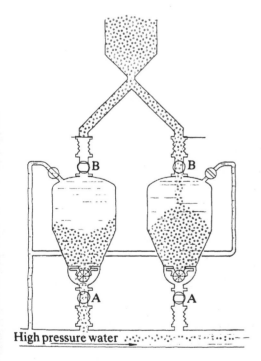

High pressure water

**Figure 15.** Two vessel lock hopper system

They have the advantage of continuous feed, but may have pressure and wear limitations
(8). One example is the Kamyr feeder (Figure 16) originating in the USA, for which the
previous standard design was limited to 25 bar (360 p.s.i.), although higher pressure units
are likely to be developed; pilot plant trials in the USA were carried out on this type
handling < 100 mm (< 4 inch) raw coal at relatively low pressure with reasonable success
(33). A unit was installed for trials in the coal shale slurry line at the N.C.B.'s Horden
Colliery, handling 200 tonnes/hr of shale at 20 per cent $C_w$, as proposed in Ref. 4.

## 5.3 PIPE-CHAMBER FEEDERS

This type is generally similar to the lock-hopper feeder in operating principle, but uses
pipes instead of large pressure vessels, and hence may be designed for pressures up to 160
bar (2300 p.s.i.) (30, 31). However, it is probably the most expensive system in terms of

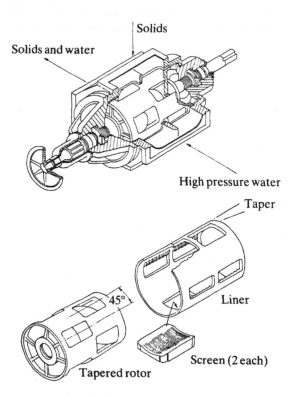

**Figure 16.** Exploded view of Kamyr rotary feeder (4)

capital cost, occupying a large space and having a complex valving and control system. Most designs use a low pressure centrifugal slurry pump to feed the slurry into the pipe-chambers, as well as a high-pressure clean water pump for emptying the chambers and driving the slurry down the line. The low-pressure slurry pump performance on very viscous slurries may limit the maximum concentration which can be handled as discussed previously in section 3.4. Pulsating flow and some dilution of the slurry is likely with a two-chamber system, so a three-chamber type may be preferable. Valves are usually of 'full-way' design, to handle large solids. Various designs of pipe-chamber feeder are described in Refs. 30, 31 and 34–37.

The Japanese Hitachi 'Hydrohoist' system has straight pipe chambers, and cam-operated control valves; the horizontal chamber version (Figure 17) can handle the largest solids. Its application for transporting stabilised slurries is described by Sakamoto *et al* (34) and the basic construction, operation and development of the vertical type, with some service experience, is given in a later work by Sakamoto *et al*. (35).

The Hungarian-designed 'Nikex' system, marketed in the UK and elsewhere by Babcock Hydraulic Handling Limited has chambers in the form of 180° bends with electronically

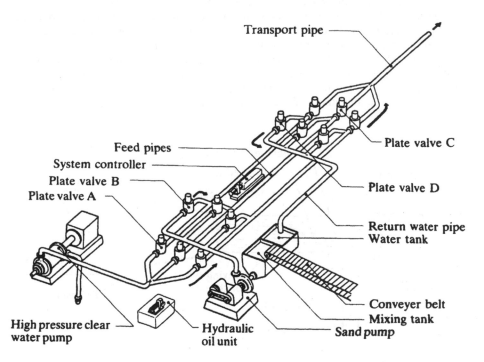

**Figure 17.** 'Hydrohoist' horizontal pipe-chamber feeder (Hitachi) (34)

controlled valve operation (31). Units are operating on coal slurries in both Hungary and West Germany. The design, construction and operation of the three-chamber pipe feeder system in the well-documented Hansa Hydromine in West Germany are described by Siebert *et al.* (36) and Kortenbusch (37); future possible applications of similar system are also discussed in Ref. 37.

## 5.4  JET PUMPS

### 5.4.1  Operating principle

A jet jump, shown diagrammatically in Figure 18, is a purely hydrodynamic device in which a jet of fluid (the driving fluid) is used to entrain more fluid. It consists of a nozzle (or series of nozzles), suction box, mixing tube (or 'throat') and a diffuser (8,38,39).

For solids-handling applications, clean high-pressure water is supplied usually by a centrifugal pump to the nozzle which converts the pressure into kinetic energy as a high-velocity jet. The jet then entrains slurry at a relatively high concentration into the suction box, and expands into the mixing tube downstream, thus accelerating the stream of entrained slurry. When mixing of the two streams is complete at the end of the throat, the more dilute mixture enters the diffuser to give a suitable discharge velocity (8,38,39). A number of different nozzle arrangements may be used (39).

A fundamental difference between a jet pump and other types is that it recognises masses rather than volumes (38); it will induce a fixed mass flow rate of mixture, irrespective of specific gravity, for a given set of operating conditions. It is, however, a relatively inefficient device, the maximum efficiency being only about 35–40 per cent (1,8).

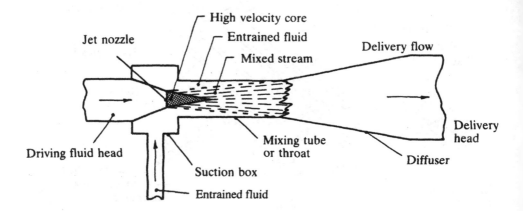

**Figure 18.**  Layout of typical jet pump

## 5.4.2   Applications

Most applications of jet pumps for handling mineral slurries are in dredging systems (38,39), where the jet unit is usually placed at the bottom end of the suction pipe. Other applications include mine hoisting, 'stowing' and drainage (39), concrete pumping, ash handling, radioactive waste disposal, etc. A jet pump may also be used in conjunction with some other type of feeder, e.g. air-lift (39) and pipe-chamber (40) types, in mine-shaft sinking or deepening.

Sizes are available for 75–500 mm (3 to 20 inches) diameter pipe, to transport up to 1520 l/s (20 000 Igpm) of combined mixture at static lifts up to 120 m (390 ft) (38). Jet pumps are also capable of handling large solids, e.g. with the annular nozzle (1,5) and side nozzle (39) types, and relatively high mixture concentrations, although some dilution will inevitably occur downstream due to the clean water jet; they can be made in abrasion-resistant materials to reduce wear. The main advantages are that abrasive solids do not have to pass through the centrifugal pump supplying the driving water (as with other types of feeder), there are no moving parts in contact with the slurry, capital and spares replacement costs are low, and the risk of pipeline blockage is reduced, due to the steepness of the jet pump head/flow characteristic. Also, its hydraulic efficiency is relatively unaffected by solids concentration; thus, if all the economic factors are considered, in cerain applications a jet pump may have an overall advantage over other types, in spite of its inherent low efficiency.

## 5.5   AIR-OPERATED SYSTEMS

Two different types of system, both using compressed air as the driving medium, are included in this section; one operates on a displacement principle ('Pneuma' pump), and in the other (air-lift), air is actually mixed with the slurry.

### 5.5.1   'Pneuma' pumping system

This Italian-designed system could be considered as a form of positive displacement pump with a compressed-air 'piston', as shown in Figure 19. The system consists of three cylinders grouped together with a common slurry discharge pipe and individual compressed-air pipes to each cylinder; it has no moving parts in contact with the mixture pumped, except for the valves (1,3). Each cylinder is filled in turn with slurry; as soon as one cylinder is filled, the inlet valve automatically closes by its own weight ('Phase 1' in Figure 18). Compressed air is supplied to the top of the cylinder and acts as a piston, thus forcing the slurry out through the delivery pipe ('Phase 2'). When the cylinder is almost empty, the air is discharged to atmosphere through the distributor and the cycle restarts ('Phase 3'). The system is available in several sizes with mixture capacities from 11–840 l/s (145–11 000 Igpm) at pressures up to 4 bar (58 p.s.i.) (1,3). The range most widely used for dredging, which is the

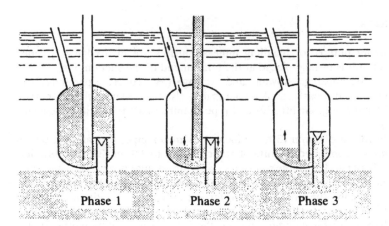

Phase 1              Phase 2              Phase 3

**Figure 19.** Compressed-air operated dredging pump system (Pneuma) (3)

main application, is from 48.5–840 l/s (640–11 000 Igpm) (1,3), and it is reported that solids up to 200 mm (8 inches) can be handled by the largest pumps (1).

## 5.5.2   Air-lift pump

An air-lift pump works on the principle that air is injected, in the form of small bubbles, into a near vertical pipe filled with liquid, so that the mixture then has a lower S.G. and upward movement results (1); if solids are entrained at the bottom of the pipe, a three-phase mixture is eventually produced (41). Refs. 41 and 42 give theoretical analyses for transporting settling slurries and homogeneous shear-thinning suspensions respectively by the air-lift principle.

Since the pipe must be vertical (or slightly inclined), the main applications are in mining, and lifting minerals, e.g. manganese modules, from the seabed (1,41). Although the air-lift pump is relatively inefficient – typically about 35–45 per cent – it is able to handle coarse solids, and also high concentration, non-Newtonian slurries (42); also, there are no moving parts in contact with the slurry. Combined jet-pump and air-lift systems have been used successfully in mine-shaft deepening and drainage applications (39).

## 6.   INTERACTION OF PUMP AND PIPELINE SYSTEM

As for any other pumping system, the pump selected for a particular duty must be matched to the characteristics of the pipeline in which it is required to work, e.g. a rotodynamic pump should operate as close as possible to its b.e.p. However, in a slurry transport application, there are more constraints which must be taken into account compared with a clean liquid system, particularly where centrifugal pumps are involved (16,18,44). First, both pump and pipeline characteristics are likely to be affected by the presence of solids; in general, pump head/flow performance will reduce (see section 3.4), and the system resistance (friction loss) increase, with increasing concentration.

Then the pipeline velocity, and hence flow rate, must always be kept above a certain critical value with a settling slurry, otherwise solids deposition will occur, with the risk of eventual pipeline blockage (8,9,16,18,43,44). Head losses on the suction side are also important from the N.P.S.H. aspect, for all types of pump (9,16,18,44). Hence, the operating 'envelope' for a constant-speed solids-handling centrifugal pump in a particular system may be relatively small.

These effects, all of which depend on the properties of the slurry being pumped, are illustrated in Figure 20, which shows how the possible operating range of a constant-speed

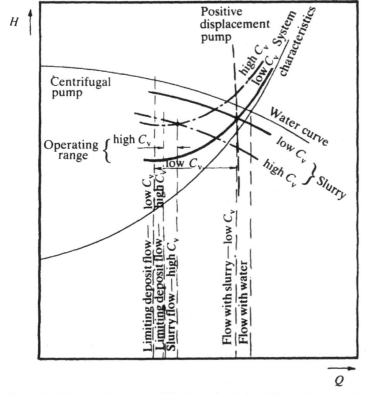

**Figure 20.** Effect of mixture concentration on pump head/flow and system characteristics with a settling slurry, showing possible operating range

centrifugal pump will be reduced as concentration increases. Thus, the $H$–$Q$. characteristic should be as steep as possible to give maximum operating stability, the ultimate being with a positive displacement pump where flow remains almost constant, independent of head, for constant speed; alternatively, variable-speed drive may be considered for a centrifugal pump, particularly where difficulty in controlling concentration is expected (8,9,16,18,43,44). The problem may be even more difficult with a non-Newtonian slurry, as shown in Figure 21, when the $H$–$Q$. characteristic of a centrifugal pump may become unstable at low flow rates as concentration increases (18,22) (see section 3.4).

Pumps may be connected in series (usually centrifugal) for high head applications, and in parallel to cover a wide flow range (8,14–16). Refs. 8, 9, 16, 18, 22, 43 and 44 all discuss the various factors affecting pump and system interaction in more detail; Ref. 9 deals specifically with dredging applications. Refs. 18 and 44 also describe the development of a computer program for selecting solids-handling pumps from a manufacturer's range, given the pipeline and slurry details.

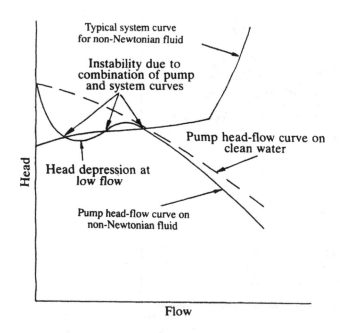

**Figure 21.**   Effect of viscosity on centrifugal pump head/flow and system characteristics with a non-Newtonian slurry, showing possible operating instability (22)

# 7. GENERAL COMMENTS

The following topics on which there would appear to be a need for further R & D work, either in terms of developing new types of pump and pump components, or providing more information to improve existing designs, are suggested:

1. Pumps for non-Newtonian 'stabilised' coarse particle slurries need to be developed. At present, the concrete-handling and rotary ram pumps appear to be the only ones which are suitable for this duty, but are limited particularly in terms of capacity/unit; higher pressures might also be desirable.
2. More compact, simpler, cheaper pumps for mine hoisting applications, particularly for handling run of mine coal. Pipe-chamber feeders, as currently used for this duty, are generally bulky, complex and expensive. Multi-stage centrifugal pumps might be considered.
3. Improved methods of gland sealing. Packed glands, which are still generally preferred for slurry applications, require considerable maintenance. So far, mechanical seals have not been too successful with abrasive slurries, but it is known that certain seal manufacturers are interested in further development.
4. Improved valve design for reciprocating positive-displacement pumps, to handle larger solids.
5. Effect of solids on centrifugal pump performance. There is a need for more data when handling coarse particle, 'stabilised' and fine particle non-Newtonian slurries, and also for better correlation of existing data.
6. Abrasive wear. As noted in the chapter 'Wear in pumps and pipelines', further evaluation of the newer construction materials, both metals and plastics, and investigation of wear patterns in centrifugal pumps is desirable.

# 8. REFERENCES

1. Odrowaz-Pieniazek, S. (February 1979) 'Solids-handling pumps – a guide to selection', *The Chemical Engineer*.
2. Thompson, T. L., Frey, R. J., Cowper, N. T. and Wasp, E. J. (September 1972) 'Slurry Pumps – a survey', *Proc. Hydrotransport 2 Conf.*, Paper H1, BHRA, Cranfield.
3. Willis, D. J., and Truscott, G. F. (May 1976) 'A survey of solids-handling pumps and systems, Part II – Main survey', BHRA, TN 1463.
4. Paterson, A. C. and Watson, N. (September 1979) 'The N.C.B. pilot plant for solids pumping at Horden Colliery', *Proc. Hydrotransport 6 Conf.*, Paper H1, BHRA.
5. Graham, J. D. and Odrowaz-Pieniazek, S. (May 1978) 'The design and operation of a slurry pumping system from 1000 m underground to surface', *Proc. Hydrotransport 5 Conf.*, Paper J2, BHRA, Cranfield.
6. Wilson, G. (September 1972) 'The design aspects of centrifugal pumps for abrasive slurries', *Proc. Hydrotransport 2 Conf.*, Paper H2, BHRA, Cranfield.
7. Warman, C. H. (June 1965) 'The pumping of abrasive slurries', *Proc. 1st Pumping Exhibition and Conf.*, Earls Court, London, 15 pp, including 12 Figures.

8.   Bain, A. G. and Bonnington, S. T. (1970) 'The hydraulic transport of solids by pipeline', (book) 1st edition, Pergamon Press.
9.   de Bree, S. E. M. (1977) 'Centrifugal dredge punps', Nos. 1–10, IHC. Dredger Division, Holland.
10.  White, J. F. C. and Seal, M. E. J. (August 1982) 'Seasand handling and pumping at Cockburn Cement', *Proc. Hydrotransport 8*, Paper A5 pp. 63–76, BHRA, Cranfield.
11.  Read, E. N. (August 1982) 'Experience gained with large dredge pumps in a sand mining operation', *Proc. Hydrotransport 8 Conf.*, Paper F1 pp. 263–78, BHRA, Cranfield.
12.  Kratzer, A. (March 1979) 'Some aspects on the design and selection of centrifugal pumps for sewage and abrasive fluids', *Proc. Pumps '79*, 6th Technical Conf., BPMA., Paper D3, BHRA, Cranfield.
13.  Burnett, M., Harvey, A. C. and Rubin, L. S. (September 1979) 'A self-controlling slurry pump developed for the US Department of Energy', *Proc. Hydrotransport 6 Conf.*, Paper G2 pp. 305–14, BHRA, Cranfield.
14.  Sabbagha, C. M. (August 1982) 'Practical experiences in pumping slurries at ERGO', *Proc. Hydrotransport 8 Conf.*, Paper A1 pp. 1–16, BHRA, Cranfield.
15.  Guzman, A., Beale C. O. and Vernon, P. N. (August 1982) 'The design and operation of the Rossing tailings pumping system', *Proc. Hydrotransport 8 Conf.*, Paper A2 pp. 17–36, BHRA, Cranfield.
16.  Crisswell, J. W. (August 1982) 'Practical problems associated with selection and operation of slurry pumps', *Proc. Hydrotransport 8 Conf.*, Paper F5 pp. 317–38, BHRA, Cranfield.
17.  Ahmad, K., Baker, R. C. and Goulas, A. (March 1981) 'Performance characteristics of granular solids – handling centrifugal pumps – a literature review', BHRA, TN 1666.
18.  Walker, C. I. (September 1980) 'Pumping solid-liquid mixtures: performance characteristics of centrifugal pumps when handling fine granular solids in suspension, and a computer-aided pump selection procedure for slurry applications', *M.Sc Thesis, S.M.E.* Cranfield Institute of Technology.
19.  Cave, I. (May 1976) 'Effects of suspended solids on the performance of centrifugal pumps', *Proc. Hydrotransport 4 Conf.*, Paper H3, BHRA, Cranfield.
20.  Sellgren, A. (September 1979) 'Performance of centrifugal pump when pumping ores and industrial minerals'. *Proc. Hydrotransport 6 Conf.*, Paper G1 pp. 291–304, BHRA, Cranfield.
21.  Gillies, R. *et al.* (August 1982) 'A system to determine single pass particle degradation by pumps', *Proc. Hydrotransport 8 Conf.*, Paper J1 pp. 415–32, BHRA, Cranfield.
22.  Johnson, M. (August 1982) 'Non-Newtonian fluid system design – some problems and their solutions', *Proc. Hydrotransport 8 Conf.*, Paper F3 pp. 291–306, BHRA, Cranfield.
23.  Rouse, W. R. (August 1982) 'Operating experience with a residue disposal system', *Proc. Hydrotransport 8 Conf.*, Paper A6 pp. 77–90, BHRA, Cranfield.
24.  Holthuis, C. H. and Simons, P. W. H. (November 1980) 'The GEHO diaphragm pump – a new generation of high-pressure slurry pumps', *Proc. Hydrotransport 7 Conf.*, Paper A3 pp. 17–32, BHRA, Cranfield.
25.  Zoborowski, M. E. (August 1982) 'African experiences with pipeline transportation of rutile and sand by hydraulic exchange cylindrical diaphragm pump', *Proc. Hydrotransport 8 Conf.*, Paper P3 pp. 515–26, BHRA, Cranfield.
26.  Eriksson, B. and Sellgren, A. (May 1978) 'Development of slurry transportation technology in Sweden', *Proc. Hydrotransport 5 Conf.*, Paper J5, BHRA, Cranfield.
27.  Sauermann, H. B. and Webber, C. E. (November 1980) 'High pressure pinch valves in hydraulic transport pipelines', *Proc. Hydrotransport 7 Conf.*, Paper A4 pp. 33–40, BHRA, Cranfield.
28.  Verkerk, C. G. (August 1982) 'Transport of fly ash slurries', *Proc. Hydrotransport 8 Conf.*, Paper F4, pp. 307–16, BHRA, Cranfield.

29.  Childs, W. D. (September 1979) 'Pulseless high pressure and flow slurry pumps', *Proc. Hydrotransport 6 Conf.*, Paper G5, BHRA, Cranfield.
30.  Kuhn, M. (May 1978) 'Feeding of solid matter into hydraulic conveying systems', *Proc. Hydrotransport 5 Conf.*, Paper F1, BHRA, Cranfield.
31.  Szivak, A., Illes, K. and Varga, L. (May 1978) 'Up-to-date hydraulic transport systems for the delivery of industrial wastes', *Proc. Hydrotransport 5 Conf.*, Paper F2, BHRA, Cranfield.
32.  Tianyu, W. *et al.* (November 1980) 'Study on high-pressure coal feeding techniques by using bin-type feeder', *Proc. Hydrotransport 7 Conf.*, Paper B3 pp. 55–70, BHRA, Cranfield.
33.  Funk, E. D., Barrett, M. D., and Hunter, D. W. (May 1978) 'Pilot experiences with run-of-mine coal injection and pipelining', *Proc. Hydrotransport 5 Conf.*, Paper F3, BHRA, Cranfield.
34.  Sakamoto, M. *et al.* (May 1978) 'A hydraulic transport study of coarse materials including fine particles with hydrohoist', *Proc. Hydrotransport 5 Conf.*, Paper D6, BHRA, Cranfield.
35.  Sakomoto, M. *et al.* (September 1979) 'Vertical type hydrohoist for hydraulic transportation of fine slurry', *Proc. Hydrotransport 6 Conf.*, Paper F1 pp. 257–68, BHRA, Cranfield.
36.  Siebert, H. *et al.* (November 1980) 'Further experience with horizontal and vertical hydraulic hoisting of coarse run-of-mine coal at Hansa Hydromine', *Proc. Hydrotransport 7 Conf.*, Paper B2 pp. 41–54, BHRA, Cranfield.
37.  Kortenbusch, W. (August 1982) 'Latest experience with hydraulic shaft transportation at the Hansa Hydromine', *Proc. Hydrotransport 8 Conf.*, Paper J5 pp. 471–84, BHRA, Cranfield.
38.  Wakefield, A. W. (November/December 1971) 'Jet pumps for dredging and transportation Parts 1 and 2', *Cement, Lime and Gravel*.
39.  Debreczeni, E. *et al.* (September 1979) 'Hydraulic transport systems in the mining industry using jet slurry pumps', *Proc. Hydrotransport 6 Conf.*, Paper G3 pp. 315–28, BHRA, Cranfield.
40.  Kuhn, M. (September 1979) 'Developments for an improved system of vertical solids transportation', *Proc. Hydrotransport 6 Conf.*, Paper F3 pp. 283–90, BHRA, Cranfield.
41.  Weber, M. and Dedegil, Y. (May 1976) 'Transport of solids according to the air-lift principle', *Proc. Hydrotransport 4 Conf.*, Paper H1, BHRA, Cranfield.
42.  Heywood, N. I. and Charles, M. E. (May 1978) 'The pumping of pseudo-homogeneous, shear-thinning suspensions using the air-lift principle', *Proc. Hydrotransport 5 Conf.*, Paper F5, BHRA, Cranfield.
43.  Tarjan, I. and Debreczeni, E. (May 1976) 'Determination of hydraulic transport velocity for pumps with various characteristics', *Proc. Hydrotransport 4 Conf.*, Paper H2, BHRA, Cranfield.
44.  Walker, C. I. and Goulas, A. (March 1981) 'Computer-aided slurry pump selection', *Proc. Pumps – the developing needs, 7th Tech. Conf.* BPMA., Paper 15, BHRA, Cranfield.
45.  Boyle, B. E. and Boyle, L. A. (May 1976) 'Hydraulic transportation of coal from collieries to coasts', *The Australian Institute of Mining and Metallurgy Conf.*, pp. 303–11, Illawara, N.S.W.

# 3.  WEAR IN PUMPS AND PIPELINES

## 1.  INTRODUCTION

Wear is a very important consideration in the design and operation of slurry systems, as it affects both the initial capital costs and the life of components. It may be defined as the progressive volume loss of material from a surface, due to all causes.

There are two main causes of wear:

1. erosion (in the most general sense), involving mechanical action by:
   (a)  solid particles – abrasive wear
   (b)  cavitation
2. corrosion, involving chemical or electro-chemical action.

Both effects can take place simultaneously – 'erosion-corrosion' – the relative proportions depending on the particular application as described by Bain and Bonnington (1). Hence the choice of construction materials is usually a compromise between abrasion resistance, corrosion resistance, mechanical strength and cost.

The information given below comes mainly from BHRA literature surveys on abrasive wear in rotodynamic machines (4) and in pipelines (7), a 'Pipe Protection' Review (8), carried out jointly by BHRA and the Paint Research Association, and the more useful papers from BHRA 'Hydrotransport' Conferences. Turchaninov's book (9) on pipeline wear includes an extensive bibliography, almost exclusively on Russian work.

The more general aspects of wear, applicable to both pumps and pipelines, are considered first, followed by a discussion of those aspects relating specifically to either pumps or pipes. There is, however, a disturbing, though consistent theme running throughout: that is, in spite of a considerable amount of literature, the general lack of reliable quantitative data on many of the factors affecting wear, from which wear rates for new systems might be predicted with reasonable certainty.

## 2.  FACTORS AFFECTING WEAR: TYPES OF WEAR

In general, these apply to both pumps and pipes. Many of the references deal with these topics in varying detail.

## 2.1  BASIC FACTORS AFFECTING ABRASIVE WEAR

There are three main components in the basic wear process: the abrasive particles, the material against which they impinge, and the relative impact velocity (e.g. 1–5,7,9,10,21, 22,24–26). Thus one may divide the relevant factors into three main groups, comprising the properties relating to:

1. abrasive slurry – particle hardness, size, shape (i.e. angularity or sharpness and relative density; solids concentration in mixture;
2. construction materials – composition, structure, hardness;
3. flow – speed, direction (i.e. impact angle); pipe flow regime (i.e. type of particle motion).

## 2.2  TYPES OF ABRASIVE WEAR

It has been suggested that wear terminology tends to be rather loose – the terms 'wear', 'erosion', 'abrasion', 'abrasive wear' are often applied to any wear situation. The various types of 'abrasive wear' are defined and discussed below.

Many references concerned with simulation wear tests on materials distinguish between various types of wear (2,9,21,22,24–26). In general, wear in slurry systems can be classified under three main types of abrasion: erosion, gouging and grinding (5).

### 2.2.1  Erosion abrasion

This arises from the impingement of smaller particles on the wearing surface, and probably predominates in most systems. It is usually considered to have two mechanisms, as described by Bitter (2):

1. Cutting wear is associated with the particle velocity component parallel to the surface, where stresses are due to velocity rather than impact and material is removed by a cutting or scratching type process.
2. Deformation wear is due to the normal component of velocity, where the elastic limit of the material is exceeded and plastic deformation occurs, eventually destroying the surface layer after initial work hardening.

In practice, a wide variation of impact angle usually exists, and these two mechanisms will occur simultaneously. Figures 1a and 1b show the results of Bitter's tests (2) on ductile and brittle materials, giving the relative proportions of cutting and deformation wear for a range of impact angles from 0–90°.

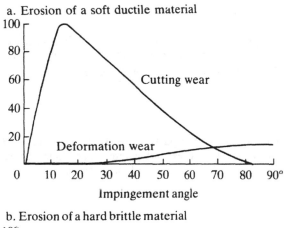

a. Erosion of a soft ductile material

Cutting wear

Deformation wear

Impingement angle

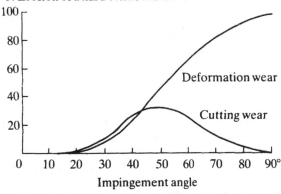

b. Erosion of a hard brittle material

Deformation wear

Cutting wear

Impingement angle

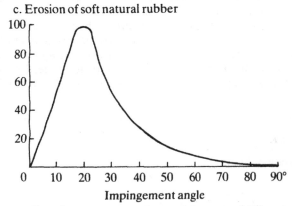

c. Erosion of soft natural rubber

Impingement angle

Erosion wear rate as a percentage of maximum

**Figure 1.** Effect of impingement angle on erosion wear of different materials (2,5)

## 2.2.2   Gouging abrasion

This occurs when large particles impinge with sufficient force that high impact stresses are imposed, resulting in the tearing out of sizeable fragments from the wearing surface. This type of wear would be significant in dredging and gravel-handling systems.

## 2.2.3   Grinding wear

This involves the introduction and crushing of fine particles between two surfaces in close proximity, e.g. pump internal leakage clearances and glands. Very high stresses are involved, and wear occurs by cutting, local plastic flow and micro-cracking.

## 2.3   WEAR THEORY

A number of simple expressions are given in the literature (e.g. 1,4,5,7,10,21,22,25,32,34, 37,39), based on wear test results, for wear rate as a function of velocity, material hardness, grain size or solids concentration. The Japanese review paper (10), gives a summary of such expressions. The one most often quoted for both pump and pipe wear is:

$$\text{wear} \propto (\text{vel})^n$$

where index $n$ may vary depending on the material and the other factors involved. For pump wear (4,5,51), the most common value of $n$ appears to be 3, but for pipe wear (1,7,10) there is much more variation in the value quoted, lying between 0.85 and 4.5, though the range is mainly between 1.4 and 3.0 (10). However, in practice it is doubtful if wear will vary with velocity as a simple power law.

Bitter's fundamental study of erosion phenomena (2), strictly for dry conditions, develops separate expressions for cutting and deformation wear, based on particle mass, velocity and impact angle, and the various properties of the surface being eroded.

Other detailed analyses relating to wear in hydraulic machines consider wear as affected by forces and velocities acting on a particle in liquid flow. Bergeron (11) attempts to predict wear rates in similar centrifugal pumps handling solids with varying properties, with simplified assumptions such as pure sliding of the particles over the surface, from the initial expression:

$$\text{wear} \propto \frac{U^3}{D}\,(\rho_s - \rho_l)\,d^3 p K$$

where  $U$ = characteristic velocity of liquid
$\rho_s$ = density of particles
$\rho_l$ = density of liquid

$d$ = diameter of particles (assumed spherical)
$D$ = characteristic dimension of machine
$p$ = number of particles/unit surface area
$K$ = experimental coefficient depending on particle abrasivity.

In a much more involved analysis in a later paper (12), but starting with the same basic assumption, he develops a complicated expression based on the statement:

Wear $\propto$ solid-liquid density difference $\times$ acceleration of main flow $\times$ coefficient of friction $\times$ thickness of particle layer $\times$ flow velocity.

This takes account of the difference between solid and liquid velocities.

More recently, both in the UK and West Germany, computer methods have been developed for analysing flow patterns, particle concentrations (13) and trajectories (14,15) through a pump impeller, from which wear patterns may be predicted. (See section 6.1.1.)

Some authors, mostly from the Soviet Union and Eastern Europe (4,11,30) also develop expressions for pump service life, either in terms of pump total head and the other properties of slurry and construction materials which affect wear, or based on a statistical analysis of pump wear tests.

Regarding pipelines, Turchaninov (9) suggests that wear is a function of the hydraulic drag of the slurry on the pipe, and gives some analysis; he also develops expressions for predicting pipeline life, but all are empirical, relying on experimental data. Another Soviet author (16) gives a method, also empirical, for predicting particle degradation in a pipeline (see section 3.2.1). Other papers (25,32) give empirical equations for wear as a function of concentration (25) (see section 3.3) or particle size and velocity (32) (see section 5.1).

It is, perhaps, debatable whether these more complex theories can be used to predict absolute wear rates with any certainty; most involve coefficients which have to be determined experimentally in any case. However, a theoretical treatment may help in predicting trends when only one or two of the relevant factors are altered, or in the correlation of results.

## 2.4 CORROSION

Corrosion may be defined as the destructive attack on metal by chemical or electro-chemical reaction with its environment (8).

### 2.4.1 Basic mechanism

Corrosion of metals by liquids is normally electro-chemical, and involves the two processes concerned with basic galvanic cell type reactions. As stated in Refs. 1 and 8, the metal shows a tendency to go into solution, forming positively charged ions. This departure of metal ions leaves the surface with an excess of electrons. A closed electrical circuit can then be set up in which the electrons travel from one area (anodic) of the surface through the metal to other areas (cathodic), where they are absorbed in a number of ways. As this process continues, corrosion results. The anodic and cathodic reactions occur simultaneously and at the same rate.

Corrosion in slurry pipelines is generally due to the presence of dissolved oxygen in the slurry, although other factors, such as pH value, are also involved (33). Therefore, corrosion control is usually attempted with oxygen inhibitors which combine with the oxygen to prevent it from reacting with the pipe. This type of corrosion control generally works best in slurries which do not have strong ionising potentials, such as coal or limestone slurries.

## 2.4.2   Types of corrosion

Probably the most common type of attack in slurry systems is 'erosion-corrosion', where the rate of corrosion is accelerated by scouring of the surface and removal of protective oxide or scale films by the impacting solids. This is discussed fully in Refs. 1 and 8.

Apart from general corrosion, certain types of localised corrosion may also occur in pumps and pipelines, depending on the construction materials and the particular working environment. These types could include crevice corrosion, bimetallic corrosion ('electrolytic action'), intergranular corrosion, weld decay and pitting, as well as local erosion-corrosion. Localised attack linked to some mechanical factor may take the form of fretting corrosion, impingement attack, cavitation damage, stress corrosion cracking, hydrogen cracking and corrosion fatigue. These types are all discussed in some detail in the Pipe Protection Review (8).

## 2.4.3   Evaluation of corrosion effect

It is often advisable to carry out field or laboratory tests to evaluate the corrosion effect for particular operating conditions. Care is needed in the selection of a method to simulate actual conditions as close as possible. Among the factors to be considered when testing corrosion resistance are temperature, composition of the solution, velocity, concentration, galvanic effect, type of corrosion and susceptibility to localised attack, equipment design and stress effects (33). Hassan *et al.* (8) give the more common methods of evaluation; those observing changes in weight and physical properties are most often used. Ferrini *et al.* (33) describe a device for the direct measurement of internal pipe corrosion, based on an electrochemical measuring method; it gives test results for fresh and salt water, and coal slurry under both static and flowing conditions, with and without oxygenation. It also shows the extent to which the elimination of oxygen reduces corrosivity.

## 3. EFFECTS OF SLURRY PROPERTIES

A number of references (e.g. 1, 7–10, 17, 24, 25, 32, 35) discuss the different methods of pump and pipe wear testing and measurement, with their inherent advantages and limitations.

Most data have been obtained from laboratory simulation tests, as it is very difficult to measure the effects in isolation in practical systems, due to interaction between the various factors involved; hence, such data can usually be applied only in fairly general or comparative terms.

### 3.1 PARTICLE HARDNESS

Laboratory tests (e.g. 21,22) have shown that, for metals in general, wear increases rapidly once particle hardness exceeds that of the metal surface, for both scouring and impact abrasion (Figure 2). However, little or no quantitative information is available on the direct effect of grain hardness.

Rubber behaviour is more difficult to compare on a relative 'hardness' basis; tests on synthetic rubbers, e.g. 'Vulkollan' polyurethane, showed fairly constant scouring wear rates (Figure 2) and much lower than for the steels, except with the less hard abrasives.

An abrasivity index, known as the 'Miller Number', has been established for a wide range of slurries (23). The Miller number has two values; the first relates to abrasivity and represents the rate of weight loss of a test specimen, and the second to attrition (or degradation) as measured by the change of abrasivity due to particle breakdown. Although the Miller test machine was developed to investigate wear in reciprocating pumps, these results give some general indication of relative abrasiveness which may also apply to centrifugal pump and pipe wear. Coal and clays are usually regarded as having a low abrasivity (index Nos. 5–50), whereas various metal ores and concentrates, silica sand and some mine 'tailings' are highly abrasive (index Nos. 70–650). Figure 3 shows an approximate comparison of hardness values of some of the more common ores and minerals, with equivalent Brinell values for typical metals and plastics (5).

### 3.2 PARTICLE SIZE AND SHAPE

It has been found that, in general, wear rate increases with grain size and sharpness, (e.g. 1,4,5,7,9,10,12,22,24,25,32,34,49) angular particles will cause more wear than rounded ones. Some authors suggest that wear is proportional to size for cutting wear, but is independent of size for direct impact (4,7,21). Large particles can also cause gouging of the surface on impact.

More recent Dutch simulation tests (24,25) related to the wear in dredging systems also studied the effect of particle size, as well as other factors. Although the earlier paper (24) suggested some dependence on impact angle, the later one (25) indicates that the size effect

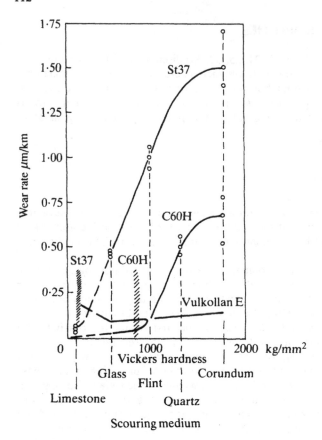

Water/solids mixture ratio by volume $= 1\cdot1$
Velocity of test specimen 6.4 m/sec
The steel hardness range is shown cross-hatched
$H_v = 110$ kg/mm$^2$ for St37, $H_v = 750$ kg/mm$^2$ for C6011

**Figure 2.** Effect of grain hardness of abrasive media on steels and Vulkollan from scouring wear tests (22)

depends more strongly on velocity and type of steel eroded (see Figure 4); however, both show the same general trend of wear increasing with grain size.

Wear tests on a small dredge pump impeller (3) showed differences in blade wear patterns depending on grain size (e.g. sand or gravel), but no quantitative results are given on wear rate. The work at Cranfield Institute of Technology (14,15) enables trajectories of

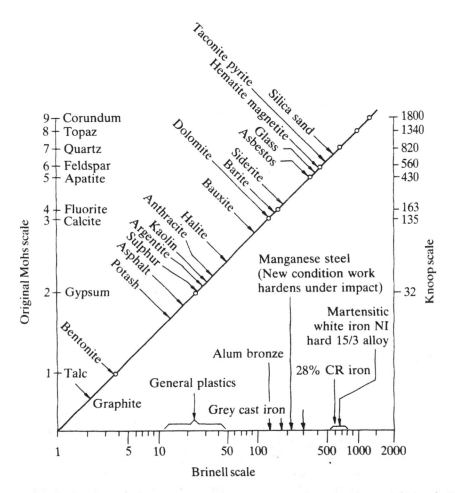

**Figure 3.** Approximate comparison of hardness values of various common ores and minerals (5)

different sized particles, and hence wear patterns, to be predicted. Figures 5a and 5b show the general decrease in pump impeller and volute wear (or increase in life) with decreasing particle size, from actual operating experience in an Australian alumina refinery (49).

There is a wide variation in results from different pipe wear simulation rigs (e.g. *rotating pipe toroid, rotating pipe specimens, etc.*) as reported by Truscott (7), although the Japanese review paper (10) suggests that wear is proportional to (grain size)$^{0.75-1.0}$ from various laboratory tests. Figure 6 shows the effect of grain size from actual pipe wear tests on a closed-circuit rig at BHRA (61). At mean velocities of 4 and 6 m/s the wear rate was

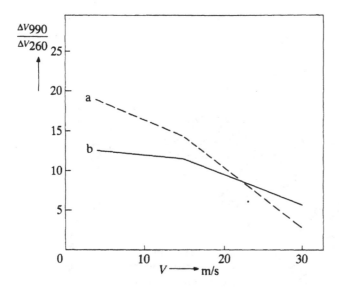

$\Delta V_{990}$ = Vol. wear for 990$\mu$m mean particle size
$\Delta V_{260}$ = Vol. wear for 260$\mu$m mean particle size

**Figure 4.** Effect of slurry velocity V, and particle size on volume wear rate for two steel types with sand slurry from rotating specimen simulation tests (25)

approximately proportional to particle size. However, at the lower velocity of 2 m/s the wear rate increased less rapidly with size. The effect of degradation of the larger particles on the wear rate is substantial at the higher velocities. Results from American pipe wear tests (32), suggest that wear rate (mm/yr) is proportional to (mean particle size)$^{2.15}$, but this effect may become less significant with increasing size.

Size and shape effects are more critical for rubber linings of both pumps and pipes than for metals; large, sharp particles are likely to cause severe damage (4,5,7,20,43,44,46,48, 51,56). There is some variation in the size limits quoted, depending on the types of abrasive and rubber, but approximately 5–6 mm ($\frac{3}{16}$ to $\frac{1}{4}$ inch) is the most generally accepted. Table 1 gives á classification of solids-handling centrifugal pumps according to particle size (5).

## 3.2.1  Degradation (attrition)

Degradation or attrition of particles (i.e. reduction of size and/or sharpness) will occur as solids pass along a pipeline, causing a reduction of wear along its length, or when passing through a

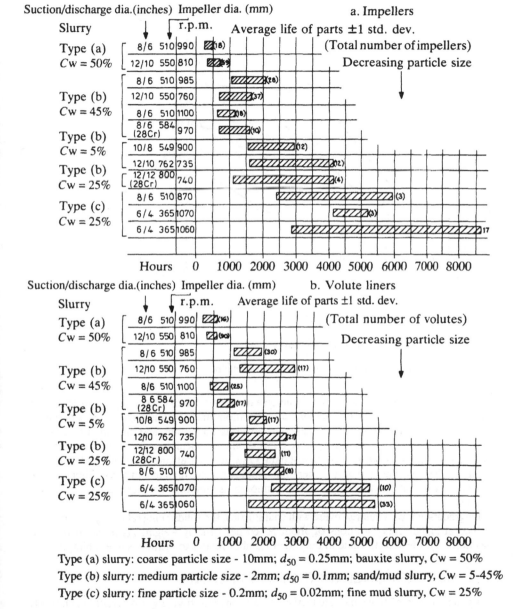

Type (a) slurry: coarse particle size - 10mm; $d_{50}$ = 0.25mm; bauxite slurry, $Cw$ = 50%
Type (b) slurry: medium particle size - 2mm; $d_{50}$ = 0.1mm; sand/mud slurry, $Cw$ = 5-45%
Type (c) slurry: fine particle size - 0.2mm; $d_{50}$ = 0.02mm; fine mud slurry, $Cw$ = 25%

**Figure 5.**   Impeller and casing life with various slurries in an alumina refinery (49)

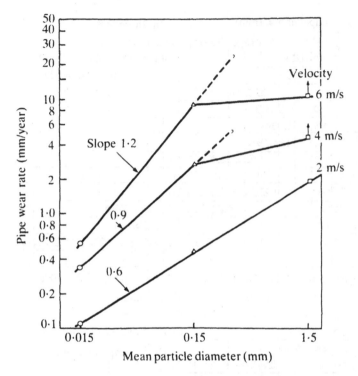

**Figure 6.** Effect of particle size on wear rate of mild steel pipe

pump. The rate of degradation will depend on a number of factors, including the structure and hardness of particles, velocity, mixture concentration and pipeline length. It is a particular problem in recirculation type slurry test rigs, affecting both pump and pipe wear results.

A few authors (e.g. Trainis (16)) give expressions for calculating degradation in pipelines using empirical coefficients based on test data. Only limited data are available on degradation in pumps, which is likely to be an order of magnitude greater than that in pipes. There is some evidence (28) that positive displacement pumps cause uniform attrition, unlike centrifugal pumps.

Degradation is particularly severe with coal slurries, the solids being relatively soft, but may reduce at higher concentrations of about 50–60 per cent by weight. As mentioned earlier (section 3.1), the second value in the Miller number relates to attrition susceptibility.

## 3.3   MIXTURE CONCENTRATION AND DENSITY

It is generally agreed that wear rate increases with concentration (e.g. 1,4,5,7,10,12,30,34,49), but there is some doubt as to the actual relationship; again, relatively little quantitative data are

**Table 1.** Classification of solids-handling centrifugal pumps according to particle size (5)

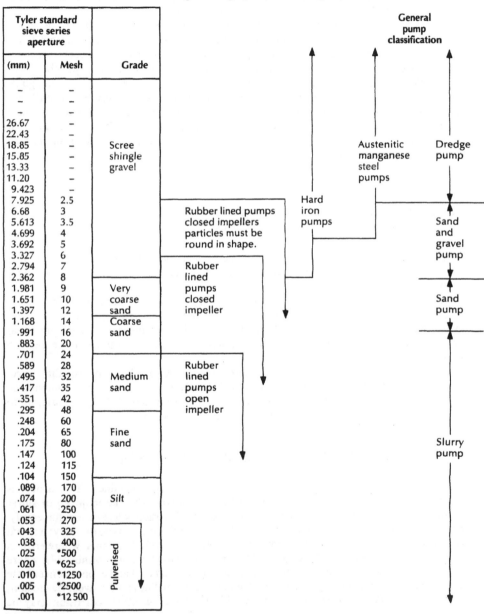

| Tyler standard sieve series aperture | | | | | | General pump classification |
|---|---|---|---|---|---|---|
| (mm) | Mesh | Grade | | | | |
| – | – | | | | | |
| – | – | | | | | |
| – | – | | | | | |
| 26.67 | – | | | | | |
| 22.43 | – | | | | | |
| 18.85 | – | Scree | | | Austenitic manganese steel pumps | Dredge pump |
| 15.85 | – | shingle | | | | |
| 13.33 | – | gravel | | | | |
| 11.20 | – | | | | | |
| 9.423 | – | | | | | |
| 7.925 | 2.5 | | | Hard iron pumps | | |
| 6.68 | 3 | | Rubber lined pumps closed impellers particles must be round in shape. | | | Sand and gravel pump |
| 5.613 | 3.5 | | | | | |
| 4.699 | 4 | | | | | |
| 3.692 | 5 | | | | | |
| 3.327 | 6 | | | | | |
| 2.794 | 7 | | Rubber lined pumps closed impeller | | | |
| 2.362 | 8 | | | | | |
| 1.981 | 9 | Very coarse sand | | | | Sand pump |
| 1.651 | 10 | | | | | |
| 1.397 | 12 | | | | | |
| 1.168 | 14 | Coarse sand | | | | |
| .991 | 16 | | | | | |
| .883 | 20 | | | | | |
| .701 | 24 | | | | | |
| .589 | 28 | Medium sand | Rubber lined pumps open impeller | | | |
| .495 | 32 | | | | | |
| .417 | 35 | | | | | |
| .351 | 42 | | | | | |
| .295 | 48 | | | | | |
| .248 | 60 | | | | | |
| .204 | 65 | Fine sand | | | | Slurry pump |
| .175 | 80 | | | | | |
| .147 | 100 | | | | | |
| .124 | 115 | | | | | |
| .104 | 150 | | | | | |
| .089 | 170 | | | | | |
| .074 | 200 | Silt | | | | |
| .061 | 250 | | | | | |
| .053 | 270 | | | | | |
| .043 | 325 | | | | | |
| .038 | 400 | | | | | |
| .025 | *500 | Pulverised | | | | |
| .020 | *625 | | | | | |
| .010 | *1250 | | | | | |
| .005 | *2500 | | | | | |
| .001 | *12 500 | | | | | |

*Theoretical values

available. Kawashima *et al.* (10) indicated that wear is proportional (volume concentration, $C_v)^{0.82-2.0}$ from the review of various laboratory test results.

It is reasonable to assume that wear will depend on the number of particle impacts on the surface, which in turn depends on the concentration. However, at higher concentrations, mutual interference between particles will tend to reduce the frequency of impacts (1). It has been suggested that wear is approximately proportional to concentration for the lower concentrations, but the rate of increase reduces at the higher values (4,5,25,34). Although de Bree states that volume wear was virtually independent of concentraton between 14 and 33 per cent $C_v$ from his earlier high-speed simulation tests at 30 m/s (24), his subsequent low-speed tests at 4 m/s (25) intended to apply mainly to wear of pipe junctions and fittings, indicated that volumetric wear, $\Delta V$, increases with concentration, $C_v$, according to a 'law of diminishing returns' (see Figures 7a and 7b):

$$\Delta V = k \ (1 - e^{-\beta C_v})$$

Where $k$ = factor depending on material properties
 $\beta$ = exponent depending on test conditions.

However, if the specific wear rate (i.e. volumetric wear/weight of solids transported) is considered, then this reduced with increasing $C_v$ (see Figures 8a and 8b) for both the high- and low-speed tests.

Figures 5a and 5b also show the general decrease in pump wear (in terms of an increase in life) as concentration decreases, from alumina plant service experience (49).

Most of the published data relates to pipe wear (7). Figure 9 shows results of a pipe wear test at BHRA on an emery slurry (61); the wear rate showed a linear increase with concentration.

Some theoretical expressions show wear depending on the solid/liquid density difference, varying either directly – if other factors remain constant – or as a more complicated function. The flow profile and particle trajectory analyses given in Refs. 13–15 attempt to predict the variation in concentration through a pump impeller.

## 3.4   CORROSIVE SLURRIES; CORROSION CONTROL

Most available information on corrosion is concerned with pipelines. This aspect is particularly important is unlined steel pipelines, which comprise the majority of slurry systems, and where both erosion-corrosion and mechanical erosion will exist simultaneously. Corrosion may be more of a problem than erosion, depending on the type of slurry; it is said to be the only wear problem with the Savage River iron ore line in Tasmania, where erosion was claimed to be negligible. Pipe wear tests at the Colorado School of Mines (34), where corrosion and erosion effects could be separated, showed that coal and copper concentrate slurries caused wear mainly by corrosion, whereas phosphate lines tend to wear by erosion, as shown in Table 2. However, alkalinity will also have an effect; corrosion may be

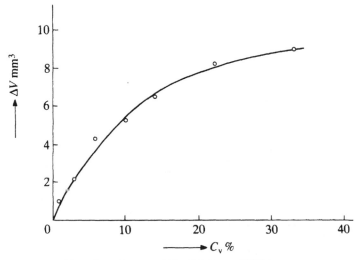

Equation of curve: $\Delta V = 9.8 \, (1 - e^{-0.08\,C_v})$

a. Low carbon steel, grade 42

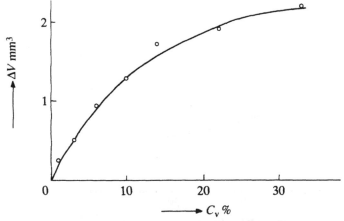

Equation of curve: $\Delta V = 2.35 \, (1 - e^{-0.08\,C_v})$

b. 15/3 CrMo Alloy steel, grade CBR315.3

**Figure 7.** Effect of mixture concentration, $C_v$ on volume wear $\Delta V$ for two steel types with sand slurry from rotating specimen simulation tests
Slurry velocity = 4 m/s
Mean particle size = 990 $\mu$m (25)

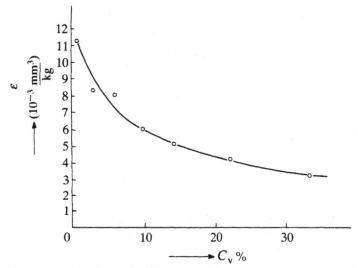

a. Low carbon steel, grade 42

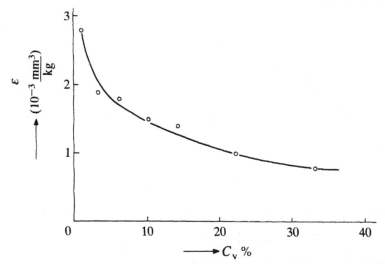

b. 15/3 CrMo Alloy steel, grade CBR315.3

**Figure 8.**  Effect of mixture concentration $C_v$, on specific wear rate ($\epsilon = \frac{\text{Vol. wear}}{\text{Wt. of impinging solids}}$) for two steel types with sand slurry from rotating specimen simulation tests
Slurry velocity = 4 m/s
Mean particle size = 990 $\mu$m (25)

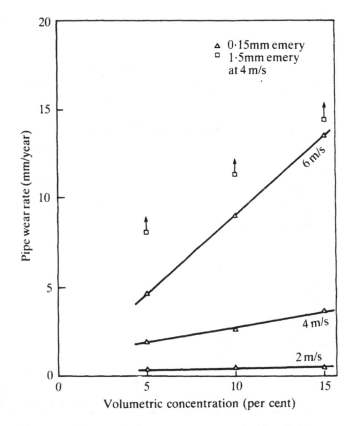

**Figure 9.** Effect of concentration on wear rate of mild steel pipe

negligible with a coal slurry if the pH value is sufficiently high. Combined corrosion and erosion effects caused fairly rapid failure of the original unlined steel pipe handling a micaceous residue slurry in a china clay process (41) with pH values in the range 2.5–5.5. Other pipe wear tests (33,36) also showed high wear rates under erosion/corrosion conditions, and with a large oxygen content (33).

Corrosion may be controlled in long slurry lines either by 'solution conditioning' involving oxygen removal, e.g. by de-aeration or the addition of sodium sulphite and pH control, or by adding oxidising inhibitors, such as high concentrations of nitrites or chromates (1,8). Corrosion is found to be more severe at the beginning of some coal, iron ore and sand pipelines (7,42). Some authors suggest that chromates, particularly if used at too low a concentration, will promote local pitting, which may be avoided by the addition of

**Table 2.** Wear rates of steel pipe with various slurries, from pipe test loop data (7)

| Reference | Type of solids | Specific gravity | | Slurry concentration (%) | | Mean velocity |
|---|---|---|---|---|---|---|
| | | Solids | Slurry | by wt. | by vol. | |
| | Hematite iron ore – | | | | | m/sec |
| | dry ⎰ | – | 1.45 | – | – | 2.5 |
| Tavares | ground ⎱ | – | 1.45 | – | – | 2.5 |
| *et al.* | ⎰ after | – | 1.74 | – | – | 2.2 |
| | ⎱ degradn | – | 1.74 | – | – | 2.2 |
| | wet ⎰ after | – | 1.72 | – | – | 2.2 |
| | ground ⎱ degradn | – | 1.72 | – | – | 2.2 |
| | | | | | | m/s |
| | Limestone | 2.68 | – | 37–68 | | 0.9–1.0 |
| Smith | Limestone | 2.70–2.62 | – | 53.7 | | 2.73 |
| and | and clay | | | | | |
| Link and Tuason | Magnetite | 4.73 | – | 30–66 | | 1.1–2.2 |
| | Teconite tailings | 2.89 | – | 60 | | 2.3 |
| | Phosphate | 2.87 | – | 40 | | 1.2 |
| | Phosphate | 2.5 | – | 25.5 | | 1.9 |
| | Phosphate | 2.5 | – | 47 | | 2.23 |
| | Copper concentrate | 5.01 | – | 45–52 | | 5.64 |
| | Anhydrite | 2.74 | – | 36.5 | | 2.05 |
| | Coal refuse | 1.73 | – | 36.4 | | 3.41 |
| (Wear test results from studies performed at Colorado School of Mines Research Institute) | | | | | | |
| Elliott and Gliddon | Crushed coal | – | – | 50–60 | | Up to 3.05m/s |
| | | | | | | m/sec. |
| Postlethwaite | Sand | – | – | – | 21–31.5 | 1.7–3.3 |
| and Tinker | Sand | – | – | – | 15–25 | 1.7–3.3 |
| | Iron ore | – | – | – | 12–20.4 | 2.3–3.3 |
| | Iron concentrate | – | – | – | 15–25 | 1.5–2.8 |
| | | | | | | m/sec |
| Barker and | Magnetite | 4.9 | – | – | 5.0 | 2.0 |
| Truscott | concentrate | | – | – | 5.0 | 3.0 |
| | | | – | – | 5.0 | 4.0 |
| (Wear test results from BHRA studies – preliminary) | | | | | | |

| Particle size | | | | Pipe dia. | Wear rate | | |
|---|---|---|---|---|---|---|---|
| | | | | | Corrosion | Erosion | Total |
| Distribution % | | | | | | | |
| −200 mesh | −325 mesh | | | mm. | | | mm/yr |
| 90 | 30 | | | 29.5 | | | 0.400 |
| 98 | 90 | | | 29.5 | | | 0.260 |
| 98 | 90 | | | 27.1 | | | 0.140 |
| 98 | 90 | | | 27.1 | | | 0.044 |
| 93 | 80 | | | 27.1 | | | 0.050 |
| 93 | 80 | | | 27.1 | | | 0.040 |

| Distribution % | | | | mm | mm/M. dry tons | | |
|---|---|---|---|---|---|---|---|
| +100 | −100 +200 | −200 +325 | −325 mesh | | Corrosion | Erosion | Total |
| 2.8 | 21.5 | (←75.7→) | | 154 | ND | ND | ND |
| 31.83 | 8.17 | (←60 →) | | 154 | − | − | 0.89 |
| 0 | 0 | 1.3 | 98.7 | 270 | − | − | 0.25 |
| 49.0 | 4.3 | 3.0 | 43.7 | 154 | − | − | 4.47 |
| 25.6 | 50.0 | (←24.4→) | | 154 | 0.84 | 1.1 | 1.98 |
| (←48.0→) | | (←52.0→) | | 52.5 | − | 41.9 | 41.9 |
| 43.3 | 37.5 | (←19.2→) | | 52.5 | − | 30.0 | 30.0 |
| 0 | 0 | 2.4 | 97.6 | 102 | 12.2 | 3.6 | 15.8 |
| 56.8 | 16.7 | 10.9 | 15.6 | 154 | 4.65 | − | 4.65 |
| 40 | 20 | 20 | 20 | 154 | 0.076 | − | 0.076 |

| Particle size | Pipe dia. mm | Corrosion | Erosion | Total |
|---|---|---|---|---|
| 0–12.5 mm | 63.5 | − | 'Extremely low' if corrosion prevented. | − |
| | 102 | | | |
| | 254 | | | |

| mesh | mm. mean | Pipe dia. | mm/yr | | |
|---|---|---|---|---|---|
| 60–100 | 0.21 | | 3–9 | | |
| 30–50 | 0.54 | 49.3 | 4–33 | | |
| | 0.21 | | 13–24 | | |
| | < 0.04 | | 20–25 | | |

| Particle size | Pipe dia. | Corrosion | Erosion | Total mm/yr |
|---|---|---|---|---|
| | | | | 0.1 |
| 1.7mm. max., 0.16mm | 38.1 | − | − | 1.25 |
| mean | | − | − | 3.1 |

hexametaphosphate ('Calgon'); however, for the Ohio coal pipeline, it was found that sodium dichromate alone could be used without causing pitting (1,7). It is recommended that laboratory tests should be carried out to determine the most effective inhibitor for a particular slurry (1,8) (see section 2.4.3).

Other possible methods of controlling or preventing corrosion, but which may be economic only for the shorter lines, are cathodic protection and the use of coatings or linings. The main advantage of rubber and polyurethane linings is that they are also abrasion-resistant (see section 4.1), whereas the thinner coatings and paints tend to give corrosion protection only (1,7).

## 4.   EFFECTS OF CONSTRUCTION MATERIAL PROPERTIES FOR PUMPS AND PIPES

As noted in the previous section, most of the quantitative data comes from laboratory simulation tests; only very limited data are available from actual pump or pipe tests, though the choice of pipe materials in commercial use is limited in any case. Otherwise, information from service experience is usually either qualitative or related to component life in fairly general terms.

## 4.1   TYPE: COMPOSITION, STRUCTURE

This section considers mainly abrasion resistance aspects.

### 4.1.1   Metals

Chemical composition, microstructure and work-hardening ability all play an important part in determining the wear resistance of metals, austenitic and martensitic steels being notably better than ferritic.

The most comprehensive set of data applicable to rotodynamic pump wear comes from Stauffer's material tests (22) in Switzerland (Escher Wyss). The data are compared on a basis of 'resistance factor', R (= volume wear of reference steel/volume wear of test material). Some Polish pump wear tests (29) also give useful comparative data on wear resistance. In general, 'Ni-hard' (4 per cent Ni, 2 per cent Cr) white iron and high chrome (12–26 per cent) cast irons and cast steel alloys were found to have the highest resistance of the cast materials, followed by 12 per cent Mn austenitic cast steel. However, 18/8 austenitic stainless steel gave only moderate resistance. Spheroidal Graphite irons were somewhat better than ordinary grey cast irons. Almost all the non-ferrous metals in Stauffer's tests were less resistant than mild steel and cast iron; tin bronzes tended to be slightly better than aluminium bronzes for the cast alloys, although 30 per cent Ni 2.5 per cent. Al bronze gave the best result for the wrought alloys. The most wear-resistant materials of all were sintered tungsten carbide, followed by hard chrome plating and the hard Cr–Co–W alloy weld overlays. More recent simulation tests on dredge pump materials (24) showed that a 15/3 CrMo alloy steel was generally better than a 3 per cent Cr steel, but was also more sensitive to the amount of impact wear (see section 5.2).

Although having not very high abrasion resistance, aluminium bronze, 18/8 and some Mn stainless steels all have good cavitation resistance. Hard Cr plating can give excellent resistance to both, provided surface preparation of the base metal is adequate. The 13 per cent Cr stainless steel is also reported as giving good abrasion and cavitation resistance, from water-turbine experience.

Service experience with centrifugal pumps (e.g. 4–6, 48–50) suggest that 'Ni-hard' and high Cr cast irons are most commonly used for general abrasive solids-handling duties with moderate-sized particles. However, for gravel and dredging applications, where the solids are relatively large, high Mn steels are often preferred, being work-hardened by impact (4,5); 'Ni-hard' and high Cr cast iron tend to be brittle and hence prone to shock damage, although still offered by reputable manufacturers for impellers handling coarse abrasives (50). Want (49) states that, although laboratory tests indicated a 27 per cent increase in wear resistance for a 28 per cent Cr cast iron over 'Ni-hard 4', pump wear tests did not show the same improvement for casing liners; some were slightly better, others worse.

Newer pump materials which are claimed to have better wear resistance than 'Ni-hard 4' are 15/3 Cr Mo alloy cast iron (e.g. 'Norihard' NH 153) (18,49) when hardened, and a 26 per cent (minimum) Cr carbide alloy (Pettibone Corpn. 'Diamond Alloy'); the latter is also claimed to have better shock resistance than most hard Ni alloy steels, though still slightly brittle.

Information relating to metal pipe materials is somewhat limited. Simulation tests, all using different techniques, include data from a Japanese report (19), where 1 per cent and 18 per cent Cr steels gave much better resistance than mild steel, and from Dutch (25) and German (26) sources, both intended to apply to wear of bends, fittings, pipe joints, etc. Some Hungarian pipe wear tests gave specific wear rates for cast iron and aluminium pipes as about 35 per cent and 55 per cent greater, respectively, than that for steel. However, the Pipe Protection Review (8) suggests that cast and spun iron pipes are more resistant than mild steel, although brittle and prone to damage during handling. There is very little published data on relative wear rates between various grades of steel pipe; pipe wear tests at the Colorado School of Mines (37) showed that a hardened steel pipe (600 HB) had about 60 per cent of the wear rate of API–5L Grade B steel pipe. Weld overlays, e.g. high Cr carbide, are sometimes used as a pipe lining (8) for very abrasive duties.

## 4.1.2  Rubbers

The ability of rubber, when vulcanised, to deform elastically under impact makes it ideally suitable to resist erosion abrasion (e.g. 2–9,43,44,51), provided that it is of sufficient thickness to avoid crushing, that bonding to the base metal is good, and that particles are not large or sharp (see sections 3.2 and 6.2.2). There is also a temperature limitation of about 80 °C for natural rubbers (4,5,7), although some synthetics, e.g. hypalon, will withstand up to 100 °C (49). However, simulation tests (21) have shown that there can be a large variation in wear rate depending on both the types of rubber and abrasive.

It is generally agreed, from both wear tests and service experience (e.g. 3–5,44,48), that soft natural rubber (Wilson 5) suggests about 35–58° Shore hardness) gives best results and is often equal or even superior to hard metal for both pump and pipe applications. However, soft rubber linings in a centrifugal pump were not very successful with a fairly coarse bauxite slurry at 60 °C, although the report (49) admits that this could have been due to the presence of 'tramp' material.

Hypalon rubber linings in a pump handling a fine bauxite slurry at 100 °C gave a life comparable with 'Ni-hard' (49). Soviet tests claim a much improved resistance for both natural and methylstyrene rubbers over butadiene styrene rubber; isoprene rubber was also very resistant (4).

Polyurethane synthetic rubber appears to be a very promising material, particularly for pipe lining where improvements over natural rubber have been reported from both wear tests (19) and service experience (7,41,56). A more recent report from English Clays Ltd (45) notes a change to a softer grade of polyurethane; it also mentions successful trials with polyurethane valve rubbers and seats in 'Mars' piston pumps. However, reports on its application to centrifugal pumps have been conflicting (48); it has been suggested that it may be more suitable for static components, such as casings, rather than impellers, but this could well depend on the grade of polyurethane and the quality of bonding. Australian mining experience with pumping cemented aggregate fill (46) indicated that polyurethane lining of the pump throat 'bush' showed least wear of a number of different elastomers tried, though none was entirely successful.

Another new proprietary gum product, 'Sellatex', has been used successfully on pump impellers and throat bushes in a South African tailings disposal system, giving much improved life over the original grade of rubber (44). Regarding pipe materials, a rubber-lined plastic pipe, comprising a core of natural rubber surrounded by layers of unsaturated polyester GRP has been introduced by a Swedish manufacturer (57). It is designed for high pressure applications, and is claimed to be suitable for 'relatively coarse' particles, wear resistant and 10–15 per cent cheaper than rubber-lined steel pipe.

## 4.1.3  Plastics and plastics coatings

There appears to be little published information so far on the use and behaviour of plastics for solids-handling pumps. Simulation jet-impact tests (e.g. 26) indicate that most, if not all, were less resistant than mild steel, with certain grades of polyethylene best, followed by nylon and 'Teflon'. The Soviets claim to have had some success in service with epoxy resins either filled with emery or granite powders, or with glass-fibre reinforcement. However, epoxy resin impellers failed rapidly in the wear tests on Polish pumps (29).

Polyvinylidine fluoride (PVDF) is a relatively new material which is claimed to be abrasion resistant. It can be used in solid form, e.g. for an impeller, with carbon fibre reinforcement or as a coating, e.g. for a cast iron casing (52).

Other types of simulation test, intended for pipe wear, appear to give conflicting results, both in terms of wear rates and order of resistance for apparently similar types of plastics. The German work noted in the literature surveys on pump and pipe wear (4,7) showed that most plastics were inferior to mild steel, except for the softest grades of PVC, in the sand blasting test. However, a list of materials given in the Pipe Protection Review (8), based in part on results from BHRA pipe-tests (35), suggests that GRP ('fibre-glass'), uPVC, ABS (Acrylonitrile-Butadiene-Styrene), high-density polyethylene and polybutylene – in increasing order of abrasion resistance – may all be superior to mild steel. Pipe wear tests in the USA (37) with a tailings slurry also showed high-density polyethylene to be more resistant than rubber or mild steel. Hence, again some doubt exists as to whether this conflicting evidence is due to differences in test methods or in grades of similar materials; errors can arise in wear tests, using weight-loss measurement, on polymers and elastomers due to water absorption (19,38), and BHRA has developed a technique to overcome this problem (38).

Hardly any service experience is available for comparison with laboratory test results; although PVC-lined mild steel pipe was included with a number of other materials in N.C.B. pilot plant wear tests at Horden Colliery (56), there was only a general comment on the problems of linings becoming detached from the pipe, or being cut by large particles. Plastics would probably be considered only for relatively short distance pipelines, from pressure limitation, thermal expansion and economic considerations (37). Although some plastics coatings and paints have better abrasion resistance than others (27), their main advantage is as a protection against corrosion (7,8,58).

## 4.1.4   Ceramics and related materials

Once again, there is relatively little and sometimes conflicting information, and probably only limited application, regarding pump and pipeline use. Although very abrasion resistant, the use of ceramics in pumps has been limited so far by their brittleness and susceptibility to thermal shock damage; there may also be particle size limitations as for rubber linings. Some makes of centrifugal pump use ceramic impeller sealing rings and ceramic coatings on shaft sleeves in way of the gland (4,48).

Cast basalt and fused alumina, which are normally regarded as being very wear resistant are used as insert linings for mild steel (7,58) and, more recently, GRP pipe (8), particularly for those sections of pipeline subjected to excessive wear, such as bends (8,46) (see section 7.2). Basalt showed fairly good wear-resistance in other simulation tests (19). Basalt linings were also tried in the pumping chambers for the pipe feeder in the Hansa 'Hydromine' in West Germany (58) as well as being used successfully in the horizontal pipelines. Fused alumina-lined steel pipe gave the best results in the N.C.B. tests (56), being reported as about nine times more resistant than basalt, although five times the cost. However, these materials are likely to be economic only for short-distance lines handling very abrasive slurries.

## 4.1.5   Concrete and cement based materials (for pipes)

Russian work on concrete pipe linings (7) claims improved abrasion resistance by 'vibro-activation', and also by the addition of a large proportion of cast iron or steel particles, or plastics resin binders ('plastic concrete') to the mix. However, all types have a much lower resistance than mild steel. Limited data from some French pipe wear tests (39) on various materials, including a urethane resin/concrete mix, give results only in terms of velocity indices for each material (see section 5.1); no actual wear rates are reported.

There is little data on asbestos cement pipes; test results suggest a performance inferior to steel (27) and even concrete (8), but one report of site experience on an Australian heavy mineral concentrate pipeline indicates a wear resistance equivalent to steel; particle size is an important factor.

## 4.2   MATERIAL HARDNESS

In very general terms, abrasion resistance for ferrous metals tends to increase with hardness (e.g. 4,5,18,21,22,24). It must be emphasised that metal hardness is not an absolute criterion of wear (4,5,21,22,24), and wear resistance is not simply proportional to hardness; other factors, including impact angle and particle properties (21,24) are also involved. Stauffer's results (22) show this general tend, but there is a large scatter, with some harder materials giving much lower resistance factors than the softer ones; the Polish pump wear tests (29) also showed similar effects. Even a trend is not apparent for the copper alloys, the resistance of tin bronzes being virtually constant, independent of hardness. It may be noted, however, that the most wear-resistant materials, such as tungsten carbide, hard chrome plating and weld overlays, were also the hardest.

These results suggest that reasonable abrasion resistance is achieved above about 300 Brinell Hardness No. (HB). Typical hardness values for high Cr alloy cast irons and 'Ni-hard' as used in abrasive duty centrifugal pumps would be in the range 550–700 HB (4,5). For the newer alloys mentioned in section 4.1.1, the Cr carbide alloy has a hardness of 500–600 HB and the hardened Cr Mo alloy cast iron (18) is about 950 Vickers Hardness (HV).

Hard materials generally are not suitable for pipe construction on economic grounds, apart from basalt and ceramic linings having hardness values equivalent to about 1000 HB, and weld overlay cladding, for very abrasive duties (8). Flame-hardened carbon steel pipe was included in the N.C.B. pilot plant wear tests (56), but again no results were given; presumably it was less resistant than zirconia alumina ceramic (ZAC) or polyurethane lining. It has been noted from German steel pipe wear tests (3) that hardness may vary across the pipe wall – a similar property could also apply to pump impeller and casing sections – and hence life will depend on the average rather than just surface hardness.

Soft rubbers are usually accepted as being more abrasion-resistant than hard ones (e.g. 3–5,44,48), but there is little or no information on the effect of hardness of plastics.

## 4.3 CORROSION AND CHEMICAL RESISTANCE

Most slurry systems are concerned with suspensions of solids in water, the main exceptions being those involved with particular chemical processes.

### 4.3.1 Corrosion resistance of metals

The corrosion resistance of mild steel is low compared with most other metals, whereas the various types of cast iron are not particularly prone to corrosive attack. Regarding pump materials, in general, the alloys of iron and steel containing Cr, Ni, Co, or Mn give improved corrosion performance over non-alloyed types. The 18/8 and 13/4 Cr/Ni stainless steels, bronzes, 'Ni-Resist' austenitic cast iron (53) and 'Hastelloy' alloys all have good corrosion resistance, although only limited abrasion resistance.

Of the more abrasion-resistant materials, 28 per cent Cr cast iron is reported to have better corrosion resistance than 'Ni-hard' or 15 per cent Cr cast iron, but is more brittle (5). 'Ferralium' (25/5 Cr/Ni alloy steel), the Pettibone Cr carbide alloy and the very hard materials, such as metallic carbides, weld overlays and chrome plating, are also generally both corrosion- and abrasion-resistant.

Surface finish also affects the corrosion rate, smooth surfaces being less prone to attack than rougher ones.

### 4.3.2 Chemical resistance of non-metals

The Pipe Protection Review (8) gives a good indication of the resistance of most rubbers and plastics to chemical attack. Natural rubber has generally good chemical resistance, except when in contact with mineral oils and solvents, or strong oxidising acids. 'Neoprene' and nitrile rubbers give better resistance to oils; butyl rubbers (e.g. 'Hypalon') are also more resistant to oxidising agents, and to higher temperatures. Polyurethane also has good overall resistance. Hard rubbers tend to have better chemical resistance, but less abrasion resistance, than the softer grades.

High-density polyethylene, PVDF (52), ABS and uPVC pipe, and many plastics coatings all withstand chemical attack well, as do other non-metals, such as ceramics and asbestos cement.

## 5.  EFFECTS OF FLOW PROPERTIES

## 5.1  VELOCITY

It is generally agreed that abrasive wear increases rapidly with flow, or particle, velocity and, as indicated in section 2.3, many authors suggest a simple power law relationship for both pump and pipe wear, based either on laboratory wear tests or theoretical considerations. For pump wear, it is most often quoted that wear rate is approximately proportional to (velocity)[3] or proportional to (pump head)[3/2]; some authors (e.g. 5,51) suggest a range of index from 2.2–3. The German sandblast simulation tests (21) indicated that it may depend on the material; e.g. for steel, the index was 1.4 and for rubber, 4.6 for direct impact wear. De Bree's later simulation tests (25) gave a variation in index from 2.9–3.7 for two steels, depending on the type and also particle size (see Figure 4); low-carbon steel was more sensitive to speed variation than the harder 15/3 Cr Mo steel. Other authors suggest that there may be differences between ductile and brittle materials (4), or that there may be some critical value of velocity, below which wear increases, say linearly, but above which the increase is much more rapid. Results from the only pump wear test to investigate the effect of speed, noted by Truscott (4), agreed with the cubic exponent of velocity relationship for wear.

Although there is more variation in the values quoted for the velocity index in pipe wear, depending on the type of simulation test, Kawashima *et al.* (10) give a range of 1.4–3.0, actual wear tests on steel pipe give values mainly between 2 and 3 (7). Tests at BHRA on small diameter pipes (38 mm) with an emery slurry (61) gave indeces from 1.6–2.9, as shown in Figure 10. Karabelas (32) found values between 2.3 and 3.3 (see Figure 11) with a sand slurry, depending on the position round the pipe wall, with the maximum value at the bottom; he also compares his data with that of other investigations (see Table 3), based on his empirical expression:

$$E = 6.1 \, d_{\mathrm{m}}^{2.15} \, \bar{U}^{3.7}$$

where $E$ = max. wear rate (at bottom of pipe), mm/yr
$d_{\mathrm{m}}$ = mean particle size, mm.
$\bar{U}$ = mean slurry vel., m/s

and notes the inconsistencies between the data.

Pipe wear tests in France (39), mainly on lining materials, with a copper tailings slurry gave indices of 2.3 for steel, 2.5–2.7 for natural and synthetic rubbers, 4.0 for polyamide resin and 3.3 for urethane resin/concrete mix.

Thus, it is likely that the actual relationship between wear and velocity probably depends on most of the other factors involved in the overall wear process.

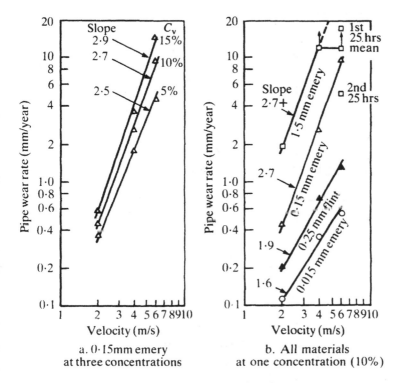

Figure 10. Effect of velocity on wear rate of mild steel pipe

## 5.2 DIRECTION (IMPACT ANGLE)

The type of material is very important in determining the effect of impact angle, as previously discussed in section 2.2.1 and shown in Figures 1a, 1b and 1c (5,22). De Bree's simulation tests (24,25) tend to confirm Bitter's results (2,4,5), with maximum wear for all three steels occurring at impact angles of about 30°–40°. Relative wear rates may also vary with angle for some steels (21,24,25), depending on test conditions (particularly particle size and velocity), and may give a large scatter of results (25). Wellinger's sand-blast tests (21) showed the relative wear rate tending to increase with angle for the steels and cast irons, reaching a maximum between 60° and 90°, whereas for rubbers, the opposite effect occurs, with cutting wear predominating at low impact angles, deformation wear being negligible (Figure 12).

In solids-handling centrifugal pumps, impact angle will vary as the solids pass through the impeller and casing passages, and will depend on the operating point on the pump head/

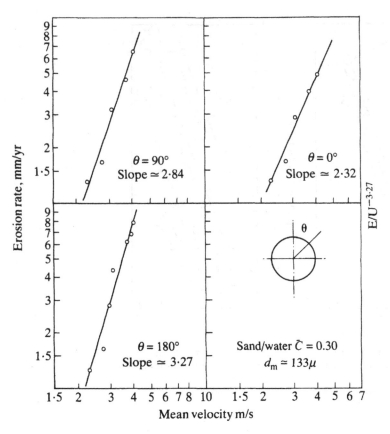

**Figure 11.** Effect of mean velocity on wear rate at various positions round pipe, from pipe wear tests with sand slurry at 30 % $C_v$ (32)

flow characteristics. Thus, hydraulic design will also have a considerable effect both on the actual wear rate and where maximum wear occurs. The computer method developed at the Cranfield Institute of Technology (C.I.T.) (14,15) for predicting wear patterns takes account of particle impact angle, as well as kinetic energy, on the impeller surfaces; typical trajectories for particles of uniform size are shown in Figure 13. In this analysis the flow field is divided into stream surfaces and the matrix through-flow analysis method is then used to calculate the streamline distribution. This enables the three components of velocity to be obtained. Next the trajectory of particles is calculated based on drag, gravitational, centrifugal and Coriolis forces.

**Table 3.** Comparison between measured and predicted pipe wear rates from other sources, based on Karabelas' empirical equation (see section 5.1) (32)

| | | | Experimental conditions | | | | |
| Investigation | Test number | Pipe I.D. (cm) | Concentration (C) | Velocity (m/s) | Mean particle diameter*, ($d_m$, $\mu$m) | Measured Erosion rate, (mm/yr) | Predicted erosion rate, (mm/yr) |
|---|---|---|---|---|---|---|---|
| Barker & Truscott | 1–2 | 3.81 | 0.05 | 2.0 | 100 | 0.10–0.15 | 0.42 |
| | 3–4 | 3.81 | 0.05 | 3.0 | 100 | 0.9–1.55 | 1.57 |
| | 5–6 | 3.81 | 0.05 | 4.0 | 100 | 2.45–4.10 | 4.02 |
| Link & Tuason | – | 5.25 | 0.12 | 1.86 | 100 | 1.60 | 0.33 |
| | – | 5.25 | 0.12 | 1.37 | 100 | 0.63 | 0.12 |
| | – | 15.41 | 0.19 | 1.22 | 130 | 0.44 | 0.15 |
| | – | 5.25 | 0.26 | 2.23 | 165 | 2.99 | 1.75 |
| Tavares *et al.* | 4 | 2.95 | 0.11 | 2.5 | 55 | 0.40 | 0.24 |
| | 6 | 2.95 | 0.11 | 2.5 | 30 | 0.26 | 0.07 |
| | 11 | 2.71 | 0.185 | 2.2 | 30 | 0.14 | 0.043 |
| | 15 | 2.71 | 0.18 | 2.2 | 35 | 0.05 | 0.06 |

*Estimates based on reported data

The wear caused by particle impacts depends on the angle of attack and kinetic energy. The method indicates the areas prone to erosion which correlate well with surface erosion tests.

Rocco *et al.* (62) considered interparticulate forces in their mathematical analysis which enabled solutions to be applied to concentrations greater than 2–5 per cent by volume which is allowable if these forces are ignored. Various numerical techniques were applied, finite element, finite volume and finite analytical method, according to their relative advantage in the flow fields in the impeller and casing. Wear was considered to be dependent on a combination of directional impact, random impact and sliding. Real wear rates were related to losses in kinetic energy by means of special laboratory tests. The analysis is complex but predictions agree well with practice and enable modifications to be made to the designs to increase pump life. Regarding pipelines, the effect of impact angle is significant primarily for flow round bends (see section 7.2) and over discontinuities, e.g. pipe joints and T-pieces, or for saltating particles (7,25).

## 5.3 FLOW REGIME (PIPE WEAR)

The type of flow regime existing in a pipeline will have an important effect on pipe wear, since it is defined by the particle dispersion in the slurry and affects the types of particle motion (sliding, rolling or saltating) along the pipe wall (1,7,10). It is possible that pump wear could also be affected indirectly, if the particle distribution across the suction pipe is changed significantly. However, the effect of flow regime alone is difficult to isolate, since

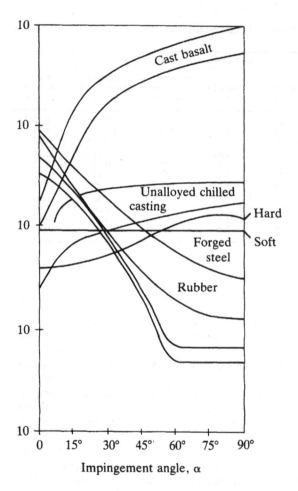

Impingement angle, α

**Figure 12.** Effect of impact angle on relative wear rates, showing wear range of different material groups, from blasting wear tests with quartz sand grain size
0.2–1.5 mm (21)

the type will be determined by flow velocity, mixture concentration, particle size and density, each of which will also affect the wear rate, and hence little test data is available. Kawashima *et al.* (10) suggest that Zandi's 'I No.' ($N_I$) describing flow regime is not adequate for general wear analysis with any type of solid particle, but could be used for individual types.

Most systems handling settling slurries operate in the heterogeneous regime with

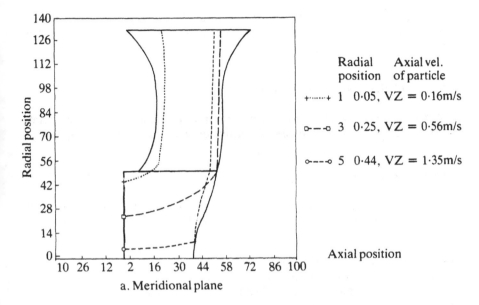

a. Meridional plane

**Figure 13.** Predicted particle trajectories in the blade passage of a slurry pump impeller, for particles released across a vertical pipe diameter (14)

turbulent flow, resulting in a solids concentration gradient across a horizontal pipe, with the highest concentration at the bottom. Hence more wear will occur in this region, due to the sliding or saltating abrasion of the larger particles (1,7,10,32,60). However, a stationary or slow-moving bed can protect the bottom of the pipe from the faster moving particles, if the increased risk of blockage with such a regime can be accepted (1).

Flow in the homogeneous regime can occur with non-settling suspensions of very fine particles, e.g. clay, chalk, etc. when the slurry behaves as a Bingham plastic and laminar flow occurs at low velocities. Wear will tend to be more uniformly distributed round a horizontal pipe wall and generally less than with a heterogeneous regime, due to the lower velocities and less turbulence (1).

Vertical pipelines also wear evenly, due to uniform distribution of solids over the pipe cross-section, though bend wear (see section 7.2) may be more severe (46).

In the special case of helically-ribbed pipe (40) although the wear rate will be higher than for smooth pipe at the same velocity, it will be similar at the limit deposit velocity condition for each case, since ribbed pipe allows lower velocities to be used.

Roco and Calder (63) have also predicted erosion rates in pipes employing similar techniques to their work in pumps (62). Included in their analysis was the lift force associated with a spinning particle in a velocity gradient. They demonstrated that due to the complexity of motion it was not possible to correlate results by simple power law equations. The analysis, however, points to ways in which scale up can be achieved from small-scale pipe loops.

## 6.  SOLIDS-HANDLING PUMP WEAR

Most published information relates to centrifugal slurry pumps, although some, mainly from service experience, is also available for positive-displacement reciprocating pumps.

## 6.1  DESIGN ASPECTS

Want (49) suggests the following criteria should be considered when selecting a slurry pump for maximum wear life, the final choice being made on the optimum combination for a given duty:

● minimum capital and replacement parts cost,
● minimum power consumption/maximum efficiency,
● minimum velocities at volute throat, and impeller blade inlet and outlet,
● minimum shaft speed (gland wear),
● minimum NPSH required (cavitation erosion).

## 6.1.1  Centrifugal pumps

Several leading authorities on solids-handling pump design mentioned in a pump wear literature survey (4) stress the importance of maintaining good hydraulic design features, as far as solids-handing considerations (e.g. unchokeability, increased section thickness) will allow, to minimise wear; in particular, rapid changes in flow direction should be avoided. Fully-shrouded impellers seem to be generally preferred, particularly for dredge pumps, though the choice between a closed or semi-open type may depend on the solids being pumped.

Figure 14 shows where high wear-rate regions would occur in a conventional centrifugal pump for clean water duty if required to pump solids; Figure 15 shows the design changes necessary to produce a satisfactory solids-handling pump and so reduce or accommodate wear (5). However, even for the latter design, the 'critical wear points' will still tend to occur in the same regions, at least for all-metal wetted parts, as follows:

Point 1.   Casing suction branch.
   2.   Impeller shrouds, near eye, especially hub side.
   3.   Impeller blade inlet edges. Blades much thicker for solids-handling design.
   4.   Impeller blade outlet edges. Blades much thicker for solids-handling design.

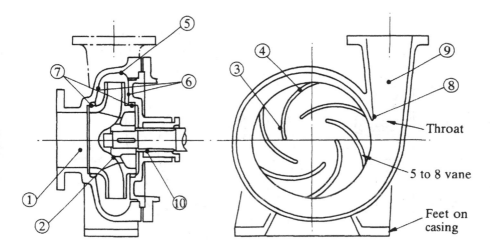

**Figure 14.**   Critical wear points in conventional centrifugal pumps (5)

5. Casing near 'cut-water'. It is suggested that a concentric or semi-concentric casing for the solids-handling design is more tolerant of 'off design' operation compared with a conventional volute, though with some sacrifice of peak efficiency; it also gives lower velocities near the cut-water, hence less wear.

6. Casing side walls and impeller outer shroud walls. Worst wear usually occurs on the suction side. Solids-handling impellers are usually fitted with 'scraper' vanes on both shrouds to reduce the flow of solids down the side spaces.

7. Impeller/casing sealing rings. Flat-faced (i.e. axial clearance) rings used in the solids-handling design are less prone to wear than the cylindrical type.

8. Casing cut-water. Concentric type casing results in a larger radius cut-water further from the impeller, giving lower velocities and less turbulence at the cut-water, hence less wear.

9. Casing discharge branch near throat.

10. Shaft seal. A solids-handling design normally has a separate clean water flush fitted to the gland, sometimes (e.g. Warman design) with the addition of an 'expeller' behind the main impeller to prevent ingress of solids.

Wear rates and patterns are likely to differ in detail from one design to another, e.g. bladed vs. 'channel' type impellers. Regarding special designs, Warman claims that his impeller shape (Figure 16a) reverses the secondary flow patterns in the casing compared with a conventional design (Figure 16b), resulting in reduced casing and impeller wear (6). However, the C.I.T. method for predicting flow patterns through an impeller (14,15) suggests that the patterns shown for the meridional plane in Figure 16 may not apply over

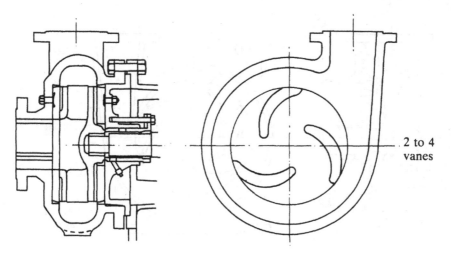

**Figure 15.** Typical centrifugal abrasive solids handling metal slurry pump liquid end construction (5)

the whole blade passage circumferentially; also, the trajectories for the larger particles (Figure 13) may not follow the flow paths. On the other hand, increased turbulence due to mixing in the volute with the Warman design may help to break any local high velocity jets from the impeller, thus reducing volute wear. A 'torque-flow' pump (e.g. Egger design) with recessed impeller is also said to suffer relatively less wear than a normal radial-flow type (4), though it must be remembered that it has to run faster to produce the same head and flow as the latter (48).

Other factors affecting wear rates and patterns are the operating point on the head-flow $(H-Q)$ characteristic, and the particle size and density. As might be expected, least wear occurs at or near the best efficiency point (5,49), when flow angles should match blade and cut-water angles, giving correspondingly low impact angles, provided that the solids trajectories follow the liquid flow paths, and also least turbulence due to flow separation. Hence wear will tend to increase both at higher and lower flow rates. Flow patterns in the volute throat above and below b.e.p. are shown in Figure 17 (49).

Local wear patterns (31) are also likely to vary for any given pump design, particularly for the impeller blades, depending on the size and density of particles passing through the passages. A typical example of impeller wear patterns is shown in Figure 18 for a rubber-lined impeller.

Figure 19 shows a comparison of wear patterns on the blade pressure surface of a small slurry pump impeller between results of paint-wear and C.I.T.'s computer prediction method; reasonably good agreement was obtained (14,15).

Several references (e.g. 4,6,43,44,46,48,51) discuss speed and/or head limitations for solids-handling centrifugal pumps, based on either manufacturers' recommendations or

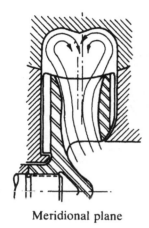

Meridional plane

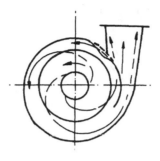

Axial plane

a. Warman design

Meridional plane

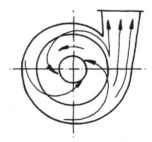

Axial plane

b. Conventional design

**Figure 16.** Comparison of suggested impeller and volute flow patterns between Warman and coventional pump designs (shrouded impellers) (6)

users' experience, and particularly for rubber-lined pumps (4,43,44,46). Crisswell (51) suggests limits for impeller eye velocity of 1.5 m/s (5 ft/s), though this may need to be increased to 2.5 m/s (8.2 ft/s) for coarse slurries to avoid settlement, impeller top speed of 25–28 m/s (82–92 ft/s) and head/stage of 30 m (100 ft) for pumps handling coarse slurries.

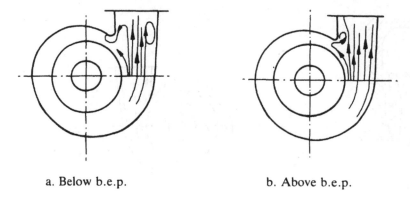

a. Below b.e.p.                                    b. Above b.e.p.

**Figure 17.**   Flow patterns in a volute throat at off-design conditions (49)

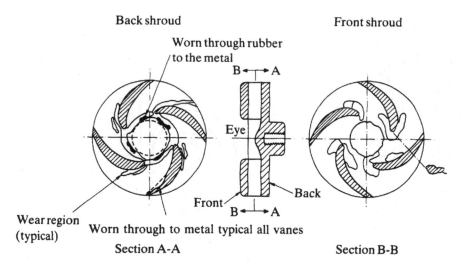

**Figure 18.**   *Typical wear patterns in a rubber-lined slurry pump impeller (31)*

However, the effect of specific speed ($N_s$) is also important (5,6,49) in designing a pump for a given duty to achieve least wear, and the common practice of limiting impeller tip speed alone can be misleading (5). Wilson (5) shows that a higher $N_s$ design, which would be cheaper in both capital and running costs, would also be likely to have a higher wear rate;

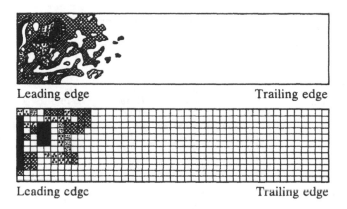

Leading edge                    Trailing edge

Leading edge                    Trailing edge

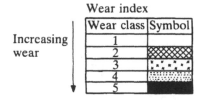

**Figure 19.** Comparison of predicted and experimental wear patterns on the blade pressure surface of a slurry pump impeller running at b.e.p., with sand slurry ($d_{50}$ = 1.8 mm, $C_v$ = 3.4 – 12 %)

although the tip speed would be only slightly increased, there would be a significantly larger increase in both peripheral and axial velocities at the impeller eye. Warman (6) also points out the advantage of the lower $N_s$ design, but suggests this is due to a lower tip speed. Table 4 gives a comparison of four different slurry pump designs for the same nominal duty (49), with design data for each, including impeller velocities, and also actual lives achieved; Pump 'D', with the largest impeller, least blades and slowest running speed, gave the longest life but had the highest running cost.

## 6.1.2  Reciprocating pumps

For piston and plunger pumps where the whole of the 'fluid end' is in contact with the slurry, parts subjected to the most severe wear are valves and seats, piston and piston rod or plunger packings and bushings (42,48). Plunger pumps are normally used for the more abrasive duties, having only one sealing point, with a clean water flush to the packing (47).

**Table 4.** Comparative details of four centrifugal slurry pumps for the same duty, handling bauxite slurry (49)

| Type (a) slurry SG 1.7<br>System flow 450 m³ h<br>System head 26 m slurry<br><br>Suction/discharge diameter (in.)<br>Type of volute | A<br><br><br><br>10/8<br>spiral | B<br><br><br><br>10/10<br>circular | C<br><br><br><br>12/10<br>spiral | D<br><br><br><br>12/10<br>spiral |
|---|---|---|---|---|
| Relative capital cost | 1.0 | 1.14 | 1.07 | 1.08 |
| Relative wear part cost | 1.0 | 1.60 | 1.29 | 1.88 |
| Efficiency at duty point (%) | 64 | 56 | 60 | 62 |
| Diameter of impeller (mm) | 686 | 890 | 762 | 914 |
| Number of vanes | 5 | 5 | 8 | 4 |
| Shaft speed (rpm) | 620 | 480 | 540 | 440 |
| Velocity of discharge (m/s) | 3.86 | 2.54 | 2.54 | 2.54 |
| Impeller exit tip vel. (m/s) | 22.3 | 22.4 | 21.5 | 21.1 |
| Impeller entry tip vel. (m/s) | 8.1 | 6.3 | 8.5 | 6.9 |
| NPSH required (m) | 2.8 | 2.6 | 1.5 | 1.0 |
| Initial Choice | | | *** | |
| Actual average life (h) | 920 | 1000 | 900 | 1100 |
| Relative operating costs<br>(wear parts only) | 1.0 | 1.47 | 1.32 | 1.57 |

Some form of stem-guided 'mushroom' valve with rubber or polyurethane disc insert, and with a hardened metal seat, appears to give best results (42,48), although rubber or PU-coated ball valves have also been used successfully for larger particle slurries (3 to 6 mm ($\frac{1}{8}$ to $\frac{1}{4}$ inch) size) at lower pressures (150 p.s.i. maximum) (48). Normally the particle size limit for reciprocating pumps is about 2 mm, although White and Seal (42) state that particles up to 10 mm could be handled by 'National' piston pumps without damaging the PU valve inserts, from operating experience with an Australian sand dredging system. Holthuis *et al.* (54) also mention the development of valve designs for handling particles of 10–15 mm for the 'GEHO' pump.

Special designs, e.g. the Japanese 'Mars' pump with an oil buffer 'piston' (41,45), the Dutch 'GEHO' piston diaphragm pump (54) and the USA 'Zimpro' hydraulic exchange cylindrical diaphragm pump (55), have been developed to isolate the fluid end parts, with the exception of the valves, from direct contact with the slurry. Valve wear can still be a problem, and Rouse (45) gives details of various modifications to the valve design and materials for 'Mars' pumps operating on micaceous residue slurry; best results were achieved with PU valve rubbers and seats, and valve bodies hard-faced with 'Admig 60' 9.5 per cent Cr alloy. Makers of the 'GEHO' (54) and 'Zimpro' (55) pumps claim longer valve life due to lower velocities through the valves compared with other types of reciprocating pump.

## 6.2 EFFECTS OF WEAR ON PERFORMANCE, COMPONENT LIFE AND SEALING

### 6.2.1 Performance

It is normally assumed that pump performance will be reduced by wear, first due to increased internal and external leakage and eventually, in the case of centrifugal pumps, by the reduction of impeller diameter and general passage roughening. There is, however, very little quantitative data available from service experience on the effect of wear on performance. Want (44) notes that it is often reported that the initial wear rate is high, followed by a gradual reduction, but one test series actually showed an increase in performance over the first half of the pump's life. It was suggested this was due to an initial polishing action by the slurry on the wetted surfaces.

### 6.2.2 Component life: metal vs. rubber lining

As noted in section 2.3, some theoretical expressions have been developed for estimating pump life, but all involve empirical coefficients based on previous test data or service experience.

Although the survey of solids-handling pumps (48) gives limited component life data for various applications, it must be emphasised that these refer strictly to the particular slurry conditions involved. Hence, no life figures can be reliably regarded as 'typical', and no attempt has been made to correlate them. In fact, some pump manufacturers suggest that it is impossible to make a realistic correlation of data, even if it is more extensive, due to the wide variation in site conditions experienced, and thus they tend to be reluctant to give information on life, in case of misinterpretation. A Polish paper (30) quotes service lives of high Cr steel parts varying from 84 hours pumping sand to 20 000 hours with coal. In general, pump impellers tend to wear faster than casings, only about a third to a half of the life being obtained in some cases.

Figures 5a and 5b (49) show the mean life of impellers and casings for five different slurries in an alumina refinery. Read (50) also gives component wear details for a 24/20 inch Warman gravel pump with high Cr cast iron wetted parts handling coarse beach sand. Other papers, mainly from Australian and South African sources, give more general life data on centrifugal pumps handling various types of mine back-fill (20,46), beach sand (42), gold slime (43) and uranium mine tailings (44).

There are many reports (e.g. 4,5,48,51) of the longer life of rubber over metal parts for centrifugal pumps, and provided that the particles are not large or sharp, bonding is good and heads and temperatures are relatively low – about 45 m (150 ft)/stage and 80 °C (180 °F) respectively. Improvements by factors of 2 to 20 over various grades of steel and cast iron and for different duties have been recorded. However, a number of papers mention problems on site with 'tramp' materials in slurries damaging rubber linings (5,20,43,44,46,48,49,51); Crisswell (51) advocates protection against tramp metal for such pumps, and suggests a horizonal discharge to prevent this material from falling back into the pump. Linings may also become detached from casings due to poor bonding, or under severe cavitation conditions (46,51). Guzman *et al.* (44) note a much improved component life for modified A–S–H. pumps, installed in series, with a lower impeller tip speed compared to that for the original design, and a new rubber material (see section 4.1.2).

For reciprocating positive displacement pumps, valve wear tends to be the limiting factor, although piston, cylinder liner and piston rod wear may be equally significant for some duties. Tables 5a and 5b (47) give comparative lives for a Wilson-Snyder triplex horizontal piston pump, handling abrasive mud and limestone slurries. The operating speed was higher than normal for this type of pump. For the abrasive mud duty, the pump was converted to a flushed plunger type; hard-chromed carbon steel plungers lasted over 600 hours. However, the makers of the 'GEHO' piston diaphragm pump claim a life of up to 3000 hours, with an iron ore slurry, compared with 300–500 hours for plunger pumps (54). Some other papers give life data on various components of 'OWECO' pumps handling beach sand (42) (about 3500 hours for the modified valve seats), 'Mars' pumps with micaceous residue (45), and a 'Zimpro' pump with a rutile/sand slurry (55).

**Table 5.** Life of single-acting triplex horizontal piston pump components (47)

**(a)** with abrasive mud/slime slurry (very arduous)

| Component | Anticipated (h) | First 840 hours of operation (h) | Next 1156 hours of operation (h) |
|---|---|---|---|
| Valve | 100 | 60 | 30 |
| Valve Seat | 133 | 105 | 64 |
| Valve Spring | – | 200 | 180 |
| Piston | 43 | 70 | 46 |
| Piston Rod | – | 42 | 292 |
| Cylinder Liner | 30 | 84 | 89 |

**(b)** with limestone slurry

| Component | Life (h) |
|---|---|
| Valves | 1500 |
| Valve Seats | 2000 |
| Piston | 650 |
| Piston Rods | 800 |
| Liners | 450 |

## 6.2.3   Gland sealing

This is generally regarded as a major problem area for virtually all types of pump, and throughout the range of applications (e.g. 4–6,42–44,46–48,50,51).

Soft-packed glands are normally preferred for both rotating and reciprocating shafts, with clean water flushing; hard-surfaced shafts or sleeves also assist in reducing wear. Wilson (5) discusses various gland and sealing sytem designs in some detail, and Sabbagha (43) and Guzman *et al.* (44) mention modifications to gland flushing systems after initial problems. Read (50) reports that stainless steel (Grade 316) shaft sleeves and Teflon or P.T.F.E. gland packing gave the best results in a Warman gravel pump on a beach sand slurry; impregnated Teflon packing also performed best in piston rod glands in 'National' reciprocating pumps, again with a beach sand slurry (42), though a moulded rubber packing looked promising on trial.

Impeller 'scraper' (or 'back') vanes and centrifugal seals (e.g. Warman 'expeller' seal) can effectively reduce the pressure at the gland, needing only grease lubrication of the packing (5,6,48). Chrisswell (51) discusses the application and limitations of water-flushed glands vs. expeller seals and back vanes; in one series pumping system (46), the first stage pump was fitted with an expeller seal, which was trouble-free, and the second stage had a separate water-flushing system.

User-experience suggests that mechanical seals tend to be unsatisfactory, particularly for the more abrasive applications. However, where leakage cannot be tolerated, such as in some chemical processes, double mechanical seals with tungsten carbide, 'stellite' or ceramic faces and clean liquid flushing may be the only solution (5,48). Lip seals with either clean water or grease supply have been used successfully in certain dredging applications (48).

## 7.   PIPELINE WEAR

It has been suggested by different authors that slurry pipelines could be designed for a mean life in the range of 10 to 30 years, if the pipe material is well-chosen to suit the particular application (1,7), although certain applications may require only a relatively short life (59).

## 7.1   STRAIGHT PIPES: GENERAL AND LOCALISED WEAR

### 7.1.1   Metal (unlined) pipes

A number of references in the pipe wear literature survey (7) dealing with pipe wear tests and service experience give mean wear rates when transporting different slurries under various conditions. A selection of these data is given in Tables 2 and 6. Probably the most useful group is from wear tests at the Colorado School of Mines (34,37), referred to earlier, with wear due to erosion and corrosion shown separately. Postlethwaite *et al.* (36) give

**Table 6.**  Wear rates of unlined steel pipe with various slurries, from service experience (7)

| Reference | Type of solids (and location) | Specific gravity | | Slurry concentration (%) | | Mean velocity |
|---|---|---|---|---|---|---|
| | | solids | slurry | by wt. | by vol. | (m/s) |
| Davison | Heavy mineral concentrate (Queensland, Australia) | 4.0 | 1.11 | 13.2 | – | 2.53 |
| Ewing | Coal (Ohio pipeline, U.S.A.) | – | – | 60 | – | – |
| Schmidt & Limebeer | Coal (E. Transvaal, S. Africa) | – | – | – | – | 2.44 |
| McDermott | Iron concentrate (Savage River pipeline, Tasmania) | – | – | 55–60 | – | 1.37–1.83 |
| Clement & Sokoloski | Cemented 'back-fill' (Manitoba, Canada) | – | 1.7 | – | – | 2.87 |
| Kupka & Hrbek | Crushed mine stone 'tailings' (Czechoslovakia) | 2.2–2.55 | – | – | 13.6 | 3.22 |
| Munro Sambells | Micaceous residue (Cornwall, UK). | 2.65 | 1.35 | – | – | (Approx. 1.83 – design value) |

| Particle size | Pipe dia. (mm) | Wear rate | Pipeline life |
|---|---|---|---|
| 60–200 B.S. mesh | 154 | 0.25 mm mean ⎫ per<br>0.94 mm max. ⎭<br>100 000 tonnes solids | |
| Distribn. % – Tyler mesh<br>−8    −100  −200  −325<br>+100  +200  +325<br>5.78  6.0    6.2    30.0 | 254 | 'Practically negligible' in 6 yrs. using corrosion inhibitor | |
| 0–25 mm | 229 | 1.32 mm max./<br>400 000 tonnes solids | 10 yrs. estd. |
| −200 mesh | 244 | Corrosn.        Erosn.<br><0.023 mm/yr.  Nil<br>at inlet end;<br><0.1 mm/yr.<br>at outlet end | |
| 80%, 200 mesh | 152 | – | Approx. 22 months (equiv. to 1.5 M tonnes solids). |
| 0–60 mm | 150 | 1 mm/12 500–15 000 tonnes solids | |
| .002 – .25 mm | 168 | – | 9 months to failure |

comparative data for erosion alone and combined erosion/corrosion in mild steel pipe, from pipe wear tests. Slurries containing iron ores and concentrates appear to cause a wide range of erosion rates, but there is little doubt that corrosion can be a significant part of the total wear, for both these and some other slurries. It also confirms the general view (see section 3.4) that erosion with limestone and coal slurries tends to be relatively low, with corrosion predominating. A number of more recent papers reporting operating experience give wear data in terms of either pipeline life or wall thickness reduction, often by ultrasonic measurement (20,46,47,58), with various slurries; these include mine tailings and sand dredging in Japan (10), mine back-fill (20,46), beach sand (42), gold slime (43), copper mine mud/slime (47), run-of-mine coal (58) and silica sand (59). Nearly all the above literature relates to steel pipes.

As mentioned in section 5.3, the worst erosion occurs at the bottom of a horizontal pipe when transporting a settling slurry (e.g. 1,7,10,32,60). Hence an accessible short-distance pipeline may be turned through 90°–180° periodically to give more even wear and thus prolong its life (1,7,56,58–60); buried or long-distance pipelines would require an extra thickness allowance.

It is important to remember that severe local wear will occur at any discontinuity or obstruction in the pipe bore, e.g. mismatched joints (see Figure 20) or flexible couplings

**Figure 20.**  Wear caused by misalignment of pipe joints

due to increased turbulence, especially with fine particles (1,7,20, 56). It has also been reported by Bain *et al.* (1) that high local wear rates can occur at the '4 and 8 o'clock' positions round the bore of the horizontal pipe, possibly due to local turbulence with a stationary bed of solids at the bottom.

## 7.1.2 Lined pipes

As listed in section 4.1, rubber or plastics linings offer the advantage of protection against both erosion and corrosion, provided that the right grades are chosen and bonding is good (1,7,19,37,39,41,44–46,56–58). Numerous reports mention the reduction of wear with rubber lining compared with unlined steel pipes and fittings, particularly if impact wear is present and the solids not large and sharp (1,7,19,37,44,46,57). Also there are claims for the successful use of polyurethane linings (7,8,41,45,56); English China Clay's service experience (41,45) reports improvements of 8 to 20 times the life for a mild steel pipe, and about twice the life of rubber lining. Both Guzman *et al.* (44) and Rouse (45) note improvements in life with softer grades of rubber (44) and polyurethane (45) lining; they also stress the need for special joint design to prevent local distortion of the lining, which could cause turbulence and hence wear.

A few reports from West Germany and the Soviet Union (7–9,58) note considerable improvements in life with basalt lining, factors of 15 to 20 being claimed compared with steel pipe. However, Zoborovski (56) reports poor results with basalt lining, due to severe local wear of the segments, but good performance from fused alumina; although much more expensive, this could still be an economic proposition.

## 7.2  PIPE BENDS

Pipe bends will wear much more rapidly than straight pipe, particularly at the outer radius, due to the increased impingement angle of the solids and more turbulence (1,7,20,26,43,47, 58,60). Bend geometry also affects the wear rate; although some authors merely suggest that wear will increase as the bend radius/pipe diameter ratio reduces (7), Brauer's comprehensive series of bend wear tests (26) show that wear increases significantly for bends of 2–3.5 $R/D$ ratio, rising to a maximum at $R/D = 2.8$. Hence bends in this $R/D$ range should be avoided if possible. Bend angle, as well as $R/D$ ratio, also affects the position and depth of the wear cavity.

Further references (20,47,58) giving data suggest that bend wear may be of the order of twice that for straight pipe, although one report (46) shows similar rates for horizontal steel pipe. However, the same report mentions severe bend wear in vertical pipelines. Bends may be thickened locally (60) or protected by a lining (1,43,46) to give increased life.

## 7.3  UNITS OF WEAR RATE

Pipe wear rates have been expressed in many different units, both absolute and derived (or 'specific'). It would be desirable to have a more generally agreed type of unit, to enable results from different sources to be more readily correlated. The Japanese review paper (10) suggests that much more data is needed for a proper analysis of all the various wear factors.

## 8.  CONCLUSIONS AND RECOMMENDATIONS

Owing to the large number of interdependent factors affecting wear in both pumps and pipelines, and the limited and often conflicting data available from both tests and service experience, it is not possible to lay down a comprehensive set of design rules to give reliable prediction of wear rates; there is still considerable scope for further research. In particular:

1. There is a need for further analysis of existing data, and/or experimental work, to attempt to resolve some of the apparently conflicting data, e.g. on the effects of particle size, concentration, material type (both metals and non-metals), etc., between (a) different types of simulation test and (b) laboratory tests and service experience. Recommendations for the best types of simulation rig, if any, for both pump and pipe wear would be desirable.
   Although progress will be made on a relative wear basis, it is unlikely that any correlation will be sufficiently complete to enable absolute wear rates to be predicted with any certainty for some considerable period.
2. More research is needed to investigate claims for the wear resistance of some of the newer construction materials, e.g. high Cr and CrMo cast iron alloys vs. 'Ni-hard', different grades of polyurethane and plastics vs. natural rubber, for both pumps and pipes as appropriate.
3. Further work on the computer prediction of wear patterns in centrifugal pumps, covering different designs and extending the ranges of particle size and concentration, are required. At least one pump manufacturer is actively pursuing this approach at present.

## 9.  REFERENCES

1. Bain, A. G,. and Bonnington, S. T. (1970) 'The hydraulic transport of solids by pipeline', (Book), pp. 131–6, 1st Edn. Pergamon Press.
2. Bitter, J. G. A. (January/February and May/June 1963) 'A study of erosion phenomena, Parts 1 and 2', *Wear*, 6.
3. Wiedenroth, W. (September 1970) 'The influence of sand and gravel on the characteristics of centrifugal pumps; some aspects of wear in hydraulic transportation installations', *Proc. 'Hydrotransport 1' Conf.*, Paper E1., BHRA, Cranfield.

4. Truscott, G. F. (October 1970) 'A literature survey on abrasive wear on hydraulic machinery', BHRA. Publn. TN 1079.
5. Wilson, G. (September 1972) 'The design aspects of centrifugal pumps for abrasive slurries', *Proc. 'Hydrotransport 2' Conf.*, Paper H2, BHRA, Cranfield.
6. Warman, C. H. (June 1965) 'The pumping of abrasive slurries', *Proc. 1st Pumping Exhibn. and Conf.*, 15 pp, includ. 12 Figs., Earls Court, London.
7. Truscott, G. F. (May 1975) 'A literature survey on wear in pipe-lines', Publn. TN 1295, BHRA.
8. Hassan, U., Jewsbury, C. E. and Yates, A. P. J. (1978) 'Pipe protection: a review of current practice', Joint report by BHRA/Paint R.A. Publ. by BHRA.
9. Turchaninov, S. P. (June 1979) 'The life of hydrotransport pipelines', (Book). Nedra Press, Moscow (1973). Transln. By Terraspace Inc., Rockville, MD., USA.
10. Kawashima, T. *et al.* (May 1978) 'Wear of pipes for hydraulic transport of solids'. *Proc. 'Hydrotransport 5' Conf.*, Paper E3, BHRA, Cranfield.
11. Bergeron, P. (November 1950) 'Similarity conditions for erosion caused by liquids carrying solids in suspension. Application to centrifugal pump impellers', *La Houille Blanche, 5*, Spec. No. 2, pp. 716–29, (In French), Transln. T 408, (1950), BHRA.
12. Bergeron, P. (June 1952) 'Consideration of the factors influencing wear due to hydraulic transport of solid materials', *Proc. 2nd Hyd. Conf.* 'Hyd. Transport and Separation of Solid Materials', *Soc. Hydrotech. de France*, (In French).
13. Roco, M. and Reinhart, E. (November 1980) 'Calculation of solid particles concentration in centrifugal pump impellers using finite element technique'. *Proc. 'Hydrotransport 7' Conf.*, Paper J3, pp. 359–76, BHRA, Cranfield.
14. Ahmad, K. (November 1982) 'A theoretical and experimental investigation of erosion prone areas on the blade surfaces of a centrifugal impeller handling granular solids', Ph.D. Thesis, Fluid Engineering Unit, Cranfield Inst. of Tech.
15. Goulas, A. (June 1983) 'The calculation of velocity profiles in a slurry pump impeller'. *Fluid Engineering Unit Report*, No. 82-AG-53, Cranfield Inst. of Tech.
16. Trainis, V. V. (1963) 'A method for computing coal comminution in a pipeline during hydraulic transport', *Ugol*, No. 9, pp. 37–41, (In Russian).
17. Baker, P. J. and Jacobs, B. E. A. (May 1976) 'The measurement of wear in pumps and pipelines'. *Proc. 'Hydrotransport 4' Conf.*, Paper J1, BHRA, Cranfield.
18. Kratzer, A. (March 1979) 'Some aspects on the design and selection of centrifugal pumps for sewage and abrasive fluids', *Proc. 'Pumps '79', 6th. Tech. Conf.*, BPMA, Paper D3, pp. 169–80, Publn. by BHRA, Cranfield.
19. Murakami, S. *et al.* (November 1980) 'Wear test of pipe linings for hydraulic transport of dam deposit', *Proc. 'Hydrotransport 7' Conf.*, Paper H2, BHRA, Cranfield.
20. Thomas, E. G. (April 1977) 'H.M.S. float fill development – Stage I – Pumping', Mount Isa Mines Ltd., Tech. Rep. No. RES MIN 48.
21. Wellinger, K. and Uetz, H. (1955) 'Sliding, scouring and blasting wear under the influence of granular solids'. *VDI – Forschungsheft, 21*, 449, B, 40 pp, 85 Figs, (In German).
22. Stauffer, W. A. (1958) 'The abrasion of hydraulic plant by sandy water', *Schweizer Archiv für Angewandte Wiss. u. Technik, 24*, 7/8, pp. 3–30. (In German), Transln. by C.E.G.B. No. 1799, 13 pp. 20 Figs.
23. Prudhomme, R. J., Rizzone, M. L., Schiemann, T. H. and Miller, J. E. (September 1970) 'Reciprocating pumps for long-distance slurry pipelines', *Proc. 'Hydrotransport 1' Conf.*, Paper E4, BHRA, Cranfield.
24. De Bree, S. E. M., Begelinger, A. and de Gee, A. W. J. (March 1980) 'A study of the wear behaviour of materials for dredge parts in water-sand mixtures', *Proc. 3rd. Int. Symp. Dredging Tech.*, Paper E1 pp. 161–80, BHRA, Cranfield.
25. De Bree, S. E. M., Rosenbrand, W. F. and de Gee, A. W. J. (August 1982) 'On the erosion

resistance in water-sand mixtures of steels for application in slurry pipelines'. *Proc. 'Hydro-transport 8' Conf.*, Paper C3, BHRA, Cranfield.
26.   Brauer, H. and Kriegel, E. (October 1963) 'Investigations on wear of plastics and metals', *Che. Ing. Tech.*, **35**, 10, pp. 697–707 (In German).
27.   Johns, H. (June 1963) 'Erosion studies of pipe lining materials – Third Progress Report', *US. Dept. of Intr., Bureau of Reclam.*, Chem. Eng. Lab. Report No. P-93, 27 pp., D.D.C. No. AD 428514 (unclass.).
28.   Schriek, W., Smith, L. G., Haas, D. B. and Husband, W. H. W. (October 1973) 'Experimental studies on the hydraulic transport of coal', *Report V*, Saskatchewan Res. Council.
29.   Zarzycki, M. (1969) 'Influence of the pump material on service life of the impellers of rotodynamic pumps in transport of mechanically impure fluids', *Proc. 3rd Conf. on Fluid Mechs. and Fluid Machy.*, Budapest.
30.   Bak, E. (December 1966) 'Construction materials and testing results of the wear of pumps for transporting solid media', *Biuletyn Glownego Instytuta Gornictwa 12*, (In Polish), BHRA translm. available.
31.   Wong, G. S. (September 1978) 'Coal slurry feed pump for coal liquefaction', *Elect. Power Res. Inst.*, Proj. 775-1, Final Report EPRI AF-853.
32.   *Karabelas, A. J. (May 1978) 'An experimental study of pipe erosion by turbulent slurry flow', Proc. 'Hydrotransport 5' Conf.*, Paper E2, BHRA, Cranfield.
33.   Ferrini, F., Giommi, C. and Ercolani, D. (August, 1982) 'Corrosion and wear measurements in slurry pipeline', *Proc. 'Hydrotransport 8' Conf.*, Paper C1 pp. 133–44, BHRA, Cranfield.
34.   Link, J. M. and Tuason, C. O. (July 1972) 'Pipe wear in hydraulic transport of solids', *Mining Congress Jnl.*, pp. 38–44.
35.   Boothroyde, J. and Jacobs, B. E. A. (December 1977) 'Pipe wear testing, 1976–1977', BHRA Publn. PR 1448.
36.   Postlethwaite, J., Brady, B. J. and Tinker, E. B. (May 1976) 'Studies of erosion-corrosion wear patterns in pilot plant slurry pipe-lines', *Proc. 'Hydrotransport 4' Conf.*, Paper J2, BHRA, Cranfield
37.   Tczap, A., Pontius, J. and Raffel, D. N. (May 1978) 'Description, characteristics and operation of the Grizzly Gulch tailings transport system', *Proc. 'Hydrotransport 5' Conf.*, Paper J4, BHRA, Cranfield.
38.   Jacobs, B. E. A. (August 1982) 'The measurement of wear rates in pipes carrying abrasive slurries', *Proc. 'Hydrotransport 8' Conf.*, Paper C2, pp. 145–60, BHRA, Cranfield.
39.   Nguyen, V. T. and Saez, F. (September 1979) 'Design of a pipeline for a highly abrasive and corrosive slurry'. *Proc. 'Hydrotransport 6' Conf.*, Paper H2, pp. 367–8, BHRA, Cranfield.
40.   Smith, L. G. *et al.* (May 1978) 'Pressure drop and wall erosion for helically-ribbed pipes', *Proc. 'Hydrotransport 5' Conf.*, Paper B2, BHRA, Cranfield.
41.   Sambells, D. F. (May 1974) 'A practical solution to pumping an abrasive slurry', *Proc. 'Hydrotransport 3' Conf.*, Paper J1, BHRA, Cranfield.
42.   White, J. F. C. and Seal, M. E. J. (August 1982) 'Seasand handling and pumping at Cockburn Cement', *Proc. 'Hydrotransport 8'*, Paper A5, pp. 63–86, BHRA, Cranfield.
43.   Sabbagha, C. M. (August 1982) 'Practical experience in pumping slurries at ERGO', *Proc. 'Hydrotransport 8' Conf.*, Paper A1, pp. 1–16, BHRA, Cranfield.
44.   Guzman, A., Beale, C. O. and Vernon, P. N. (August 1982) 'The design and operation of the Rössing tailings pumping system', *Proc. 'Hydrotransport 8' Conf.*, Paper A2, pp. 17–36, BHRA, Cranfield.
45.   Rouse, W. R. (August 1982) 'Operating experience with a residue disposal system', *Proc. 'Hydrotransport 8' Conf.*, Paper A6, pp. 77–90, BHRA, Cranfield.
46.   Davidson, C. W. (November 1978) 'Cemented aggregate fill – underground placement', *Mount Isa Mines Ltd., Tech. Rep.* No. RES MIN 53.
47.   Graham, J. D. and Odrowaz-Pieniazek, S. (May 1978) 'The design and operation of a slurry

pumping system from 1000 m underground to surface', *Proc. 'Hydrotransport 5' Conf.*, Paper J2, BHRA, Cranfield.
48. Willis, D. J. and Truscott, G. F. (May 1976) 'A survey of solids-handling pumps and systems, Part II – Main survey'. Publn. TN 1463, BHRA, Cranfield.
49. Want, F. M. (November 1980) 'Centrifugal slurry pump wear – plant experience', *Proc. 'Hydrotransport 7' Conf.*, Paper H1, pp. 301–14, BHRA, Cranfield.
50. Read, E. N. (August 1982) 'Experience gained with large dredge pumps in a sand mining operation', *Proc. 'Hydrotransport 8' Conf.*, Paper F1, pp. 263–78, BHRA, Cranfield.
51. Crisswell, J. W. (August 1982) 'Practical problems associated with selection and operation of slurry pumps', *Proc. 'Hydrotransport 8' Conf.*, Paper F5, pp. 317–38, BHRA, Cranfield.
52. Demillecamps, E. and Lambillote, M. (March 1979) 'The use of P.V.D.F. in pumps handling corrosive media', *Proc. 'Pumps '79', 6th Tech. Conf. BPMA.*, Paper D4, pp. 181–216. Publ. by BHRA, Cranfield.
53. Morrison, J. C. (March 1983) 'Materials and construction for contaminated water and sea-water pumps', *Proc. 'Pumps – the Heart of the Matter', 8th Tech. Conf., BPMA*, Paper 11, pp. 295–308, Publ. by BHRA, Cranfield.
54. Holthuis, C. H. and Simons, P. W. H. (November 1980) 'The GEHO diaphragm pump – a new generation of high-pressure slurry pumps', *Proc. Hydrotransport 7' Conf.*, Paper A3, pp. 17–32, BHRA, Cranfield.
55. Zoborowski, M. E. (August 1982) 'African experiences with pipeline transportation of rutile and sand by hydraulic exchange cylindrical diaphragm pump', *Proc. 'Hydrotransport 8' Conf.*, Paper P3, pp. 515–26, BHRA, Cranfield.
56. Paterson, A. C. and Watson, N. (September 1979) 'The N.C.B. pilot plant for solids pumping at Horden Colliery', *Proc. 'Hydrotransport 6' Conf.*, Paper H1, BHRA.
57. Eriksson, B. and Sellgren, A. (May 1978) 'Development of slurry transportation technology in Sweden', *Proc. 'Hydrotransport 5' Conf.*, Paper J5, BHRA, Cranfield.
58. Siebert, H., Kortenbusch, W. and Harzer, H. (November 1980) 'Further experience with horizontal and vertical hoisting of coarse run-of-mine coal at Hansa Hydromine', *Proc. 'Hydrotransport 7' Conf.*, Paper B2, pp. 41–54, BHRA, Cranfield.
59. McElvain, R. E. and Hardy, T. H. (May 1978) 'Operational report on a 5 mile (8 km) pit to flotation plant transport line with a simplified approach to slurry transport'. *Proc. 'Hydrotransport 5' Conf.*, Paper J3, BHRA, Cranfield.
60. Wright, D. and Brown, J. (September 1979) 'Hydraulic disposal of P.F. ash from C.E.G.B. Midlands Region Power Stations', *Proc. 'Hydrotransport 6' Conf.*, Paper H3, pp. 379–88, BHRA, Cranfield.
61. Jacobs, B. E. A. and James, J. G. (October 1984) 'The wear rates of some abrasion resistant materials', *'Hydrotransport 9' Rome, Italy*, pp. 331–44, BHRA, Cranfield.
62. Roco, M. C., Addie, B. R., Dennis, J. and Nair, P. (October 1984) 'Modelling errosion wear in centrifugal slurry pumps', *Proc. 'Hydrotransport 9' Rome, Italy*, pp. 291–316, Cranfield.
63. Roco, M. C. and Cader, T. (September 1987) 'Numerical method to predict wear in pipes and channels', *7th International Conference on the Internal and External Protection of Pipes, London, UK*, pp. 137–48, BHRA, Cranfield.

# 4. MATERIAL PREPARATION AND DEWATERING

## 1. MATERIAL PREPARATION

### 1.1 INTRODUCTION

The general techniques applicable to material preparation for slurry transport are the same as those used for size reduction in the mineral processing industry.

Dewatering is very important for materials such as coal, in terms of both capital and operating costs of the transport system and, insofar as the transporting water is being released to the environment, dewatering may have important repercussions. In some circumstances the inevitable loss of transported material in the separation process may be unacceptable and the difficulty and cost of dewatering fine material could well influence the choice of whether to transport the solids in a coarse or finely ground state.

### 1.2 CLASSIFICATION

Normally a system of screens is used to segregate the feed into two streams; one which may be passed to a later stage of the comminution process and the other which is fed to the first stage crusher. Screens may be either stationary or vibratory. Much development has taken place over many years and reliable systems of screening may be achieved by using manufacturers' recommended equipment.

An interesting method of screening has been devised by the National Coal Board, in which sizing is brought about by dynamic means. Coal is dropped onto a rotating wheel of spokes, somewhat resembling a bicycle wheel. The fine material falls through the gaps between the spokes while the coarser lumps are thrown to one side by the spinning action. The advantage of this type of screen is that it can deal with damp solids, separating out fines from the 2–6 mm range downwards. Previously, coventional screens became blinded by the sticky fines.

### 1.3 CRUSHING AND GRINDING

An important factor in size reduction is the power requirement and maintenance costs of the equipment. Work by Bond (1) has shown that the specific power requirement is given by:

$$W = 9.83 \ W_i \left( \frac{1}{\sqrt{p}} - \frac{1}{\sqrt{F}} \right)$$

where $W$ = grinding power, KWh/tonne
$W_i$ = work index
$p$ = size at which 80 per cent of product passes, $\mu$m
$F$ = size at which 80 per cent of feed passes, $\mu$m

The work index varies from about 10 to 20 depending on the material. The value reduces to about 10 for wet grinding and would be increased if a large proportion of material were less than 70 $\mu$m.

For the case of coal transport, in which the coal is to be used for firing of boilers, it could be considered that the size reduction would be required anyway and that this expense should not be regarded as part of the transport cost. The size reduction could be undertaken in two stages. One to achieve the optimum size distribution for a fine particle slurry, as in the Black Mesa pipeline, and a second stage to render the coal suitable for pulverised fuel combustion. For the case of the so-called 'coal-water mixtures' the size reduction might be undertaken in one stage with the coal being transported as a very fine slurry followed by direct firing with no separation from the water. Large scale systems, however, need to be optimised as well as possible and the above formula would only be used in preliminary feasibility calculations. Actual power and an assessment of remaining costs would need to be made by reference to manufacturer and perhaps by means of pilot plant trials as has been done at the ETSI coal evaluation plant (2).

Perry and Chilton (1) stated that plant efficiencies may range from 25–60 per cent, relative to the power used in a laboratory mill, which itself considerably exceeds the theoretical work needed for grinding. The crushing or grinding equipment may be used in a once through process or incorporated in closed-circuit systems with arrangements generally as in Figure 1. The object of closed circuit grinding is to permit the mill to operate

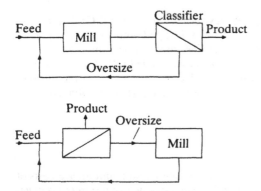

**Figure 1.** Alternative arrangements for closed circuit grinding

effectively with a relatively coarse product and consequently a relatively high throughput. The delivered product obtained by classification represents the fine component of this coarse crushing at a recirculation rate which may be up to 10 times the product rate. For a given top size the product of a closed-circuit system will have a more uniform size than the corresponding product of a once through process. In the coarse-crushing stage, the classifiers will be screens but in progressing to the fine grinding stage inertia separation will be employed, either built into the air/solids transport system within the mill, or using external cyclones. The choice of grinding equipment and the associated system will depend very much on the nature of the feed material (hardness, stickiness or 'soapiness') and on whether the hydraulic transport or the final use of the material demands control of size distribution rather than only top size. While the coarse-crushing stage will generally be dry, the finer grinding in certain designs of mill may be dry or as a slurry. However, it will be realised that closed-circuit wet grinding, using hydrocyclones instead of air cyclones, will not necessarily give the same product characteristics as would the dry mills.

## 1.3.1 Coarse crushing

This is generally done by jaw, gyratory or roll crushers, (Figure 2). In the jaw crusher the solids fall between a fixed and a swinging jaw driven by a heavy flywheel mechanism, and this type of action is best suited to hard materials. As a very approximate order of magnitude, jaw crushers set to a size of 51 mm will have a power consumption in the region of 0.75 kWh/tonne.

The gyratory crusher comprises two cones, one of which is oscillated vertically and both being free to rotate about their axes. Primary crushing employs steep cone angles and a small size reduction ratio, while secondary crushing can be performed using wide cone angles. Some adjustment in performance is possible in the sense that a short oscillating stroke gives an evenly crushed product while a long stroke gives greater capacity but with a

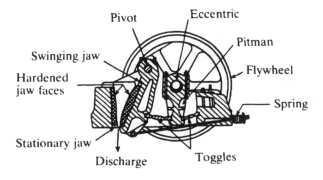

**Figure 2a.** *Blake jaw crusher (4)*

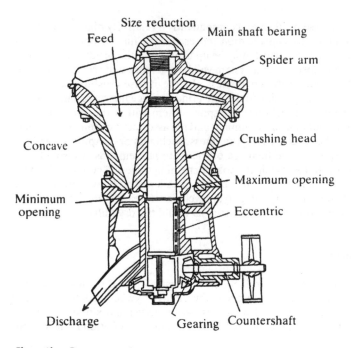

**Figure 2b.** Gyratory crusher (8)

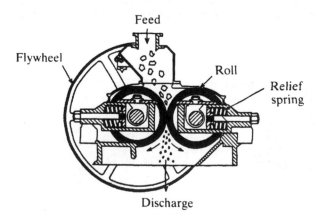

**Figure 2c.** Smooth roll crusher (8)

wide size distribution. For a given size setting the gyratory crusher will have a specific power consumption, perhaps two-thirds of that of the jaw crusher, but from the above remarks about control of size distribution, generalisations may be misleading.

The roll crusher suffers from greater wear than the oscillating types but is well-suited to wet, sticky materials containing clay. Crushing may be by a single roller against a fixed plate or between two rollers, and the rollers may be smooth or fitted with teeth. Adjustment in performance in this case is by the choice of choked or free crushing. By feeding at a high rate a heap of solids form over the rolls and crushing is partly by one particle against another. This 'choked' crushing produces a greater proportion of fines than the free crushing which occurs when each feed lump is nipped individually by the rollers. Power consumption has been quoted as 0.3 kWh/tonne for soft rock and 0.75 kWh/tonne for hard rock.

## 1.3.2 Secondary crushing

The gyratory and smooth-faced roll crushers mentioned above can also be arranged for secondary crushing, where reduction ratios of 8:1 and 15:1 respectively may be achieved. The gyratory machine is more likely to be operated dry and the presence of high moisture can lead to loss of grinding efficiency. The roll crushers operate with either wet or dry material. In secondary crushing it is more likely that the size distribution of the product will become important, because of its effect on the next stage of material preparation, or because of the hydraulic-transport characteristics if it is being pumped as a coarse slurry. The roll crusher may be operated in the choked or free grinding mode, and may be incorporated in a closed circuit with an external classifier. The hammer mill can be fitted with screen bars beneath the impacting hammers, which results in oversize material being retained until crushing is complete. If the material has a high moisture content, oscillation of the breaker plate is sometimes employed and the screen bars may be located external to the machine so that means of reducing the risk of clogging (such as vibration) can be employed. While the hammer mill is particularly suited to soft friable materials such as coal and limestone it will also accommodate clay, unlike the machines employing simple crushing.

Where the material is suitable, the impact principle can be extended by arranging for vigorous circulation within the machine, the particles then breaking by impact with other particles or fixed surfaces and grid screens. Substantial size reduction is then possible with a relatively high machine throughput at an energy cost below 0.37 kWh/tonne but rising rapidly to 11 kWh/tonne when reducing soft materials to < 30 mesh (500 $\mu$m) or 45 kWh/tonne for 'tough' material at the same size. Product sizes less than 200 mesh (75$\mu$m) are possible with hammer mills handling suitable materials, but it will be realised that the rapid rise in energy consumption as the size is reduced may make it worthwhile to leave the final stage to a type of plant better suited to the duty.

## 1.3.3   Grinding

In those mineral processing applications which require a fine dry product a choice exists
between grinding by the action of balls or rollers on a platen, or essentially by tumbling the
material with heavy balls or rods in a cylinder. The former, 'vertical-spindle' machines feed
the material onto the grinding ring and the ground product is transported away by an
upward air current. This air is then usually passed through an inertia type of classifier which
enables the oversize to be recirculated to the grinding ring. It will be realised that this action
requires that the feed material and product shall be free-flowing and dry. In the context of
hydraulic transport it may not be economic to supply the heating air necessary to ensure
satisfactory operation of this type of mill and it is, therefore, unlikely to be found in this
type of application.

The rod or ball tube mill is, in fact, no more tolerant to the moisture levels of say 20 per
cent which tend to make materials 'sticky' but it offers the facility for adding further water
without losing the capacity to give a fine product. The general arrangement is a hollow
horizontal cylinder which may be of uniform diameter or conical at one end, (Figure 3)
lined with replaceable segments of wear-resistant materials such as manganese steel or
Nihard. The linings will be smooth when handling fine soft material, but may be fitted with
'lifters' which produce a partial impact action suited to certain applications. A charge of
balls or rods within the tube is tumbled by rotation at a speed some 75 per cent of that
'critical' speed which would retain them by centrifugal force throughout a revolution. The
ball charge is usually about half of the mill volume and after a period of time the charge will
assume a wide range of sizes: it is possible to arrange to add new large balls and extract the
worn small balls as a continuous process. Material is fed to the cylinder through a trunnion
and, in the case of wet grinding, the slurry will flow to a larger diameter trunnion at the

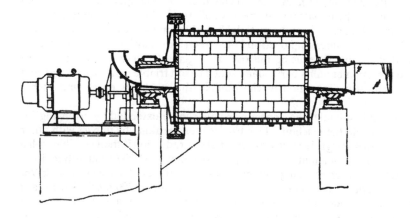

**Figure 3.**   Tube mill (G.E.C. Ltd)

discharge end carrying the fine product. For dry grinding it is usual to provide sweeping air to carry the fines to the discharge trunnion, simultaneously drying the material if necessary. Conical ball mills have been developed because of the observation that the larger balls tend to migrate towards the discharge end which is contrary to the requirement that they should be available for the breakage of large material at the feed end. By arranging the larger diameter of the tube at the feed end, the large grinding balls will migrate towards this point.

The choice of balls or rods as the grinding medium depends on the type of material and product required. The rod mill is fairly tolerant to the presence of moisture in the feed, and its action involves almost continuous line contact between rods for the whole length of the tube. The nature of rod packing effectively prevents oversize material being discharged and the rod mill may be used in open circuit to give a product which contains relatively little fine material.

The ball mill has the characteristic of giving a relatively fine product but efficient use of a given size of mill will usually require closed-circuit operations with recirculation of the oversize. The choice between fully dry and fully wet grinding will involve many considerations which can probably be decided only after comparative tests on the material. If the feed material is relatively dry and if closed-circuit grinding is indicated, operation as dry grinding may be preferable because a wide choice is available of solids classifiers which operate relatively effectively because of the density difference between solids and air. In comparison, classification of solids in a slurry by the hydrocyclone (see section 2.3 on drying) is less versatile and, in fact, a build-up of very fine material may be impossible to avoid. This may lead to a need to bleed off a portion of the recirculating slurry or to use large settling tanks.

It is not possible to generalise on the relative effectiveness of wet and dry grinding because of the wide difference in optimum operating setting of a given mill in the two situations. For example, Perry and Chilton (1) quote tests showing that with an intermediate ore charge the selectivity of grinding was about the same in wet as in dry operation. However, selectivity could be increased in wet grinding by increasing the ore charge and by decreasing the charge for dry grinding. A comparison of the methods in open circuit grinding suggested that wet processing gave the higher capacity and efficiency. This advantage might well be offset by a greater rate of wear of the grinding media and liners resulting from the combined action of corrosion and erosion or the alternative cost of adding corrosion inhibitors. In assessing the alternatives available it is worth noting that loss of grinding material in wet grinding, or fan power for transport and classifying in dry milling, may each involve an operating cost comparable with that of the mill itself. The close relationship between product size and power consumption was mentioned in the introduction in connection with the Bond Work Index formula which was derived for the tube mill type of operation. For example, a ball mill grinding a wet slurry of metallurgical ore has been quoted as consuming 3.5 kWh/tonne when grinding to 90 per cent < 10 mesh (1.7 mm) and 11–15 kWh/tonne at 90 per cent < 100 mesh (150 $\mu$m) which is in approximate agreement with the Bond formula. Inspection of tabulated data by manufacturers for a given type of mill suggests that the specific power consumption for a constant product size is likely to reduce substantially as the mill size is increased.

## 1.4  SOME PLANT ARRANGEMENTS

The considerations involved in slurry preparation can be illustrated by reference to some typical installations. The Ohio coal pumping line was based on the availability of fine tailings from a coal washery, to be supplemented by run of mine coal. Because of this, wet screening and secondary crushing were employed and it was possible to produce economically a high-fines mixture (maximum size 14 mesh, 1.2 mm), with a minimum tendency to settle. Intermediate storage was adopted to permit control and to avoid the segregation which might have been expected if mixing was done prior to storage. It is demonstrated by Bain and Bonnington (3) that production of a high fines mixture by grinding involves considerable plant complication.

In the Black Mesa development it has been adequately demonstrated that a slurry without excess fines can satisfactorily be re-started after prolonged shutdown and there is considerable incentive to limit the fines to 20 per cent less than 325 mesh (50 $\mu$m) so as to give the best possible conditions for dewatering by centrifuge. The top size is 14 mesh (1.2 mm). Thus, it will be seen in Figure 4 that the dry coal is screened and the oversize crushed from 51 mm to 10 mm (2 inch to $\frac{3}{8}$ inch) prior to the addition of water in the final stage of grinding by rod mills, which have provision for closed cycle operation using simple

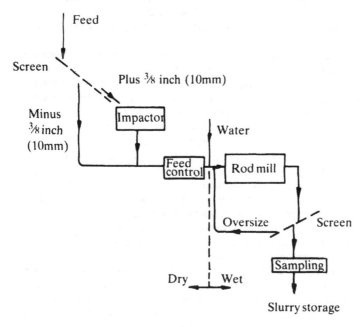

**Figure 4.**  Flow diagram of Black Mesa preparation plant

screens for separation of the oversize. Variable speed belts and belt weighers control the coal feed rate and this, together with water supply control, ensures the uniformity of slurry properties. Properly designed and maintained belt weighers are capable of an accuracy of the order of 1 per cent.

It is considered by some authors including Bernstrom (4) that the final stage of grinding by rod mills could be carried out as effectively and cheaper by a further stage of crushing.

Although there is provision for routine laboratory sampling, a continuous check on slurry consistency is made by measuring the pressure differential in a small pipeline loop. Continuous agitation is provided in the three final storage tanks each holding a two-hour supply at normal pumping rate (a further standby tank is also installed).

In the Trinidad limestone pipeline it was decided to reduce the material before pumping to a size less than 52 mesh (300 $\mu$m) which led to the choice of a ball mill. Closed circuit wet milling was adopted and was shown to increase mill capacity by over 50 per cent (5). It was originally intended to use woven wire screens for the classification in the mill circuit, but tests showed the screens to have unacceptably short life and a 'sieve bend' developed by the Dutch State Mines was substituted. (This is discussed in section 2 on dewatering.) This sieve bend itself calls for a recirculation system as its classification performance demands a constant slurry flow rate across it.

## 2. DEWATERING AND SOLIDS RECOVERY

In some cases treatment may be simple or unnecessary. At one extreme, coarse material may be left to drain naturally and at the other extreme hydraulically-transported coal in the form of CWM may be burned without further treatment.

The three major processes involved in dewatering are:

1. particle sedimentation, either natural or in a centrifugal field;
2. filtration, in which water drains through a cake of solid either naturally, with the assistance of a centrifugal action or assisted by pressure or vacuum;
3. thermal drying which, because of its cost, would be avoided if possible.

All of these principles may be employed in any one dewatering plant and the skill involved in devising the process flow sheet lies in making the most economical use of the individual elements. Essentially each individual separation process is best suited to a limited range of duty in respect of particle size range and input concentration and, especially for the more sophisticated centrifuges, there is a limit to the throughput of a single machine. Similarly each process will characteristically produce solids discharges at particular moisture contents, although the variable nature of solids results in a fairly wide range in performance.

As an illustration of the general intention, Figure 5 shows the sort of flow sheet which could be used to dewater a slurry of coarse material. The presence of fine material in the slurry would prevent it being fed to the basket centrifuge in the original low concentration,

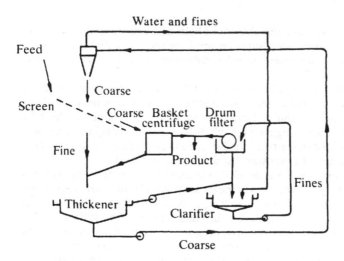

**Figure 5.** The principles of dewatering a coarse slurry

because the fines would be carried directly through the coarse apertures of the centrifuge. However, by initially separating the fines by a screen and reducing the water content first in the thickener and then in the hydrocyclone the fines can be remixed with the coarse materials for feeding to the centrifuge. The clarifier is necessary to give an acceptably clean effluent and the vacuum filter recovers the finest solids component. Slurries in the non-Newtonian category are likely to be at high volumetric concentration and there is little scope for dewatering by any other means than high-performance centrifuges followed by thermal drying.

## 2.1 SCREENS

If large single-size particles were being transported, the screen would act as an effective dewatering method. In more usual applications it serves to separate large particles with relatively little water from the fines component carrying most of the liquid. Indeed, in some circumstances the size separation may be improved by spraying further water, especially if the fines component is relatively sticky. The design and selection of screens is well established and manufacturers will advise on the most suitable screening medium and whether the duty is better achieved by vibrating the screen. Multiple deck screens may be used if it is desired to segregate the incoming slurry into more than two size fractions.

A design particularly mentioned in relation to dewatering is the Dutch State Mines sieve bend. This design is compact and less likely to give excessive carry over of fines with the coarse component. The curved screen, usually comprising wedge bars transverse to the flow and rounding of the upstream edge of the bar, results in the unusual feature that the size of particles screened out decreases as wear occurs. It is, however, relatively simple to reverse the screen at intervals so presenting a new sharp edge and sharpening the previously worn edge. Although this design was originally developed as a stationary screen, rapping mechanisms have been developed more recently. Differing material characteristics are accommodated by varying the angle turned through. For minerals and ores of 2000–200 $\mu$m the angle would be about 50°, whereas fine material (300–40 $\mu$m) at high solids concentration would require an increase to about 270°.

The 'Elliptex' dewatering screen has the feature that by a sophisticated vibration pattern on a horizontal frame a partial squeezing of the cake is achieved, with a greatly improved dewatering action. It has been reported (6) that a coal feed from 3 mm down to zero was dewatered to 12.5 per cent surface moisture even after being sprayed with wash water at the feed end.

## 2.2  THICKENERS AND CLARIFIERS

The thickener is the basic applicaton of sedimentation, being a tank in which sufficient retention time is available for solids to settle while relatively clear water is removed at the top. In order to operate continuously, the bottom of the tank is of conical shape and a series of revolving rakes scrape the settled material towards a well in the centre from which the sludge is pumped by a diaphragm or centrifugal pump. Slurry feed is usually to the centre of the tank, and liquid discharges over weirs at the tank walls.

It is generally necessary to ensure that the feed to a thickener is relatively low in solids concentration, typically no more than 20 per cent by weight, except for the heavier metallurgical ores. The solids content of the underflow will be of the order of 20–50 per cent by weight for coal and 60–70 per cent for heavy ores.

High molecular weight flocculants are often used to enhance settling rates and thereby reduce the tank size required. Considerable care is necessary to ensure good mixing of the flocculant and it is, of course, essential to guard against the risk that the flocculant may be detrimental to any further intended treatment of the settled sludge.

The clarifier performs essentially the same function as a thickener but as the intention here is to produce a clear liquid effluent the tank proportions are different and the rake mechanism much lighter. Because the feed material is of small size it is much more likely that flocculation will be necessary to ensure that particles will not be carried out with the effluent.

Thickeners and clarifiers are economical in power demand but, being limited by gravity settling, involve a relatively high capital investment. For this reason, methods of increasing the gravitational effect are beneficial in many circumstances.

## 2.3  HYDROCYCLONES

The most usual design of hydrocyclone takes the form shown in Figure 6, in which the slurry is fed into the cyclone tangentially under pressures of typically about 100 kN/m² (15 p.s.i.)

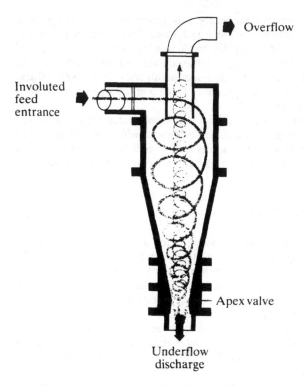

Overflow

Involuted
feed
entrance

Apex valve

Underflow
discharge

**Figure 6.** Hydrocyclone (Kreb Engineers)

although this may reach 350 kN/m$^2$ (50 p.s.i.) and the resulting swirl subjects the particles to a high centrifugal acceleration. The feed may leave the cyclone in two alternative streams, the outlet tube or vortex finder carrying predominantly clear water and the underflow orifice at the cone apex discharging the remaining liquid and coarse solids. Because particles are subject to both the centrifugal action and fluid drag, the prediction of the degree of separation in a given cyclone is difficult, although several theories have been developed. It has been demonstrated that a fraction of the particles, whatever their size, must be carried with the liquid to the underflow and consequently a sharp cut-off in size separation is not possible. Two measures of performances are used, one being the 'cut size', i.e. the particle size at which equal proportions of this size will emerge from the overflow and the underflow. However, considerably more useful information is derived from the grade efficiency curve (Figure 7) which enables a calculation to be made of the proportions of each component of the incoming size distribution which will emerge in the two directions. This is especially important where the emerging streams are to be treated by

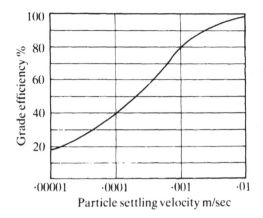

*Figure 7.* Typical grade efficiency curve for a hydrocyclone

further processes which are sensitive to size distribution. The important design variable is the diameter of the cylindrical section of the cyclone. For a given flow the cut size increases with cyclone diameter to the power of about 1.5 while the pressure drop decreases as the fifth power of diameter. The aim of getting as low a cut size as feasible is, therefore, best met by multiple small cyclones and, for a given duty, there will be an optimum combination of size and number (i.e. capital cost) and the cost of pumping power. The cyclone thickener is most usefully applied in the particle size range 5 to 200 $\mu$m and many variations are found in individual features of design and in the process flow sheet, for instance by combining cyclones in series and/or recirculating the flows. Several manufacturers have developed multiple-cyclone assemblies which share a common feed and discharge arrangement and which can accommodate wear resistant construction materials such as resin and rubber.

## 2.4 CENTRIFUGES

### 2.4.1 Solid bowl

The general principle of the centrifuge shown in Figure 8 is that the slurry is fed through a central tube, and centrifugal sedimentation in the rotating bowl causes the solids to form an outer layer at the wall of the bowl, while the liquid forms the inner layer. A weir is fitted at the cylindrical end of the bowl and the liquid leaves over this weir. In order to remove solids continuously, the inner rotor carries a scroll and rotates at a slightly different speed from that of the bowl so progressively pushing the solids towards the conical ('beach') section and the solids discharge, equivalent to the underflow of a normal thickener.

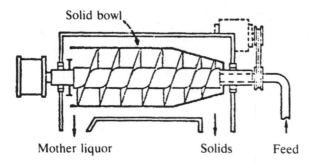

Figure 8.  Solid bowl centrifuge

The performance of any particular centrifuge can be adjusted to suit the characteristics of the feed material by variations of rotational speed of the drum and differential speed of the scroll and by adjusting the height of the overflow weir. For example, an increase in weir height will increase the residence time of the slurry and give a clearer water effluent, but at the expense of reducing the time available for solids to drain in the conical section with the result that the solids will contain more moisture. Heating of the slurry will generally give a much reduced loss of solids in the effluent and a reduction in moisture content in the discharged solids. Similar benefits may be gained by adding flocculants either before the centrifuge or within the bowl, depending on the type of additive. The dewatering effectiveness of a solid bowl centrifuge is very dependent on the proportion of fine material in the feed but in many applications the solids discharge will contain about 20 per cent moisture by weight.

## 2.4.2   The disc centrifuge

This type, shown in Figure 9 does not employ a scroll and the centre stack of discs rotates at the same speed as the bowl. The relatively narrow passages for slurry flow give conditions in which particles will readily reach the underside of the upper disc, from which they are progressively ejected by centrifugal action, emerging from holes in the bowl periphery.

## 2.4.3   The peeler centrifuge

These are of batch type and of relatively low throughput and high power consumption. Their use would, therefore, be confined to special application to the very finest fraction of slurry. The slurry is fed to a horizontal cylindrical bowl having an overflow weir. Some of the separated water will leave over the weir and the remainder may be removed at the end of the cycle by inserting a skimming device. The solids cake is then removed, at lower

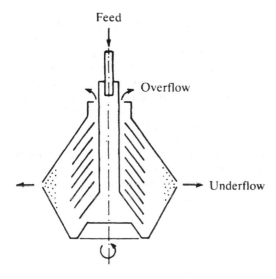

Feed

Overflow

Underflow

**Figure 9.** The disc centrifuge (6)

speed, by inserting a plough. The high centrifugal field possible with these machines gives a degree of dewatering of fine material which is not possible with any other type of centrifuge but at a considerable additional cost.

## 2.4.4 Basket centrifuges

By using a perforated bowl the mode of action of the centrifuge changes from sedimentation to filtration and it becomes more difficult to prevent fine material being carried through with the liquid. In respect of the liquid component, therefore, this type of machine will be midway along the flowline.

A basket centrifuge with a conical bowl can be designed such that solids fed to the apex will slide towards the outlet without assistance. However this method lacks flexibility as the cone angle may need to be changed to suit differing materials, and there is no independent control of retention time. More usually, therefore, solids are transported by an internal scroll or by giving an axial oscillation to the bowl.

The alternative cylindrical bowl machine transports the solids by a 'pusher' having a relatively short axial oscillation applied by an eccentric or, in large machines by hydraulic action. Each backward movement of the pusher allows a new ring of solids to form and the forward motion then pushes the whole cake towards the outlet.

The perforated basket is formed usually from wedge wire bars which for the conical machine are more conveniently arranged circumferentially but in the cylindrical machine may be circumferential or axial. The minimum practicable spacing is about $100 \cdot \mu$m and this represents an ultimate limit on the particle size which can be retained, but in any particular application the spacing will be chosen to give the best compromise between the moisture content in the delivered solids and the proportion of fine material carried over. When handling coarse material the moisture content of the solids can be in the range 5–10 per cent by weight with a loss of perhaps 3 per cent of the solids with the effluent.

The selection of centrifuges of all types requires considerable experience and laboratory-scale tests are only of limited value. For a given range of existing machines the best choice will embrace the throughput loading (i.e. number of machines required) and the dewatering performance. The interaction of throughput and performance and the internal design adjustment possible can be found only by pilot test on the largest feasible scale.

## 2.4.5   Screen bowl centrifuge

In the screen bowl centrifuge the slurry is first centrifuged in a solid bowl section. The sedimented solid is then transferred via a scroll to a second part of the bowl constructed in the form of a screen. Here the remaining liquid is centrifuged through the screen openings. There are two effluent flows from the machine, one from the bowl and one from the screen. Some fines from the solid will pass through the screen but the system allows a lower moisture content to be achieved than by using a bowl alone.

In tests carried out by the National Coal Board (7) the moisture content was found to be lower than could be achieved by vacuum filtration. An important advantage of centrifuges over conventional vacuum filters is that they take up much less space. Measured on a volume basis the ratio is approximately 1:10. This type of centrifuge has been considered for some of the long distance pipelines planned for the USA.

## 2.5   VACUUM FILTRATION

Although the driving pressure available in vacuum filtration is limited, this type of machine has a logical place in the choice of dewatering technique. The high forces involved in centrifuges limit the choice of filter medium and the compaction of the cake may be detrimental to filtration. At the expense of providing a large filter area and low pressure differential the vacuum filter offers a more flexible choice of filter medium and mode of operation which makes it the most economical choice in appropriate cases. Most commonly, the continuous filter takes the form of a drum or discs rotating on a horizontal axis (8) the drum being partially submerged in a tank containing the feed slurry. The vacuum applied inside the drum causes a cake to form on the filter cloth which becomes progressively dryer as it rotates. Depending on the individual design and material characteristics the dewatered cake is then discharged by scraping with a fixed knife, by running the filter cloth over secondary rollers of small diameter or by applying a blow-back pressure to a segment of the drum. The drum speed, submergence and vacuum each have a significance in determining performance, depending on the type of material and the choice of filter cloth. Filtration rate may also be improved by applying heat via a steam hood.

Vacuum filters operate continuously and automatically and are, therefore, low in operating cost, but capital cost is high. They can successfully dewater a wide variety of materials, including those of fibrous nature, but are not suitable for quick-settling slurries. The capital cost is reduced by using the disc filter which provides a much larger filter surface and this type is generally used for handling large volumes of relatively free-filtering solids in the size 40–200 mesh (350–75 $\mu$m). The cake produced by a disc filter will, however, have a higher moisture content than that from a drum filter. Reported maximum capacities of vacuum filtration range from 0.1 kgs$^{-1}$ m$^{-2}$ for finely ground minerals under a vacuum of 85 kNm$^{-1}$ (25 in Hg) to 4 kgs$^{-1}$ m$^{-2}$ for gravity concentrates and sand under a vaccum of 20 kNm$^{-2}$ (6 in Hg). Much lower capacities are, however, reported.

## 2.6 PRESSURE FILTERS

The limitation on dewatering performance due to the available pressure across a vacuum filter can be overcome by applying pressure, but only at the expense of losing the facility for continuous operation. Automatic pressure filters are available, however, which move the filter cloth over small diameter rollers causing the cake to separate from the cloth. This improved operation enables batch filtering to be linked more readily to continuous processes. These presses, now built with filter areas of between 100 m$^2$ and 200 m$^2$, are intended for coal preparation, cement production and effluent treatment. Pressure is almost invariably supplied by pumping of the slurry, although in some cases the final part of the dewatering cycle may be the application of air pressure which gives a low cake moisture content. The usual form of the filter press comprises multiple cells held together in a frame by hydraulic pressure. The design of the supporting plate is such that the filter cloth is adequately supported against the differential pressure which is usually of the order of 700 kNm$^{-2}$ (100 p.s.i.) but may reach 2 MNm$^{-2}$ (300 p.s.i.). Compressed air can be applied to the slurry chamber as the final stage to assist dewatering and cake discharge or an alternative design makes use of a rubber membrane to enable the cake to be compressed by external air pressure. Considerably reduced cake moisture content is possible by this method.

For most applications of solids pumping it is probably that a vacuum filter will be adequate and the less convenient pressure filter would not be necessary except for colloidal and slimy material such as sewage sludge.

## 2.7 THERMAL DRYING

In cases where delivered solids must be completely dry there is no alternative but to incorporate a final thermal drying stage. However, it will almost invariably be found that it is most economical to remove the majority of the liquid by mechanical means as the cost of heat to evaporate the remaining moisture will be high in those situations where the heat is supplied solely for this purpose. In an integrated process such as a power station the availability of low-grade heat from the boiler flue gas means that thermal drying is relatively inexpensive and normal coal deliveries at 15–20 per cent moisture are dried while being milled.

Where heat is supplied by firing in a separate oil or coal burner the gases will probably leave the drier at a relatively high temperature and recovery of the sensible heat in the gas and certainly the latent heat of the moisture is possible only by considerable sophistication. The flash dryer is so-called because the particles are exposed to the high velocity, high temperature, gas stream for a very short time and are thereby protected from overheating, which may be detrimental to some materials. This feature of design means that only surface moisture will readily be removed, and absorbed moisture is retained in the particle. The aerodynamics of the device also make it difficult to reconcile the relative settling rates of the large and small fractions of the feed material. The fluidised bed dryer operates by bubbling the drying air through a bed of the solids which can result in greatly increased residence time, giving more complete removal of absorbed moisture. The velocity necessary to fluidize the material will be such that some fine particles will be carried over and, as with the flash dryer, air cleaning cyclones are necessary both to give acceptable air quality for environmental reasons and to prevent excessive loss of solid material. Fluidised beds have the property of good heat transfer and drying heat may be supplied via steam tubes in the bed if relatively cheap process steam is available.

A further arrangement of air dryer is the rotary machine in which solids are tumbled within a sloping cylinder passing from one end to the other by the action of internal baffles. Hot air in co-current or counter flow will then carry out the function of drying without being required to transport or fluidise the particles. There is inevitably some carryover of fine material and this design does not avoid the need for gas cleaning. An advanced dewatering scheme was devised for the ETSI project. Fluidised bed dryers were used together with cyclones and filters to remove the particulate matter from the exhaust gases. Pre-heating of the centrifuge feed was employed to obtain the best water removal.

## 2.8  OIL AGGLOMERATION TECHNIQUES

These recently developed techniques illustrate an integrated approach to the problems of dewatering and product suitability.

### 2.8.1  Handleability of pipeline product

Coal from a pipeline of the Black Mesa type is very cohesive at moisture levels above 10 per cent, although the coarser fraction, above 250 $\mu$m, is easy flowing at this moisture level. This cohesiveness makes the coal unsuitable for use in conventional coal handling equipment and this limits its marketability and hence the potential application of slurry pipelines. To avoid costly thermal dewatering, Shell (9) have developed the Pellitiser Separator. The basic principle is to split the stream into two with the split being at 250 $\mu$m. A hydrocarbon binder added to the fine coal water slurry preferentially wets the coal particles which can then be agglomerated by agitation. The free ash is not agglomerated so the process also acts as a coal ash separator.

The +250 $\mu$m stream is dewatered in a screenbowl centrifuge and the pellets removed from the water-ash flow by screening. Overall water content of the product is approximately 11 per cent. Tests indicate that the system works and that the upgraded coal can be stored and handled in conventional equipment.

## 2.8.2   Suitability of pipelined coal for coke manufacture

Coal supplied by pipeline is likely to be unsuitable for good quality coke manufacture. This arises from the particle size being too small and the fact that the clay content coats the particles, resulting in a poorly bonded coke unsuitable for blast furnace operation. This has been noted particularly with a range of Australian bituminous coals.

Under suitable conditions it has been established by BHP of Australia (10) that separation of the coal from the clays can be achieved by adding oil to the coal slurry in a stirred tank. The agglomerated particles can be dewatered by use of a static screen. Tests have shown that coke made from the agglomerated particles had a superior strength to that made from the normal pipelined coal.

As a result of a joint-venture between BHP and BP Australia, further developments have taken place resulting in the 'Integrated Pipeline Transportation and Coal Cleaning System' (IPTACS), (11). In this process coal is reduced in size to a suitable particle size distribution and mixed with oil and water to form a fine slurry. Whilst the slurry travels through the pipeline agglomerates are formed. These agglomerates, up to 3–4 mm in diameter, are removed from the slurry by pumping it over screens. The underflow from the screens contains the mineral reject, which can be dewatered using screen bowl centrifuges. Very little coal finds its way into the waste matter and is thus a considerable improvement on conventional methods.

## 2.9   OPERATIONAL EXPERIENCE AT THE MOHAVE GENERATING PLANT

It is interesting to note the developments which have been necessary to give satisfactory performance at the receiving end of the Black Mesa pipeline. The delivered solids were expected to contain about 20 per cent less than 325 mesh (approx. 50 $\mu$m) and it was intended that the centrifuges would deliver coal to the mills at about 25 per cent moisture. The solids carried over with the water were to be removed by settling and then returned to the centrifuge.

After some development it was found necessary, in order not to overload the mill-drying capability, to direct the settled fines straight to the boiler burners and to heat the slurry to 140 °F (60 °C) to give a moisture content of 20 per cent. The steam used for slurry heating entails a loss of station efficiency. Some development work was also carried out to increase the life of the centrifuge filters, using a ceramic coating, and the conical beach using hardened stainless steel. Replacement intervals exceeding 15 000 hours have been achieved.

## 2.10  DISPOSAL OF WASTE MATERIAL

Where it is necessary to recover the solids from the liquid, such as with mine tailings or fly ash, a method of disposal is to pump the slurry to local areas for land reclamation. The normal method for fine solids is to distribute the material across low lying areas and to build up the region gradually. Alternative disposal methods are to pump the waste back down mine shafts or out to sea but the latter method although usually cheap causes the greatest environmental damage.

The disposal of water slurry effluents may include water evaporation, power plant and other industrial use, irrigation, recycle to the slurry system and discharge (12). For discharged water any standards relating to concentrations of dissolved materials will have to be met. In the USA the operation would require a permit from the Environmental Protection Agency. As there are no standards for slurry discharge as such, permit issuers would need to exercise 'engineering judgement' to decide the best methods for controlling the disposal. Moore (13) considers that in many cases upgrading of the quality of slurry waste water will be required for either re-use or discharge to a surface water course. For marginal projects the additional costs could make the system uneconomic. Long-distance pipelines are not likely to be unduly influenced by these costs.

Water quality and discharge are further discussed in Ref. 14.

## 2.11  MIXING AND RECLAIMING

### 2.11.1  Mixing

Mixing is required for slurries in the pre-pumping stage and after pipelining to maintain suspension in storage tanks. Theoretical aspects are discussed in chapter 6.

Mixing is likely to be carried out by jets or direct mechanical means. In the case of mechanical mixers these may be of the propeller or paddle type. Propeller type mixers move the largest volume of slurry for a given power input and, therefore, this type is found on the largest systems. For the Black Mesa pipeline, for example, each delivery storage tank has a capacity of 30 280 m$^3$ and each agitator is driven by a 375 kW motor. For smaller systems a variety of mechanical mixing arrangements can be considered which mainly concern the position at which the shaft enters the tank. The shaft may be inclined to the main tank axis to enhance the mixing effect.

Mixers do not necessarily maintain all the slurry in suspension, there are normally fillets of slurry remaining on the tank bottom at the junction between the tank walls and floor. Extra power is required to improve the homogeneity of a mixture and excess power may be needed to remove all fillets. This further points to a method of reducing power consumption since instead of trying to maintain a nearly homogeneous slurry in one large tank the storage could be split between two or more tanks (15). The tank which supplies slurry to the next stage of treatment, for which a closely controlled concentration may be required, could be fully mixed. Preceeding tanks, only carrying out a storage function, could be allowed to have more marked concentration gradients, thus reducing their power requirement.

The design of effective and reliable mixers is a technology of its own and best results are likely to be achieved by consulting the relevant manufacturers. One solution for solving both mixing and mechanical problems is to use a conical vessel. The mixer, a long rotating scroll, is traversed around the conical wall.

Large changes of scale can introduce problems and a more effective solution may be to use a number of mixers of established design and size rather than to risk too large a change of scale.

Further aspects of practical mixer design are discussed by Harrah (16).

## 2.11.2 Slurry reclaim

If slurry is allowed to settle it will form a relatively high concentration mixture of solids and water. The flow characteristics of this material will depend to a large extent on its particle size distribution and specific gravity. If it is required for a further stage in the process it may need to be re-slurried. This can be carried out by the Marconaflo System.

The method operates by using a high velocity jet, which is rotated or oscillated to cut into the settled material. The slurry thus formed is drawn into a solids handling centrifugal pump to be transferred to the next stage in the treatment process.

The system can be mounted in tanks with the jet feed either passing through the tank or supplied from beneath. Dynajet systems are mobile units combining the moveable jet with the solids handling pump. They can be suspended from a crane, boom or gantry and lowered into the material to be recovered, thus enabling solids recovery to be made from settling ponds.

Application to coarse coal handling is discussed by Sims (17) but the method has been successfully applied to many different types of slurry.

## 3. GENERAL COMMENTS

'The hydraulic transport of solids' is often considered only from the point-of-view of the flow of mixtures through pipes. When the overall economics is considered, it is essential to take account of the material preparation and dewatering aspects. On short pipelines preparation costs can completely change the overall costs of transportation. For coal pipelines in particular the cost of dewatering can also be significant. The best way of dewatering a coal slurry may not yet have been developed. Certainly considerable expense has been incurred in the construction of pilot plants to develop suitable systems. This is an area in which the engineer will need to seek expert advice.

## 4. REFERENCES

1. Perry, R. H. and Chilton, C. H. (Eds) (1973) 'Chemical engineers' handbook' 5th edition. New York: McGraw Hill.
2. Derummelaere, R. H., Dina, M. L. and McEwan, P.F. (23–26 March 1982) 'ETSI coal evaluation plant'. *Proc. 7th Int. Tech. Conf. on slurry transportation*, Lake Tahoe, Nevada, USA, Washington DC, USA, pp. 27–32, Slurry Transport Assoc.
3. Bain, A. G. and Bonnington, S.T. (1970) Hydraulic transport of solids by pipeline', Oxford: Pergamon Press.
4. Bernstrom, B. (2–4 March 1977) 'An evaluation of grinding and impact crushing for reduction of coal to pipeline size', *Proc. 2nd Int. Tech.Conf. on slurry transportation*, Las Vegas, Nevada, USA, Washington DC, USA, pp. 43–9, Slurry Transport Assoc.
5. Walker, J. R. D. and Worcester, R. C. (9–13 April 1962) 'Hydraulic transport of solids: Trinidad Cement Ltd's 6 mile 2000 psi pipeline', *Pipes, Pipelines, Pumps and Valves Convention*, London.Theme 6/7 (e), 11 pp.
6. Svarovsky, K. (Ed) (1977) 'Solid-liquid separation', London, Butterworths.
7. Jenkinson, D. E. (29–31 March 1978) 'The dewatering, recovery and handling of pipeline coal', *Proc. 3rd Int. Tech. Conf. on slurry transportation*, Las Vegas, Nevada, USA, Washington DC, USA, pp. 49–57, Slurry Transport Assoc.
8. McCabe, W. L. and Smith, J. C. (Eds) (1956) 'Unit operations of chemical operations', New York: McGraw Hill, 945 pp.
9. Van der Toorn, L. J., van Oldenborgh and Huberts, L. J. (29–31 March 1978) 'Coal slurry pipeline end products', *Proc. 3rd Int. Conf. on slurry transportation*, Las Vegas, Nevada, USA, Washington DC, USA, pp. 69–76, Slurry Transport Assoc.
10. Rigby, G. R. and Calcott, T. G. (29–31 March 1978) 'Coking coal slurry transport: the use of oil agglomeration techniques to overcome recovery problems', *Proc. 3rd Int. Conf. on slurry transportation*, Las Vegas, Nevada, USA, Washington DC, USA, pp. 86–92, Slurry Transport Assoc.
11. Rigby, G. R., Jones, C. U. and Mainwaring, D. E. (25–27 August 1982) 'Slurry pipeline studies on the BHP-BPA 30 tonne per hour demonstration plant', *Proc. 8th Int. Conf. on hydraulic transport of solids in pipes*, Johannesburg, South Africa, pp. 181–94, BHRA, The Fluid Engineering Centre.
12. Rieber, M. and Soo, S. L. (1982) 'Coal Slurry Pipelines: A review and analysis of proposals, projects and literature', Electric Power Research Institute.
13. Moore, J. W. (26–28 March 1980) 'Quality characteristics of slurry wastewater resulting from the slurry pipelining of Eastern Coal', *Proc. 5th Int. Tech. Conf. on slurry transportation*, Lake Tahoe, Nevada, USA. Washington DC, USA, pp. 242–50, Slurry Transport Association.
14. U.S. Congress, Office of Technology Assessment (January 1978) 'Coal Slurry Pipelines', OTA-E-60.
15. Pharamond, J. C. and Olderstein, A. J. (26–28 March 1980) 'New concepts in slurry storage', *Proc. 5th Int. Conf. on slurry transportation*, Lake Tahoe, Nevada, USA. Washington DC, USA, pp. 74–83, Slurry Transport Assoc.

16. Harrah, H. W. (15–17 May 1974) 'Slurry storage agitators for slurry pumping', *Proc. 3rd Int. Conf. on hydraulic transport of solids in pipes*, Golden, Colorado, USA, pp. B4–43 to B4–52, BHRA, The Fluid Engineering Centre.
17. Sims, W. N. (15–18 March 1983) 'Coarse coal handling by Marconaflo systems', *Proc. 8th Int. Conf. on slurry transportation*, San Francisco, California, USA. Washington DC, USA, pp. 353–9, Slurry Transport Assoc.

# 5. INSTRUMENTATION

## 1. INTRODUCTION

In order to operate a solids transport pipeline efficiently it is vital that the automatic control system or the operator is adequately aware of conditions within the pipeline. Although detailed requirements will differ depending on particular applications, the main parameters which may require measurement are:

1. solids concentration (or slurry density),
2. flow rate,
3. pressure or head,
4. pump speed and power,
5. viscosity.

Of these, the first two points are the most important in industrial installations, since knowledge and control of concentration and flow rate is essential in order to avoid blockages, as well as for system evaluation. The ability to detect, for settling slurries, whether a bed of solids exists in a horizontal pipe, and if so, to what depth, is also desirable in order to minimise the risk of line blockages. For slurries having a relatively high solids content, the viscosity, measured by instruments which are 'available on site' rather than 'permanently installed', is an important parameter which affects pump performance and overall system pressure.

There appears to be very little published information on the order of accuracy required from industrial pipeline instrumentation; in fact, there is little mention of instrumentation at all in most references on existing pipeline systems. Kakka (1), in his review of meters for flow and concentration measurement in iron ore slurry transport for the UK steel industry, distinguishes between requirements for 'commercial' purposes i.e. obtaining the total tonnage of solids delivered, and for operational control; he specifies a mean accuracy of ±0.5 per cent for the former, and ±2–5 per cent for the latter, both being expressed as a genuine percentage of the true value (note that some authors consider accuracy in terms of a variation in numerical percentage concentration, either by weight or volume.

An overriding factor is that any instrumentation for use on site must be reliable, robust and easy to install, externally-mounted devices being preferred (1). Calibration, particularly if required to be done on site with the actual slurry in order to achieve adequate accuracy, should be as direct as possible, and may be seen as a major consideration for all types of instruments.

Aspects considered in the sections which follow include current applications, commercial availability and the advantages and disadvantages or limitations of the equipment discussed. Various control system philosophies are also outlined.


## 2.  SOLIDS CONCENTRATION MEASUREMENT AND METERS

The only method for direct measurement of delivered concentration for a settling slurry is by sampling, using a weigh tank, but this is not suitable for continuous recording, or for handling large tonnages. Thus, its use is usually confined to test circuits, although a large weigh tank was used in the large-scale pilot scheme for hydraulic hoisting at the Gneisenau Mine in West Germany (2). In the case of large weigh tanks, the normal procedure is the addition, after the slurry sample has been taken, of a known quantity of water up to a pre-determined mark (usually in a section of reduced area) which has been accurately positioned to indicate a known total volume. Nearly all the concentration measuring methods discussed in this section, involving meters installed in the pipeline, measure the local transport concentration (see below).

*'Delivered' vs 'transport' concentration*

It is important to recognise the difference between 'delivered' and 'transport' (or 'in-situ', 'spatial') concentration, often referred to as 'hold-up' for a heterogeneous settling slurry, due to the 'slip' effect between the solids and liquid (normally water) and also to the concentration gradient across a horizontal pipe (3). Delivered concentration refers to the slurry state at exit from the pipeline and is commonly defined (in terms of volume concentration), as:

$$C_{vd} = \frac{\text{Vol. flow rate of solids, } Q_s}{\text{Vol. flow rate of mixture, } Q_m}$$

'Transport' concentration refers to the instantaneous local conditions inside the pipeline, and is defined as:

$$C_{vt} = \frac{\text{Vol. occupied by solids}}{\text{Vol. occupied by mixture}}$$

since mass flow of mixture = mass flow of solids + mass flow of water

it can be shown that mean velocity of solids,

$$V_s = \frac{C_{vd}}{C_{vt}} \times \text{mean velocity of mixture, } V_m$$

or

$$C_{vd} = \frac{V_s}{V_m} \times C_{vt}$$

(Mean vel. of water, $V_w = \dfrac{1 - C_{vd}}{1 - C_{vt}} \times V_m$)

Thus, if 'slip' is significant, i.e. $V_s < V_m$, then $C_{vd} < C_{vt}$.

For homogeneous and pseudo-homogeneous slurries, where there is no slip, then $V_s = C_m$ and $C_{vd} = C_{vt}$.

So, for 'commercial' purposes, it is the delivered concentration which is required, whereas for control purposes, the transport concentration is more appropriate. Also, care should be taken to ensure that conditions in the measuring section are reasonably representative of the pipeline as a whole, and not affected by proximity to upstream fittings, e.g. bends.

(N.B. The terms 'delivered' and 'transport' concentration in this book are those commonly used in English-speaking countries. Note that some European papers have been translated such that the term 'transport' refers to the delivered concentrations as defined above).


## 2.1 GRAVIMETRIC TYPE (PIPE-WEIGHING DENSITY METER)

The principle of operation is that a length of pipe of known volume is weighed full of water and then compared with its weight containing slurry. The difference between the two values is proportional to the solids content. Its application to actual slurry pipeline systems has been confined so far to use in by-pass test loops (4), e.g. in the Savage River Mines iron ore pipeline in Tasmania (5). Otherwise, the method is used mainly in laboratory test circuits, and a number of different designs have been reported (6,7,8,9). There are two basic horizontal types – cantilevered U-tube, and 'straight-through' pipe – with flexible couplings joining the device to the main pipeline. The method has also been used with an inclined U-tube (8).

The horizontal U-tube version is available commercially, but only in small pipe sizes up to 40 mm (in UK). However, Baker and Hemp (7) suggest that the straight-through design, with two equal pipe lengths connected by flexible couplings (i.e. three hinge positions) and supported centrally by load cells, is to be preferred (Figure 1); this avoids the large force on a U-tube due to slurry pressure, which could be a source of error.

The method is basically simple and should give a high accuracy, being dependent on fundamental mass measurements (7). Performance is not affected by particle type or size. Calibration is reasonably straightforward requiring only dead-weight loading on the pipe. Tests on a 'straight-through' meter gave an accuracy within ±2.5 per cent with a fine sand slurry when compared with direct sampling by weigh tank.

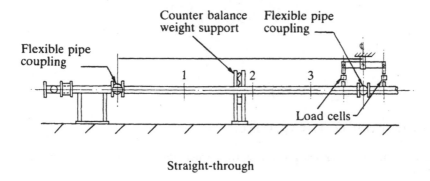

Straight-through

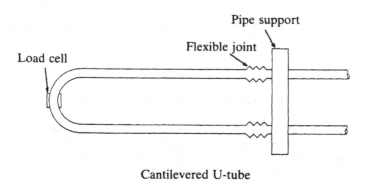

Cantilevered U-tube

**Figure 1.** Gravimetric concentration meters

There are, however, a number of possible limitations to its general use in industrial pipelines. It becomes much less sensitive for mixtures with relative densities only just greater than 1.0. Reactions at the flexible couplings affect the accuracy (7,9,10), and hence care is needed in the joint design; this could be a problem if the device is to be considered for larger pipe sizes and higher pressures. Accuracy is also affected adversely if a layer of settled solids adheres to the bottom of the pipe on shut-down, and is not removed when re-started, as was found with china clay slurry during tests at BHRA. A relatively long length of accessible horizontal pipeline is required for installation; hence, application to an existing system might be difficult in some cases. Although it is a relatively inexpensive method for smaller pipes, the costs of flexible couplings, support structure and load cells for typical slurry pipeline sizes could be quite large.

## 2.2  DIFFERENTIAL PRESSURE TYPES

### 2.2.1  Counterflow meter

This consists of a U-tube through which the slurry passes. By measuring the pressure gradient in the upward and downward lines it is possible to separate the effects of buoyancy and friction and thus determine the concentration. This meter is used mainly in laboratory test loops, although the method has also been tried in one Dutch dredging system (1,6,7,8,10). It is not generally available commercially.

The main attraction is simplicity and cheapness, as only pressure measurements, using either manometers or transducers, are required. However, in order to interpret the measurements, assumptions are made that friction losses in the vertical pipes are the same for slurry and clean water, and that 'slip' or 'hold-up' effects are negligible (1,6,7,8,10). More detailed analysis can correct for slip (7,8,10). Later work in West Germany (8,10) has developed averaging differential pressure techniques to give greater accuracy and avoid pressure tapping flushing problems, and is claimed to be simpler than the Dutch (Van der Veen) method; Kao and Kazanskij (8) quote an average error of ±4 per cent compared with the sampled concentration with a coarse coal slurry. BHRA tests using a simple manometer arrangement gave an average error of about ±10 per cent with a homogeneous china clay suspension, compared with sampling, but results with settling slurries were generally about 10–15 per cent low.

Kao and Kazanskij (8) suggest that this type of meter gives a close approximation to the delivered concentration, as 'hold-up' effects tend to be averaged out in upward and downward flow. More recent analysis by Clift and Clift (11) employs a deeper understanding of the principles involved and indicates the probable errors that can arise. Apart from doubts on the maximum achievable accuracy (7), particularly with solids having a high settling velocity, one obvious limitation is that no indication is given of conditions in any part of the system with horizontal pipe. Also, in a practical system, it might be difficult to install a special inverted U-pipe loop.

### 2.2.2  Slurry manometer

This device, designed by Warman International in Australia for test rig use, and mentioned by Baker and Kemp (7), is essentially just the upward-flowing limb of a counterflow meter and it is necessary to operate it with water and slurry in turn to calculate the concentration. There are no reports of its use on an actual pipeline; and presumably it is not available commercially. Again, it is a very simple and cheap method, using only a single manometer, but is subject to settling velocity effects and hence a correction should be applied, otherwise large errors can occur (7). Also, it was found during tests at BHRA that results with homogeneous mixtures of fine particle slurries, e.g. coal, china clay, were generally inaccurate, due to difficulty in keeping the inclined limbs connected to the manometer free of solids. The same comments regarding horizontal pipe flow as were given for the counterflow meter (section 2.2.1 above) would also apply. Thus, it appears to be of somewhat limited use and accuracy.

## 2.2.3  Venturi meter

By employing both the pressure loss across the meter and the depression in pressure at the throat it is possible to use a venturi meter to measure concentration (as well as flow) from the pressure drop across it, provided the mixture is homogeneous (7). Although an apparently simple method, only limited data are available from tests on relatively small sizes (3 inch and 4 inch inlet diameter) with small particle slurries at concentrations up to 14 per cent $C_v$ (7). Thus, extrapolation to larger pipe and particle sizes and higher concentrations, is open to doubt, and the requirement for a homogeneous mixture would restrict the application. A venturi causes some obstruction to flow, hence permanent head loss and additional risk of blockage; also, wear could be a problem with abrasive slurries (see comments in section 3.2.2 below).

## 2.2.4  Adjustable constriction device

Used in a vertical pipe, the device consists of a differential manometer connected between two tappings, the lower one being opposite a contraction produced, for example, by a standard pinch valve. The method, which is more suited to large diameter pipes, involves adjusting the constriction to give zero pressure differential for all flow rates of water by itself (in an upward direction) and is described by Bain and Bonnington (12).

## 2.3  NUCLEAR RADIATION METHODS

### 2.3.1  γ-ray absorption

This is currently the most widely used technique for concentration measurement in both industrial pipelines (1,4,5,17,18) and test loops (7,8,10,13,14,15,16), and several makes are commercially available, with claimed accuracies of ±1 per cent (1). A typical arrangement is shown in Figure 2. Since the performance can be affected by the concentration distribution across the pipe (10), the method can also be adapted to measure the distribution, and various scanning devices have been reported (7,13,14).

Developments in the USA have involved three-component concentration measurement (e.g. for coal/rock/water mixtures) using dual-beam techniques with either medium- and low-energy γ-ray sources (7,8) or γ-ray and neutron sources (15); the latter method gave an accuracy within about ±2 per cent $C_w$ (numerically) with fine particle slurries.

The main advantages would appear to be ease of installation, being a 'clamp-on' type of device, and no direct contact with the slurry. Performance is mainly unaffected by fluid

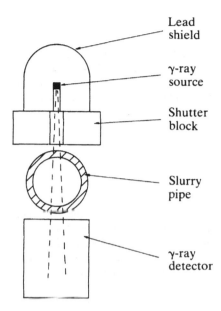

Lead
shield

γ-ray
source

Shutter
block

Slurry
pipe

γ-ray
detector

**Figure 2.** γ-ray concentration meter

properties or particle size, although Kao and Kazanskij (8) suggest that there is a critical upper limit for the latter. These meters are also generally accepted as being reasonably reliable (8). Due to concentration gradient effects, it is often recommended to install the meter in a vertical section of pipe, away from bends or other fittings (10). However, Japanese tests (16) with an iron ore concentrate slurry gave good agreement between the mean concentration given by a sampling probe and a γ-ray meter in both horizontal and vertical 150 mm (6 inch) pipes, even though a concentration gradient existed across the horizontal pipe, provided a heterogeneous suspension was obtained with no sliding bed. Considerable care is required in using this method (8,10), particularly regarding calibration; Kao and Kazanskij (8) recommend that this should be done using a test loop with the same slurry conditions, pipe size and material as for the actual pipeline to obtain reliable results, whilst admitting that this may be difficult to achieve in practice. They also suggest that there is a limitation on the maximum mixture density which can be measured for a given pipe size, e.g. S.G. of 2.5 for 400 mm (16 inch) diameter steel pipe. Response time to changes in concentration tends to be relatively long (8,10,17), although the three-component sensor reported by Verbinski *et al.* (15) gives readings at 1 second intervals. Application problems which can cause errors are entrained air in the pipeline (8,18), pipe wall deposits, or wear (8,10), diminishing intensity of γ-ray source with time (10), and over-heating of the detector head in high temperature environments (8). These meters are also relatively expensive.

## 2.3.2   Neutron moderation

Although such meters are commercially available, there is no report of their use in industrial slurry pipelines, and probably more development is needed (1). Baker and Hemp (7) suggest that the main advantage appears to be that all equipment is mounted on one side of the pipe, but this results in only a small region of the pipe section being examined.

## 2.4   ELECTRICAL METHODS

### 2.4.1   Conductivity measurement

This method has been used in certain dredging installations for some years (1,8), and meters are commercially available, with a claimed accuracy of ±2 per cent (1). More recent development work, particularly by Beck (7,19) in the UK and Koa and Kazanskij (8) in West Germany; has involved the use of conductivity measurement techniques. A conductivity meter is also included in the three-component sensor mentioned above (see section 2.3.1).

The electrodes are relatively simple to install in the pipe wall, although these have to be in contact with the slurry, and hence are subject to fouling or wear. The technique described by Beck *et al.* (19) was said to be relatively inexpensive, so may well be cheaper than, say, a nucleonic meter. Although the method can possibly be used for a wide range of slurries and sludges, individual calibration is required for each type (19). However, it may be unreliable if dissolved chemicals are present, e.g. corrosion inhibitors or drag-reducing agents (7,8), or if the solids have variable conductivity in water or tend to dissolve, such as some coals and mineral ores (8). Three or more pairs of probes arranged circumferentially are recommended for use with heterogeneous suspensions, to give representative concentration measurement (8).

Kao and Kazanskij (8) suggest that this method is much more reliable than capacitance measurement. However, Wiedenroth (10) considers that the conductivity technique is unreliable for absolute measurement, although adequate for relative values. Baker and Hemp (7) suggest that more development work is needed for both methods.

A cross-correlation conductivity technique has been used for measuring local solids velocity and concentration using traversing probes (8) although these were subject to considerable erosion. Cross-correlation may also be used for flow measurement (20) (see section 3.3.1 below), and for detecting solid deposition (21) (see section 5.1.1 below).

### 2.4.2   Capacitance measurement

Information on this method is rather limited, and its use so far seems to have been confined to experimental development, such as the Polish work mentioned by Baker and

Hemp (7) on a device to measure both the mean concentration and the distribution across the pipe.

Wiendenroth (10) suggests that the method looks promising, though more experience is needed; however, Kao and Kazanskij (8) state that it is less reliable than the conductivity technique.

## 2.5  Ultrasonic Methods

Although ultrasonic concentration meters are commercially available, there appears to be a lack of reports on their use and operating experience in industrial pipelines. Development work by Beck (7,19,22) has involved the use of ultrasonic sensors. This method has the advantage that it may be used with both electrically conducting and non-conducting liquids. Both commercial meters and those under development require calibration on the particular slurry to be pumped, for most accurate results. It is possible to use sensors either clamped to the outside of the pipe, or fitted flush within the inside in contact with the slurry. Although the 'clamp-on' type is simple to install and not susceptible to wear or fouling, the 'flush-mounted' type avoids problems with pipe wall effects, so gives better results overall (22). Balachandran and Beck (22) state that an experimental meter using an ultrasonic beam modulation technique is sensitive to both flow velocity and particle size, as well as concentration; however, cross-correlation techniques, being also applicable to flow measurement (20), can be used to compensate for velocity effects. Japanese tests (16) with an iron ore concentrate slurry showed that an ultrasonic meter in a horizontal pipe was affected to some extent by the concentration gradient across the pipe, giving an increasing value of concentration with velocity. Performance is adversely affected by air bubbles.

## 2.6  OPTICAL METHODS

Baker and Hemp (7) report that deflection and holographic techniques developed by two workers, mainly for particle size determination, may, with further development, be applicable to the industrial measurement of solids concentration.

## 3.  FLOW METERING

The most fundamental method of determining flow rate but only at the end of a pipeline and when system control is not a primary aim, is to divert the flow for a known time into a calibration tank. Alternatively, a very approximate indication of flow from the open end of a line may be gained by applying the equations of motion to the observed trajectory of the issuing slurry. Another simple type of discharge meter, the 'sifflet' eccentric nozzle, has been used successfully with various common mineral slurries. The device is formed by flattening a short length of the crown at the open end of a pipe and, according to Bain and

Bonnington (12), it has been found that the pressure drop across the resulting taper is independent of solids concentration and specific gravity when the former is less than 20 per cent by volume and when the latter is less than 2.0, for all pipe orientations. Bernouilli's equation enables the slurry velocity to be determined using the pressure drop and the pipe cross-sectional areas prior to the taper and at the exit respectively.

Before discussing the various types of in-line flow meters available, it is worthwhile outlining the factors which should be considered during selection:

1. Measurement capability –
   maximum flow and flow range;
   accuracy and repeatability;
   linearity;
   need for calibration.
2. Installation –
   pipe size range, and orientation (horizontal or vertical);
   head loss across proposed instrument;
   sensitivity to upstream flow conditions;
   convenience.
3. Application –
   slurry type and properties, i.e. viscosity, presence of entrained gas, abrasivity (wear), electrical conductivity (electromagnetic flowmeters);
   system pressure and temperature;
   steady or pulsating flow.
4. Economics –
   initial cost;
   need for maintenance/re-calibration;
   expected life.

Hayward's book on flowmeters (23) is a good aid to any selection process. Papers by Kakka (1), Debreczeni *et al.* (6) and Kao and Kazanskij (8) also review the various types of flowmeter suitable for slurry pipeline applications. Baker and Hemp (7), although primarily concerned with concentration meters, make some comments on flow metering which might be of interest to a prospective user. Irrespective of the instrument chosen for a particular application, it is vital to carry out the recommended installation and calibration procedures prior to its operational use.

## 3.1  ELECTRO-MAGNETIC METERS

This is the most common method of flow measurement, for both industrial pipelines (1,4,5,6,17,18) and test loops (6,8) and a number of makes are commercially available, with claimed accuracies of ±1 per cent (1).

These meters are generally regarded as adequate for most applications, although

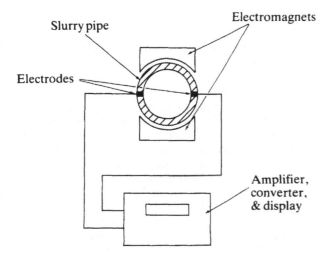

**Figure 3.** Electromagnetic flowmeter

relatively expensive (1,4). The main advantages are that the meter does not obstruct the flow, and may be used in its standard form for all types of non-magnetic slurries and sludges, provided that the liquid has sufficient conductivity (1,8). For magnetic slurries, e.g. iron ore, a compensating device is required (1), but the resulting accuracy may still be suspect. Meters do not normally require special calibration on the slurry to be handled, except possibly for non-Newtonian and metallic solid mixtures (8). Performance may be affected by a non-uniform inlet velocity distribution if the profile changes along the meter axis; hence the meter should not be installed close to an upstream obstruction to the flow (8). Also, precautions are needed to avoid exposing the meter to any external sources of magnetic radiation, and to ensure that it is properly earthed. Problems may occur in service due either to wear or contamination of the electrodes (8), but special linings are available to overcome or reduce these effects. Installation in a vertical section of pipe will help to reduce wear (Figure 3).

## 3.2 DIFFERENTIAL PRESSURE TYPES

### 3.2.1 Counterflow meter

Although the counterflow meter can be used for the measurement of flow as well as solids concentration, Clift and Clift (11) show that the reliability of velocity measurements obtained by this device is not very high, especially for large particle slurries.

## 3.2.2   Venturi meter

Few reports of venturi meters being used in industrial slurry pipelines, except possibly in the Soviet Union (8), have been discovered, although standard meters are, of course, commercially available.

Kao and Kazanskij (8) state that these instruments can give equally good results for slurry flow measurement as for single-phase flow. They quote results from Soviet tests in 1954 on a range of venturi sizes up to 300 mm (12 inch) inlet diameter on sand and coal slurries, with solids up to 6 mm size; reasonable accuracy was obtained, and performance was not affected by grain size, solids relative density or concentration with the ranges investigated. Also, although wear is often considered to be a major disadvantage, this was found not to be a problem with these slurries up to 900 mm (35 inch) pipe size. Wear can be reduced by mounting the meter vertically (6,8), and by avoiding small area ratios (6). However, Baker and Hemp (7), quoting Brook's work, observe that vertical mounting is recommended to ensure a homogeneous suspension, clean water calibration is maintained up to only 10–20 per cent $C_v$, depending on the type of solids, and abrasion-resistant materials are necessary to retain internal dimensions. Discharge coefficients may also be affected by slurries containing large, high density particles.

## 3.2.3   Bend meter

Although this type of meter has been used in slurry test loops, and does not cause any restriction to flow (6), its accuracy is only about ±2–5 per cent on clean liquids, depending on the quality of construction and installation. Also, it would be subject to considerable wear with abrasive slurries (6,8); for this reason, Kao and Kazanskij (8) consider it unsuitable for practical use.

## 3.3   ELECTRICAL METHODS

## 3.3.1   Conductivity measurement

Development work in recent years has been carried out by Beck (7,19,20) on conductivity flow meters, using cross-correlation techniques; a similar technique was used by Kao and Kazanskij to detect solids slip velocity (8). Most of the advantages and limitations of the method discussed in section 2.4.1 above for concentration measurement would also apply to flow measurement, except that, with cross-correlation, calibration on individual slurries and pipe sizes is not required (20). Beck (20) suggests that the electrodes can be very robust, the electronic circuits are simple, and hence the method has potential as a low-cost flow meter; it is likely to be much cheaper than ultrasonic systems.

## 3.3.2 Capacitance measurement

Although it may be possible to apply this method to slurry flow measurement, again using cross-correlation technique it would appear that it is more suitable for use with solids/gas mixtures, i.e. pneumatic conveying (20).

## 3.4 ULTRASONIC METHODS

This type of flow meter is commercially available in various guises, but its use in industrial pipelines would appear to be limited so far, judging by the lack of published reports. According to one UK manufacturer, the accuracy is only of the order of ±5 per cent with a single 'clamp-on' type of sensor, unless specially calibrated on site with the particular slurry, due to pipe wall effects. Faddick *et al.* (24) report results of tests on units bonded to the pipe wall with a coarse sand slurry over a range of 10–50 per cent $C_w$. Most recent development work has involved the use of ultrasonic turbulence detectors for A.C. cross-correlation techniques, directed mainly towards application in sewage sludge and similar systems.

Again, most of the comments on advantages and limitations of the method given in section 2.5 above for concentration measurement would also apply to flow measurement, except that cross-correlation techniques do not require special calibration (20). Balachandran and Beck (22) report that results were independent of particle size and concentration over the rather limited ranges investigated, using internal flush-mounted sensors; also, performance was not affected by small quantities (up to 5 per cent by volume) of gas.

The main economic advantage of commercial ultrasonic flow meters is that they tend to be considerably cheaper than, say, electro-magnetic types.

## 3.5 TRACER TECHNIQUES

The rate of progress of a 'pulse' introduced into homogeneous slurry flow via an upstream tapping point, can be determined by a suitable detection method. The pulse may take the form of, for example, a change in conductivity, colour, radioactivity or fluorescent content in the slurry and detection may be by electrical or visual means. Even if probes which produce electrical outputs are employed, the signals will not be continuous in nature, but the method can be useful for calibration of other instruments.

## 3.6 CORIOLIS FORCE MEASUREMENT

In one form of these meters the flow of slurry is passed through a U-tube. The ends of the U are held stationary while the curved part is moved cyclically. This causes the legs of the U to be bent in a plane at right angles to the plane in which they both lie. The curvature induced in the legs causes the slurry to experience a Coriolis force which acts in opposite directions on the two sides of the U, causing it to twist. Measurement of this twist is proportional to the mass flow rate of the slurry. Typically units are manufactured to be inserted into lines up to 75 mm diameter but large units are becoming available (25). Tests on an early model (26) using a coal slurry, showed it to give readings which were substantially independent of concentration, particle size and temperature.

## 4. PRESSURE

Pressure measurements are necessary in a pipeline firstly to monitor pump performance and secondly to establish the frictional head loss gradient along the line.

The measuring instrument may be mounted directly on the pipe providing it is designed to withstand the prevailing levels of vibration and its internals are not allowed to come into contact with the conveyed solids (see below).

If remotely mounted measuring instruments are employed the piezometric, or tapping, holes should be between 3 and 6 mm in diameter (depending on the solid particle size), clean-drilled through the pipe wall and at least 10 pipe diameters downstream of bends or other obstructions. For pumps with vertical deliveries a 'ring tapping' utilising four holes at 90° spacings around the pipe is desirable, but in horizontal lengths two tappings holes 45° either side of the crown is preferable to avoid problems caused by solids settlement. It is important that solid matter should not block the connecting tubes or the internals of the measuring instrument, whatever form it takes. Therefore, a flushing facility should be incorporated to purge the connecting tubes of solids prior to measurements being made. Where pressures are low translucent tubing is obviously advantageous. A catchpot can also help to prevent solids or air reaching the pressure measuring instrument and if each tapping point at a particular location in the line has a dedicated catchpot and pressure sensor then more reliable values may be obtained. Take-off tubes should ascend from the pipe to the measuring instrument and when liquid is used to transmit the pipe pressure, allowance must be made for the difference in heights between the pipeline and the instrument. In the case of a U-tube manometer, a small reservoir of water feeding down onto the column of manometer fluid in the leg furthest from the pressure tapping will keep the height correction factor effectively constant.

Although attractively simple, U-tube manometers can only be considered for relatively low pressure applications, even when measuring differential pressures, and are frequently only employed for the measurement of suction head. When this is less than one atmosphere it is preferable to keep the tubing linking the manometer to the pipe tapping hole free of water. Used with care, U-tube manometers can be very accurate and are often the first choice for research applications.

In commercial pipelines, Bourdon tube gauges are probably the most common pressure measuring instrument encountered and these can be mounted directly onto the pipeline providing the internal mechanism is isolated from the pipe fill. This may be achieved by employing a diaphragm, usually made of thin stainless steel, to act as the sensing element and this is connected to the gauge, by means of a small bore rigid tube containing a suitable liquid or gas. If gas is employed no head correction is required (see above) but pipeline pressures below atmospheric or above approximately 2 bar cannot be measured accurately. The diaphragm is normally mounted directly onto the pipe via a very short, large-bore standpipe to minimise the risk of blockages.

'Diaphragm' gauges employ a similarly mounted diaphragm but in this case the centre of the disc is connected directly to the meter pointer by a mechanical linkage.

Pressure instrument protectors have become commercially available in which a flexible impermeable sleeve isolates a captive sensing liquid from slurry flow (Figure 4). The sleeve and sensing fluid are contained within a flanged or screwed housing which is inserted into the pipeline and would seem to offer an elegant solution, since the line pressure is measured over 360° with no possibility of solids contamination or obstruction. However, a protector for larger pipe sizes can be expensive. Sleeve materials and sensing liquids are available to suit most applications and positive pressure up to 20 bar.

An inexpensive and interesting variation in devices designed to separate instruments from pipeline solids was developed during work done by Bantin *et al.* (27). A porous polythene ring, clamped between adjacent modified pipe flanges, was employed to block particles while permitting water (at line pressure) to enter an enveloping annular-enclosure into which a tapping hole had been drilled. The arrangement gave 360° sensing and a back-flushing facility was available if the porous ring became blinded with fines.

For the measurement of differential pressure a twin-tubed Bourdon gauge connected to upstream and downstream isolating diaphragms can be employed but a more sensitive alternative is the differential pressure transducer and such devices are readily linked to an

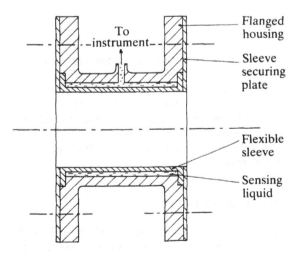

**Figure 4.** Sleeve-type pressure instrument protector

automated control system. Pressure transducers for differential or absolute pressure, may comprise a diaphragm plus some form of associated displacement detection system or, for very high pressures, a sensor made of a material the electrical characteristics of which changes with applied pressure. For measurement of pressures greater than about 10 bar the outputs from strain gauges bonded to the pipe outer wall have been successfully employed by the National Coal Board.

## 5. SOLIDS DEPOSITION DETECTION

In order to minimise energy costs it is frequently desired to operate hydraulic transport lines at the lowest velocities necessary to avoid significant solids settlement. Although the presence of a sliding or stationary bed of solids will increase the line operating pressure for a given volumetric throughput, instrumentation designed to monitor the latter two para-meters may be inadequate to detect settlement until solids build-up has reached dangerous proportions, particularly where relatively large and heavy particles are involved. Also, an increase in line pressure due to settlement may be erroneously interpreted as a change in the viscosity of the slurry caused by variations in solids concentrations or size distribution. For these reasons, an independent indication of the presence of a settled bed would be desirable. Techniques suitable for detecting a bed of settled solids in the pipeline have been investigated by Kazanskij (21), Ercolani *et al.* (28), Ferrini *et al.* (29) and Brown and Shook (30).

## 5.1 ELECTRICAL METHODS

### 5.1.1 Conductivity measurement

Kazanskij (21) has developed a D.C. cross-correlation method, using pairs of electrodes displaced axially and set in the pipe wall around the underside, to detect the height of a settled layer of solids, and to measure the solids velocity near the bottom of the pipe; the electrode diameter should be larger than the grain size. A simple version was developed for industrial use, as well as a precision type for laboratory purposes. It is claimed that this technique is preferable to Ercolani's single probe methods (28) (see section 5.1.2 below), as overall conditions in the pipe are monitored. This work appears to have been generally directed towards application in the dredging industry.

### 5.1.2 Electric charge and thermal measurements

Ercolani *et al.* (28) report the development of two types of probe, inserted in the pipe wall, both being sensitive to variation in slurry velocity. The 'electric' probe is sensitive to changes in the distribution and concentration of ionic charges round the concentric electrodes, and the thermic type, comprising two thermo-resistors, detects the variation of heat exchange between the probe and the slurry stream. In each case, the type of signal generated depends on the flow regime, and the authors claim that both types are 'simple and reliable'. However, Kazanskij (21) suggests that, since a single probe would give only a local signal, not fully representing conditions over the whole section, these methods are not suitable for industrial use, where concentration and grain size may change rapidly. Also, the thermic probe in particular would seem to be vulnerable to abrasive wear.

### 5.1.3   Measurement of voltage fluctuations in an applied transverse field

Brown and Shook (30) have developed an L-shaped probe for measuring local slurry velocities which functions in a transverse potential field applied by small electrodes recessed into the walls of a short length of 50 mm diameter acrylic plastic pipe. Voltage fluctuations produced by changes in resistivity at the probe-mounted sensors, corresponding to changes in slurry concentration, were cross-correlated. Although it could clearly be used to identify settlement, the probe described was stated to be rather fragile and so can only be regarded as a research tool. The possible limitations in the usefulness of localised flow investigations, as expressed by Ercolani (above), would also apply.

## 5.2   ULTRASONIC METHODS

Faddick *et al.* (24) suggest that an ultrasonic flow meter has the potential for monitoring the flow regime, since a sensor placed at the bottom of a pipe is affected by the presence of a sliding or stationary bed; however, more development work would be needed.

A series of ultrasonic probes mounted on the top of the pipe has been used experimentally to measure the profile of a settled solids bed in two adjoining inclined pipe sections after shut-down (29).

## 6.   PUMP POWER

The large majority of slurry pumps are electric motor-driven. Where efficiency is not of prime importance, it is generally adequate to monitor motor voltage and currents, possibly with power factor correction, mainly as a check on overall mechanical condition and any deterioration in performance. However, in applications involving the continuous operation of large numbers of relatively high-powered machines, it is usual to install power consumption (kWh) meters. It should be noted that this only enables the overall efficiency of the whole unit, e.g. pump and motor, to be determined; it is still necessary to assume the motor efficiency if the pump efficiency is to be calculated.

It is possible to measure pump input power directly by installing some proprietary type of torquemeter between pump and driver. However, this is not usually practical, particularly for standard units, unless special provision is made for extra space between the couplings to accommodate the meter.

## 7. LINE TEMPERATURES

Pipelines laid in parts of the world subject to very cold weather are either thermally wrapped or buried deeply enough to avoid frost. However, temperature sensors may still be usefully installed at several points along the line (31) in order that non-essential shutdowns are avoided when there is a significant chance that the water in the line will freeze if flow ceases.

## 8. VISCOMETRY

In general, viscosity measurements give a sufficiently accurate indication of the behaviour of a slurry. However, it is possible to extract further information about the flow properties of some non-settling slurries by using a rheogoniometer such as the Weissenberg instrument (32). The rheogoniometer is capable of measuring normal forces whereas viscometers measure shear forces.

The published literature shows that rheogonimeters are very rarely used on slurry measurements although theory suggests that normal stresses could be important in understanding such phenomena as drag reduction and yield stresses.

Van Wazer *et al.* (32) describe some sixteen rotational viscometers, eight commercially available tube or capillary viscometers and approximately ten further types, and this is by no means an exhaustive list. Many instruments are not suitable for slurries for a variety of reasons such as insufficient viscosity range, wrong shear rate range, insufficient clearances, etc.

For laboratory use in determining the shear-stress shear-rate behaviour of slurries the capillary and Couette geometries are by far the most popular. The cone and plate viscometer is sometimes used but the relatively fine clearance between the cone and plate severely restricts the slurry particle size range which can be handled.

For quality control of slurries in day-to-day running of plants, a variety of 'viscometers' are used. Equipment for these two roles will now be considered in more detail.

## 8.1 VISCOMETERS FOR LABORATORY USE

### 8.1.1 Tube or capillary viscometers

For use with slurries, these instruments are generally manufactured in-house. Typical arrangements with vertical and horizontal tubes are shown in Figure 5 and 6. Hanks and

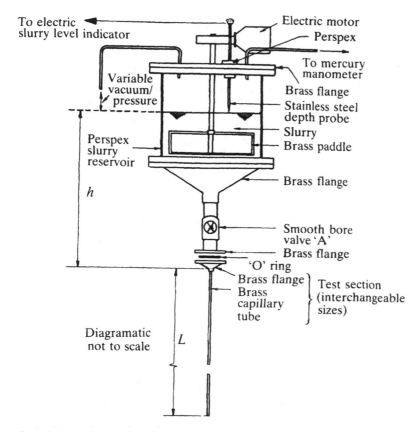

**Figure 5.** Typical vertical capillary tube viscometer

Hanks (33) describe a viscometer which is identical in principle to that shown in Figure 6 and used by BHRA. Refs 34 and 35 describe large bore (100 mm) tube viscometers for coarse particle viscous slurries ('stabilised slurries').

The range of shear rates which can be investigated using the tube instrument is dependent on tube diameter, but is usually 1000 $s^{-1}$ upwards for small tubes. For large bore instruments, it is possible to get to shear rates below 10 $s^{-1}$, although this is not easy in practice. The maximum shear rate that can be obtained is limited either by the pressure rating or capacity of the instrument or by the necessity to have laminar flow in the tube; the flow in larger diameter tubes may become turbulent.

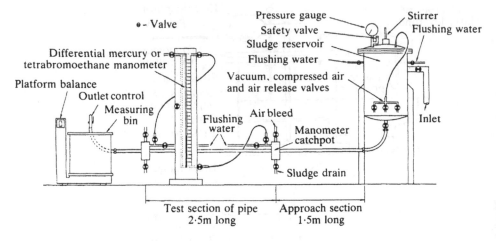

**Figure 6.** Large bore horizontal tube viscometer

## 8.1.2 Rotational instruments

Rotational instruments are generally purchased, popular models being the Contraves Rheomat, Haake Rotovisco, Ferranti and Brookfield instruments. Figure 7 shows a cross section of a typical instrument.

Rotational viscometer arrangements are generally limited to shear rates less than $1000 \text{ s}^{-1}$ and only a few instruments are capable of giving shear rates greater than $200 \text{ s}^{-1}$.

Most instruments are supplied with a range of cups and bobs to cover a range of viscosities. In practice, the cup is frequently not used, the bob being placed in a large container of slurry ('infinite sea geometry'). This practice limits the range of shear rates which can be obtained but it means that relatively coarse particles will not adversely affect results or jam in the narrow annulus between cup and bob.

An alternative to the cup and bob viscometer is the cone and plate which is sometimes used. Jones and Bullivent (36), for example, describe tests on a china clay suspension using a cone and plate rheogoniometer. These devices are limited in particle size range due to the narrow gap but are capable of high shear rates compared with cup and bob instruments.

Some manufacturers offer 'stirrers' or serrated bobs in place of the standard viscometer bobs for use with settling slurries or those that show phase separation phenomena. However, in order to obtain reliable results in terms of true shear rates and shear stresses it is necessary to undertake a lengthy calibration procedure using fluids of known non-

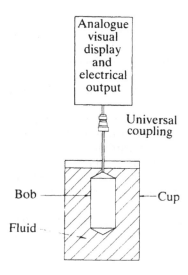

**Figure 7.** Rotational viscometer

Newtonian characteristics. Even when this procedure has been completed, it is not generally possible to assess the flow curve with the same accuracy that a true Couette geometry would give.

### 8.1.3 Other viscometers

Some other viscometer geometries do find limited use for laboratory investigations. Falling sphere viscometers can be used on slurries. Penetrometers as used in soil analysis can have some use on dense phase slurries; transverse flow plate viscometers may also be used. In general, these geometries are such that shear-stress shear-rate relationships cannot be derived and hence the methods are of limited use.

## 8.2 VISCOMETERS FOR PROCESS CONTROL

Viscometers may frequently be used for controlling a process or operation since viscosity of a slurry may be affected by concentration, temperature, chemical composition, etc.

A section of pipework within the plant may be used as a viscometer by recording the pressure drop and flow rate through it. Unless it is possible to vary the flow rate, this arrangement will only give information at one shear rate. Several manufacturers (e.g.

Contraves, Haake, Brookfield) offer in-line rotational viscometers which may be mounted in part of the plant pipework. These are generally able to operate across a range of rotational speeds (and hence shear rates) giving information about the flow curve. Units are available to cover a wide range of viscosities, temperatures and pressures. Vibrating probes may be used to obtain a viscosity measurement at a single shear rate value; vibrating spheres (37) and vibrating rods mounted inside the process pipework are both manufactured and supplied with the associated electronics.

A wide range of methods is used to check the viscosity of samples removed from a plant. The falling sphere viscometer, or a more buoyant variation, the ping-pong ball test, is used by British Coal in certain applications. The time to empty a standard funnel (short tube viscometer in principle) is used by some cement companies for controlling chalk slurry. Some manufacturers offer simplified viscometers for process control and frequently an ordinary viscometer will be used simply as a comparator to control slurry consistency from day-to-day.

All of these methods are adequate for process control since they give reproducible results and can thus be used to control a slurry within certain limits. However, they cannot be used for calculation purposes or prediction of head losses.

## 8.3   GENERAL ASPECTS OF VISCOMETRY

Two practical problems which are seldom given the attention they merit are applicability of a slurry to analysis by viscometric methods and the choice of instrument. In general, fluids which show little or no settlement will give reliable results in a viscometer. Some small vertical capillary viscometers (e.g. as in Figure 5) are capable of testing slurries which exhibit settling characteristics by using a stirrer to keep solids suspended. However, using viscosity values to predict pipeline performance of rapidly settling slurries can only be an approximate procedure. Further, with such a low viscosity slurry, pipeline flow will be turbulent and viscosity effects become much less important. The ability of some Couette viscometers to test settling slurries can be enhanced by using paddles or serrated bobs, but the same arguments apply.

At the other end of the concentration range, many slurries become granular in behaviour, the liquid phase being insufficient to fill the interstices between particles. Some slurries exhibit dilatancy at or around this concentration and others will develop such a structure that a rotating bob simply rotates in a hole in the slurry. Viscometric results under these circumstances are often unrepeatable and erratic and, although some information can be gleaned, the results must be used with extreme caution.

The choice of viscometer is never easy. Some workers advocate the use of several instruments in parallel to ensure good results. Cheng (38) has proposed an 'upper bound flow curve' based on results from several viscometers. Provided that the properties of a given slurry are well understood it has been shown that good agreement can be obtained between rotational and tube instruments, e.g. Ref 38. However, this does require a thorough experimental technique and close attention to the correct data reduction methods.

It must also be emphasised that incorrect use of any instrument will give erroneous results.
Summarising the practical aspects of using tube and rotational viscometers:

1. Rotational viscometers
   - small sample needed (2 litres maximum),
   - easy to operate,
   - may be purchased complete,
   - shear rate range up to approximately 300 s$^{-1}$,
   - may be used to characterise thixotropy,
   - may be used to characterise wall slip,
   - different bobs and cups give wide viscosity range,
   - different length bobs used to estimate end effects,
   - scale up calculations can be complex and lengthy,

2. Tube viscometers
   - larger sample needed (5 litres to 500 litres depending on type),
   - may be difficult to operate,
   - generally constructed in-house,
   - shear rate range up $5 \times 10^4$ s$^{-1}$,
   - difficult to characterise thixotropy,
   - may be used to characterise wall slip,
   - different tube diameters give wide viscosity range,
   - different length tubes used to estimate end effects,
   - scale up calculations from tube to tube may be relatively simple.

Both tube and rotational instruments are limited in the maximum particle size which can be passed. Both types require many tests to be conducted on a single fluid if end effects and wall slippage are to be eliminated.

A reasonable method of obtaining the best from both instruments is to use a rotational instrument to scan through a range of slurry types (e.g. particle size, pH, temperature, concentration, additive, etc.), to select the most promising mix. A tube instrument can then be used to check flow properties on these slurries and also give pointers on scale up procedures.

## 9. ABRASIVITY

Developed about ten years ago, the Miller Number (39) provides a relative measure of a slurry's abrasivity and attrition (the rate at which abrasivity changes with time). The two-part number, which may be determined with a 'rubbing block' device, is useful for estimating the service life of pumps, valves, etc. The weight loss data are fitted to the equation:

$$w = At^B$$

where $w$ is weight loss in g, $t$ is time in hours and $A$ and $B$ are constants.

The abrasivity is the rate of change of $w$ with time two hours after the start of the test. An additional constant is introduced to make the abrasivity of 220 mesh corundum equal to 1000. The attrition is found by dividing the second differential coefficient of the above, equation by the first differential coefficient, again evaluated two hours after the start.

It has been suggested that ultimately the attrition factor could become important in predicting rheological changes that occur in a long pipeline. For the present, however, the Miller Number's main advantage is that it allows pipeline operators to assess the probable effect, on equipment life, of types and grinds of solids different from those currently transported.

## 10. CONCLUSIONS

The $\gamma$-ray type of density meter is perhaps the most common instrument used in large mineral slurry systems for monitoring solids concentration, arguably the most important operating parameter.

Flow measurement and control is usually accomplished with electromagnetic flow meters, although ultrasonic velocity meters can have certain advantages and may therefore gain more acceptance in the future.

Although the Bourdon-type dial gauge is still the most common instrument used for measuring suction and discharge pressure, other types of pressure transducer provide direct electrical output for system monitoring and control. In both cases it is important that pipeline solid cannot clog the instrument used and various isolating and/or purging devices are available to prevent this happening.

Significant effort appears to be being directed towards the development of an instrument that will instantly detect solids deposition, since this prompt information would allow reduced normal operating velocities without excessive risk of blockages occurring.

Certain additional instruments are available, some of which are separate from the pipeline but kept 'on-site', which enable slurry viscosity, abrasivity and corrosiveness to be monitored. Use of these instruments allows the required values of the main system parameters to be adjusted in order to optimise overall operational efficiency.

## 11. REFERENCES

1.  Kakka, R. S. (May 1974) 'Review of instruments for measuring flow rate and solids concentration in steelworks slurry pipeline', *Proc. 3rd Int. Conf. on the Hydraulic Transport of Solids in Pipes*, Colorado, USA, Paper F6, BHRA.
2.  Kuhn, M. (1976) 'New possibilities and current development of hydraulic transport in the mining industry of the Federal Republic of Germany', *Proc. 4th Int. Conf. on the Hydraulic Transport of Solids in Pipes*, Alberta, Canada, Paper E4, BHRA.
3.  Gandhi, R. L. (1976) 'An analysis of 'hold-up' phenomena in slurry pipelines', *Proc. 4th Int. Conf. on the Hydraulic Transport of Solids in Pipes*, Alberta, Canada, Paper A3, BHRA.

4.  Aude, T. C., Cowper, N. T., Thompson, T. L. and Wasp, E. J. (28 June 1971) 'Slurry piping systems: trends, design methods, guidelines'. *Chem. Engineering.*
5.  Peterson, A. C. and Watson, N. (1979) 'The N.C.B.'s pilot plant for solids pumping at Horden Colliery', *Proc. 6th Int. Conf. on the Hydraulic Transport of Solids in Pipes*, Canterbury, England, Paper H1, pp. 353–66, BHRA.
6.  Debreczeni, E., Meggyes, T. and Tarjan, I. (May 1978) 'Measurement methods in an experimental rig for hydraulic transport', *Proc. 5th Int. Conf. on the Hydraulic Transport of Solids in Pipes*, Hanover, Federal Republic of Germany, Paper G1, BHRA.
7.  Baker, R. C. and Hemp, J. (May 1978) 'A review of concentration meters for granular slurries', *Proc. 5th Int. Conf. on the Hydraulic Transport of Solids in Pipes*, Hanover, Federal Republic of Germany, Paper B1, pp. 75–88, BHRA.
8.  Kao, D. T. and Kazanskij, I. (March 1979) 'On slurry flow velocity and solids concentration measuring techniques', *Proc. 4th Int. Conf. on Slurry Transport*, Slurry Transport Association.
9.  Okude, T. and Yagi, T. (May 1974) 'The spatial solid concentration and the critical velocity', *Proc. 3rd Int. Conf. on the Hydraulic Transport of Solids in Pipes*, Colorado, USA, Paper E2, BHRA.
10. Wiedenroth, W. (September 1979) 'Methods for the determination of transport concentration and some problems associated with the use of radiometric density meters', *Proc. 6th Int. Conf. on the Hydraulic Transport of Solids in Pipes*, Canterbury, England, Paper B2, BHRA.
11  Clift, R. and Clift, D. H. M. (1981) 'Continuous measurement of the density of flowing slurries', Brief Communication, *Int. J. Multiphase Flow*, Vol. 7, No. 5, pp. 555–61.
12. Bain, A. G. and Bonnington, S. T. (1970) 'The Hydraulic Transport of Solids by Pipeline', Pergamon Press, Book.
13. Carleton, A. J., French, R. J., James, J. G., Broad, B. A. and Streat, M. (May 1978) 'Hydraulic transport of large particles using conventional and high concentration conveying', *Proc. 5th Int. Conf. on the Hydraulic Transport of Solids in Pipes*, Hanover, Federal Republic of Germany, Paper D2, BHRA.
14. Przewlocki, K., Michalik, A., Korbel, K., Wolski, K., Parzonka, W., Sobota, J. and Miss Pac-Pomarnacka (September 1979) 'A radiometric device for the determination of solids concentration distribution in a pipeline', *Proc. 6th Int. Conf. on the Hydraulic Transport of Solids in Pipes*, Canterbury, England, Paper B3, pp. 105–12, BHRA.
15. Verbinski, V. V., Cassapakis, C. G., de Lesdernier, D. L. and Wang, R. C. (September 1979) 'Three component coal slurry sensor for coal/rock/water concentrations in underground mining operations', *Proc. 6th Int. Conf. on the Hydraulic Transport of Solids in Pipes*, Canterbury, England, Paper C4, pp. 161–8, BHRA.
16. Hayashi, H., Sampei, T., Oda, S. and Ohtomo, S. (November 1980) 'Some experimental studies on iron concentrate slurry transport in pilot plant', *Proc. 7th Int. Conf. on the Hydraulic Transport of Solids in Pipes*, Sendai, Japan, Paper D2, pp. 149–62, BHRA.
17. Jordan, D. and Wagner, R. (May 1978) 'Supervision and control of the hydraulic conveyance of raw coal by modern measuring equipment at the Hansa Hydromine', *Proc. 5th Int. Conf. on the Hydraulic Transport of Solids in Pipes*, Hanover, Federal Republic of Germany, Paper G2, BHRA.
18. Schriek, W., Smith, L. G., Haas, D. B., Husband, W.H.W. and Shook, C. A. (May 1974) 'Experimental studies on the hydraulic transport of coal', *Proc. 3rd Int. Conf. on the Hydraulic Transport of Solids in Pipes*, Colorado, USA, Paper B1, BHRA.
19. Beck, M. S., Mendies, P. J., Walecki, T. and Gatland, H. B. (May 1974) 'Measurement and control in hydraulic transport systems using cross-correlation measurement systems and fluidic diverters', *Proc. 3rd Int. Conf. on the Hydraulic Transport of Solids in Pipes*, Colorado, USA, Paper F5, BHRA.
20. Beck, M. S. (1981) 'Correlation in instruments: cross-correlation flow meters', *J. Phys. E.: Sci. Instrum.*, Vol. 14.

21. Kazanskij, I. (November 1980) 'The anti-blockage watcher: a new measuring system for maximal solid output with minimal energy consumption and wear', *Contribution to 7th Int. Conf. on the Hydraulic Transport of Solids in Pipes*, Sendai, Japan, BHRA.
22. Balachandran, W. and Beck, M. S. 'Solids concentration measurement and flow measurement of slurries and sludges using ultrasonic sensors with random data analysis', Unpublished paper.
23. Hayward, A. T. J. (1985) 'Flow meters. A basic guide and source book for users', London, UK, Macmillan Press Ltd.
24. Faddick, R., Ponska, G., Connery, J., Di Napolis, L. and Purvis, G. (September 1979) 'Ultrasonic velocity meter', *Proc. 6th Int. Conf. on the Hydraulic Transport of Solids in Pipes*, Canterbury, England, Paper B4, pp. 113–24, BHRA.
25. Heywood, N.I. and Mehta, K. (19–21 October 1988) 'A survey of non-invasive flowmeters for pipeline flow of high concentration, non-settling slurries', *Proc. 11th Int. Conf. on the Hydraulic Transport of Solids in Pipes*, Stratford-upon-Avon, UK, pp. 131–55, Cranfield, Bedford, UK.
26. Mathur, M. P. and Ekmann, J. M. (March 1984) 'On-line instrumentation evaluation for characteristics of alternative fuels', *Proc. 9th Int. Tech. Conf. on Slurry Transportation*, Washington, DC, USA, pp. 223–30, Slurry Transport Association.
27. Bantin, R. A. and Streat, M. (September 1970) 'Dense-phase flow of solids-water mixtures in pipelines', *Proc. 1st Int. Conf. on the Hydraulic Transport of Solids in Pipes*, Coventry, England, Paper G1, BHRA.
28. Ercolani, D., Ferrini, F. and Arrigoni, V. (September 1979) 'Electric and thermic probes for measuring the limit deposit velocity', *Proc. 6th Int. Conf. on the Hydraulic Transport of Solids in Pipes*, Canterbury, England, Paper A3, pp. 27–42, BHRA.
29. Ferrini, F. and Pareschi, A. (November 1980) 'Experimental study of the solid bed profile in sloping pipe sections', *Proc. 7th Int. Conf. on the Hydraulic Transport of Solids in Pipes*, Sendai, Japan, Paper F3, pp. 283–90, BHRA.
30. Brown, N. P. and Shook, C. A. (August 1982) 'A probe for particle velocities: the effect of particle size', *Proc. 8th Int. Conf. on the Hydraulic Transport of Solids in Pipes*, Johannesburg, South Africa, Paper G1, pp. 339–48, BHRA.
31. Tczap, A., Pontins, J. and Raffel, D. N. (May 1978) 'Description, characteristics and operation of the Grizzly Gulch tailings transport system', *Proc. 5th Int. Conf. on the Hydraulic Transport of Solids in Pipes*, Hanover, Federal Republic of Germany, Paper J4, BHRA.
32. Van Wazer, J. R. *et al.* (1963) *Viscosity and Flow Measurement – A Laboratory Handbook of Rheology*, London, J. Wiley & Sons.
33. Hanks, R. W. and Hanks K. W. (23–26 March, 1982) 'A new viscometer for determining the effect of particle size distributions and concentration on slurry rheology', *Proc. 7th Int. Conf. on Slurry Transportation*, Lake Tahoe, USA, pp. 151–61, Washington DC, USA, Slurry Transport Association.
34. Anon. Hydraulic Transport of Solids Leaflet, Published BHRA, Cranfield.
35. Lawler, H. L. *et al.* (29–31 March 1978) 'Application of stabilised slurry concepts of pipeline transportation of large particle coal', *Proc. 3rd Int. Conf. on Slurry Transportation*, Las Vegas, Nevada, pp. 164–78, Washington DC, USA, Slurry Transport Association.
36. Jones, T. E. R. and Bullivent S. A. 'The elastico-viscous properties of deflocculated china clay suspensions', *J. Phys. D.: Appl. Phys.*, Vol. 8, pp. 1244–54.
37. Matusik, F. J. and Scarna, P. C. (December 1981) 'Latest instrument makes on-line viscosity control of slurries possible', *Control Engineering*, pp. 116–18.
38. Cheng, D. C.-H. and Tookey, D. J. 'The measurement of the viscosity of Barytes-Sand Pulps using a tube viscometer and co-axial cylinder viscometer', *Warren Spring Laboratory Report LR 100 (MH)*, Stevenage, UK.
39. Miller, J. E. (22 July 1974) 'Miller Number', *Chem. Engng.*, pp. 103–106.

# 6. ADDITIONAL ASPECTS OF SLURRY SYSTEMS

## 1. SYSTEM CONTROL

Successful long-distance slurry transport pipelines exist where control valves, gauges, shut-offs, pump variable drives, etc., are not incorporated (1). However, to keep overall operational costs to a minimum, and to prevent blockages, it is usually regarded as essential to keep the mixture flow parameters optimised by means of an automated control system.

It is desirable that a control system should respond to changes in mixture concentration, rather than merely provide for the selection of a given rate of solids feed. Also a steeply falling pumphead-discharge characteristic is preferable at the system operating point in order to reduce the effect, on flow stability, of solids feed rate irregularities. However, variations in the latter should be minimised: in the Hansa Hydro-Pit (2), a density gauge was installed on the suction side of the feed pump to provide the input control values for regulating the speed of the scraper-conveyor which determined the solids feed rate. Effective control of the slurry uniformity may necessitate special precautions with regards to the design of mixing hoppers and their agitators (minimum use of holding tanks and hoppers is generally recommended). This is particularly true where there is a tendency for the extremes of the size range to settle at widely differing rates.

It is possible that a controlled solids and water feed, supported by a concentration meter demonstrating uniform mixing, would be an adequate check of satisfactory operation and it would not be essential to measure slurry flow rate. The reason why it may be preferred to monitor flow rate is that it is probably the most reliable signal available to indicate potentially dangerous situations such as incipient blockage and to initiate corrective action by increasing pump speed. Although pump discharge pressure may be measured without solids interference using pressure transducers, the nature of settling-solids flow is such that the pressure signals will fluctuate quite markedly even in so-called steady conditions. Velocity cannot change except at a very slow rate due to pipeline inertia and this is likely to be a more reliable indication of the need for fine tuning of pump speed. At least two long-distance pipelines have flow meters at both ends of the transport line (3,5): in addition to providing 'built-in redundancy', the arrangement allows the control system to identify instantly a line break, whereupon alarm bells are activated in the control centre.

There are certain 'ease of handling' benefits to be gained from feeding solids materials intermittently into a continuously flowing stream of water. However, in these circumstances, the control system must not be allowed to cause the pump of an in-line series of centrifugal pumps to cease running or else blockages are possible. Also the reduced line back pressure resulting from low downstream solids concentration must not be allowed to cause the pump speed to rise excessively, since this may cause the power demand to rise above the electrical rating of the motor. These points are important since pipeline feeders may become increasingly common in the future: feeders allow higher efficiency centrifugal pumps to be used without any wear problems. Also, emergency flushing of the pipeline can be accomplished simply by halting the operation of the feeder.

Another aspect of solids flow rate control is described by Burnett *et al.* (6): a helical inducer has been developed as a self-controlling slurry pump for use at the booster stations of long pipelines. This centrifugal pumping device, which eliminates the need for sumps, level controls and variable speed drives, incorporates an air core vortex vented to atmosphere and automatically adjusts to different back pressures (i.e. to different concentrations and flow rates). The designers claim that the device, which is suitable for coarse or fine slurries, eliminates the need for 'complicated and possibly troublesome feedback control systems'.

Detailed description of slurry pipeline control systems appear to be rare, but two interesting papers are presented by Tczap *et al.* (3) and Buckwalter (4). The former of these is for a centrifugally pumped system at Grizzly Gulch, South Dakota, USA, and it is interesting to note that, although all the automatic control sequences are initiated by the operator, the validity of his commands are verified automatically prior to their execution. It has been stated that virtually all blockages of commercial pipelines are due to operator error, either by inadequate monitoring for early warning or by faulty procedures.

Design considerations for control systems suitable for slurry pipelines utilising positive displacement pumps are presented in the second of the above references (4). An override control system, described by the author, enables such a system to start and stop without any process upset. A ramp generator signal may be used to increase or decrease pump speed at a uniform rate because the rate of change of flow with speed is constant for positive displacement pumps. At the initial pumping station, flow must be the primary control parameter whereas suction and discharge pressures should be *overriding* control functions. At subsequent stations, however, suction pressure should usually be the primary control parameter, with discharge pressure providing the *override* control functions; if suction pressure increases, this is due to the input from the upstream station having risen and the pump speed would therefore be increased.

## 2.   START-UP AND SHUT-DOWN PROCEDURES

Two vital phases in the operation of a pipeline are those of start-up and shut-down. Some systems may be started and stopped quite frequently whereas others may be required to operate virtually continuously; nevertheless whatever the restart frequency reliable operation is obviously essential and must be considered at an early stage of the conception of a pipeline system. Aspects which need to be borne in mind are slurry properties such as stability on inclines and settling behaviour in the pipe, maximum allowable gradients, starting pressure requirements and start-up and shut-down procedure.

The ability to have planned shut-downs with slurry in the pipe greatly increases the flexibility of the system; however not all systems can be shut down in the pipe full condition. The major danger during starting and stopping is the formation of plugs of solids which will block the system. Plugs can be formed in a number of different ways. After shut-down solids may settle in the pipe to form a bed, in itself this would not necessarily be a problem if it can be resuspended by the passage of the remaining solids and carrier fluid across the top of the bed; unfortunately coarse particle slurries can be very difficult to restart by this means. The settled bed can slide down the pipe, if it is inclined, and cause plugging at the base of the incline. Too rapid a start, from the settled condition, may cause part of the bed to move bodily and pile up against a stationary portion; again this may result in a plug.

## 2.1   HOMOGENEOUS OR NON-SETTLING SLURRIES

Truly non-settling slurries are generally the easiest to handle in the starting and stopping phases. The slurries included in this class are fine particle, high concentration mixtures such as limestone for cement plant feed, coal-water and coal-oil mixtures and stabilised coarse particle slurries. The rheological properties of these slurries are such that downward migration of solids through the carrier fluid either does not take place or only occurs over a long period of time, thus precluding the formation of a solids bed. A full pipe containing suspended solids with little or no vertical concentration gradient avoids sliding on inclines and can usually be simply restarted by operation of the main slurry pumps. Starting pressures may be somewhat greater than normal running pressure; the magnitude of the increase will depend on mixture properties, yield stress often being an important factor.

As with many fluid systems there is always the risk of forming vapour cavities during starting and stopping if acceleration or deceleration takes place too rapidly. The slurry velocity should be increased or decreased gradually so that cavities are not able to form. Methods are available which allow the transient pressure surge conditions to be predicted (7).

## 2.2   SETTLING SLURRY

This type of slurry includes low concentration large particle mixtures such as washery tailings, mine waste, run-of-mine coal mixtures and many fine particle slurry systems; some of the better known systems which fall in this class are the Black Mesa coal pipeline in the USA, the Brazilian Samarco pipeline and Loveridge mine hoisting pipeline.

Many of these slurries, particularly the coarse particle slurries, are not considered to be restartable mixtures, i.e. the pipe line would not normally be shut down when full of slurry. In such cases, the normal procedure would be first to flush the line of solids before stopping the pumps (e.g. 8, 9, 10 and 11). A shut-down in which the slurry is replaced by water is not a swift procedure, for instance a flush velocity of 2 m/s over a distance of 20 km would take nearly three hours and systems such as the Black Mesa pipeline will take very much longer. Sometimes a slug of fines is inserted between the slurry and the flushing fluid which is normally water. This is to prevent dilution of the slurry which in some systems, particularly those containing a majority of fairly fine solids, can tend to settle at normal pumping velocities in the dilute condition; a case in point is the Black Mesa system (12), which although it can be restarted in the full pipe condition, has been flushed on occasions.

If an emergency situation arises which necessitates a complete and immediate shut-down, virtually the only action which will help to avoid plugging is to dump as much of the system contents as possible. Dump ponds can be provided at various points throughout a system for this eventuality; these are positioned, as far as possible, at the lowest points in the system allowing gravity to assist the dumping operation. The solids remaining in the pipeline can often then be restarted by pumping slowly with water at gradually increasing velocities as the finer fractions are suspended and flushed out of the system.

Some systems utilise more than one pipe or have a standby pipeline (8) and in such instances where there is only a partial stoppage, such as the failure of one or two pumps or a slurry leak on the line in use, the flow can be diverted to the standby pipe. This can avoid the need to shut down altogether or may allow a controlled shut-down to be performed.

Systems which can be allowed to stop in the line full condition are generally those pumping fine and or low density solids, e.g. the Black Mesa Pipeline USA, (12), and the Samarco Pipeline in Brazil, (13). The Black Mesa line is 439 km long and has a number of pumping stations. When a station shuts down unexpectedly the next station downstream switches suction to its water reservoir in order to keep as much slurry in motion as possible. During the early years of operation water flushing was frequent during shut-down but lately most shut-downs are of short duration and restarts with a full line are readily accomplished.

The Samarco system was originally intended for continuous operation, alternating between slurry and water batches to vary solids throughput, however during commissioning a series of shut-down tests were conducted which indicated that the slurry would restart easily and start-up and shut-down in the operating mode are now routine procedures. Batches have been shutdown in the line for periods of up to 130 hours.

As with non-setting slurries it is necessary to exercise control during shut-down to avoid vapour cavity formation. It is important, particularly in systems employing centrifugal pumps, to ensure that the pumps are clear of solids before shut-down.

If a system has been flushed free of solids then there should be little difficulty encountered during start-up. The normal procedure is to accelerate the water up to operating speeds and then gradually introduce the solids as described by Guzman *et al.* (8).

It is generally considered that the safest method of restarting a line containing fairly fine slurry (say less than 2–3 mm i.e. a restartable mixture) which has been allowed to remain in the pipe is by gradually accelerating the fluid in the top of the pipe over the settled bed; in this way the finer fractions are first suspended followed by coarser material as the velocity increases. It is important not to accelerate too quickly as this may cause the material resting on the pipe bottom to slide en masse with the risk of forming a plug. This procedure is similar to that used on the Black Mesa Line. In long pipelines with several sections it is necessary to pressurise the line initially to ensure that the slurry in all sections starts to move as nearly simultaneously as possible when pumping begins.

Systems similar to the Samarco line, which has a much finer particle slurry than the Black

Mesa system can be restarted with relative ease and once line pressurisation has been completed the pump stations are started simultaneously. If the shut-down has not been unduly long the restarting pressures are only a little above normal operating pressures.

If the system has become blocked then before remedial action can take place the plugged section of pipe must be located. The usual technique is to pressurise the line at one point and observe readings and pressurisation rates at others. There are three basic methods of blockage clearance:

- Back flush to spread out the plug followed by a rapid restart in the forward direction. This method is used to clear routine blockages on the Loveridge Mine system which pumps a dilute coarse coal slurry with a top size of about 100 mm.
- Gradual pressurisation in the forward direction, then the pressure is held until flow begins. The Samarco line has been cleared in this fashion on the only two occasions when a potential plug appeared in the line. This 'pressure massaging' technique is sometimes used to restart the Black Mesa line. Long lines are often started section by section, the upstream end being the first to run, bleed-off points at intervals down the line being opened and closed when appropriate.
- If the previous methods fail then the plug can be resuspended by high pressure water injection into the blocked section. This technique has been proved on coarse particle slurries (10).

Plugs were expelled from the Black Mesa line in the early days of operation by cutting tapping points in the pipe along the plug, injecting water into the plug and pressurising from either end. Since then the solids size distribution has been altered to ease restarting with a full line. Hard packed plugs may also be removed by cutting the pipe and jetting out the solids (9).

Many people envisage a plug as a continuous pipeline full of solids. This is not possible as most slurries which are pumped over long distances contain less than 40 per cent solids by volume. Experience indicates that plugs are generally about 15–30 m in length.

## 2.3 PARTICLE SIZE EFFECTS

The difference between the start-up and shut-down sequences for the Samarco and Black Mesa pipelines is primarily due to the different particle size distribution of the solids. The Samarco solids are fine and have a narrow size band (90 per cent between 43 and 10 $\mu$m) (13). This means that the stationary slurry settles out relatively slowly; this settlement also causes solids migration down inclines. Coal slurries similar to that pumped in the Black Mesa pipeline normally have a wider size distribution, to improve their de-watering characteristics, which makes them more susceptible to segregation during settling and start-up; consequently considerable care must be taken when starting with solids in the line. The original Black Mesa solids had a top size of about 1.5 mm to 2.0 mm and contained 14 per cent by weight less than 44 $\mu$m; this proportion of fines was increased to 19 per cent to ease starting (12). The coarser the particles comprising a slurry the more difficult it is to restart, the final particle size distribution is almost invariably a compromise between pumping power, crushing and dewatering costs and restart performance.

## 2.4   PIPE INCLINATIONS

An important concern of the pipeline designer is what inclination to specify as the maximum for construction. It is well-known that if the inclination exceeds the angle of repose of the slurry in a pipe then the solids can slide to the bottom of the slope and cause a plug. The rheological properties of this settled material can be investigated to see if it is still in a liquid, and therefore theoretically pumpable form; generally a slope limitation is imposed which will prevent sliding. The maximum slope, which may be of the order of 10 per cent (14), cannot be predicted at present and it is necessary to determine its value from test results such as those described by Olada *et al.* (15). In certain instances the cost of restricting the maximum slope can be excessive and arrangements have to be made to back flush or inject water in the event of a shut-down with slurry in the pipe. Experience gained with early pipelines has allowed slope limitations to be relaxed by as much as 50 per cent for some iron and copper concentrates (16). This has a substantial effect on capital costs by shortening pipeline length.

## 3.   VALVES

This section briefly describes common valve types and their use in slurry systems. Although there are several types of valve now used extensively on slurry handling systems, no single type is suitable for all applications. Baker and Jacobs (17) state that probably the soundest principle in the design of slurry transport systems is to avoid the need for valves wherever possible. By sensible use of gradient and system layout, the requirement for certain valves may be minimised, however, it is extremely rare to find no valves whatsoever in an operational slurry system.

There are two major problems associated with the use of valves for solids handling systems, namely those of erosion and blockage. Erosion in and around the valve will be accelerated by discontinuities in the wetted surface or significant protrusion of the closing element into the flow. Ideally, a solids handling valve should present a circular or nearly circular section to the flow of the same diameter as the adjacent pipework when fully open; this is not always possible due to the engineering and geometrical constraints associated with valve design. Any significant reduction in the duct cross-sectional area will obviously increase the risk of bridging and blockage particularly where large particle slurries are being pumped. As a flow control measure, operation of valves in the partially closed position will accelerate wear and greatly increase the chance of plugging and is not generally recommended; specially designed dissipators are available if flow control is required.

## 3.1 BALL VALVE

The ball valve utilises the same concept as the plug valve and is sometimes referred to as the 'spherical plug' valve or 'ball plug' valve. A typical ball valve (with one piece body) may be seen in Figure 1. The central ball has a circular hole through one axis with the ball/hole proportions such that when the ball is at the limit of travel either a full spherical face (closed position) or the ends of the hole (open position) are presented to the inlet and outlet. Ball valves may be supplied with either 'full-bore' or 'reduced bore' hole dimensions. Although the ball valve is usually considered for straight on/off duties, recent developments, which reduce final closure rates, have allowed flow control duties to be performed.

During slurry service, fine granular material may find its way between the ball and seat causing wear. Forced lubrication or water flushing can be applied to overcome this problem. The ball spindle is not totally isolated from the pipeline pressure and therefore a spindle gland has to be provided.

Most standard ball valves can operate at pressures of up to 50 bar and have an operational temperature range of $-30\,°C$ to $230\,°C$. Although ball valves with metal seats are available, the most commonly used seat materials are plastics or elastomers. PTFE is widely used, however other options such as nylon, graphite and PCTFE exist for special applications.

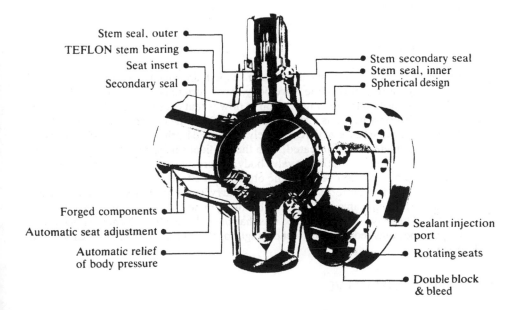

**Figure 1.** Ball valve

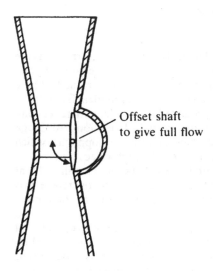

Offset shaft
to give full flow

**Figure 2.**   Butterfly valve

## 3.2   BUTTERFLY VALVE

The butterfly valve usually consists of a circular shaped disc which turns about a diametral axis within the cylindrical bore of the valve body. A 90° turn fully opens or closes the valve. In the fully open position the only restriction to flow is that caused by the thickness of the disc. However this obstruction is placed approximately in the centre of the flow and thus halves the maximum particle size that can safely be carried by the pipe. Butterfly valves are either resilient or metal to metal seated, with the resilient seated type having the flexible seating located on either the disc or the valve body. Figure 2 shows a butterfly valve with the disc axis offset to one side to overcome the obstruction effect.

It is uncommon to find butterfly valves used in slurry lines but the development of such a valve for use as a dump valve in a pipeline handling mine tailings of less than 230 $\mu$m is reported (17). The reason for this choice was that the valve could be located very close to the main pipeline, thus reducing the 'dead' length of the lateral in which a solids build-up could occur.

## 3.3   DIAPHRAGM VALVE

The diaphragm valve consists of a valve body assembly with a single flexible diaphragm which isolates the actuating mechanism from the transported slurry. There are two main diaphragm valve types available, namely 'weir' and 'straight-through'.

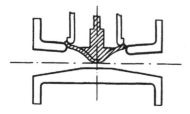

**Figure 3.** Weir type diaphragm valve

Figure 3 shows a typical weir type valve which tends to be the most commonly used design. Shut-off is obtained with a relatively low operating force and short diaphragm movement which maximises diaphragm life. Diaphragms are usually made from either an elastomeric material or PTFE with an elastomer backing. With particularly abrasive slurries, the weir design may be subject to erosive wear and the passage constriction is sometimes undesirable with large particle slurries.

The 'straight-through' diaphragm valve does not utilise a weir section so flow through the valve is in a straight line along a full bore passageway. The elimination of the weir requires a larger diaphragm movement and therefore material choice for the diaphragm is much more limited.

Diaphragm valves can normally operate within the −50 °C to 175 °C temperature range, depending on valve material choice.

## 3.4   GATE VALVE

The gate valve is possibly the most commonly used valve in industry generally, and whilst it is not specifically considered to be suitable for slurry transport, versions designed to prevent solids accumulation under the bonnet are used on slurry handling installations. The gate valve consists of a gate or disc which is moved at right angles across the line of flow between matching seats in the valve and in the fully open position provides a relatively full flow opening.

Gate valves are not intended for throttling or flow control applications since erosion of the disc seriously affects the shut-valve performance. Also, a partially open disc may suffer from damaging flow-induced vibrations.

There are numerous types of gate valve available which usually gain their name from the type of sealing disc used. For example, solid wedge, parallel slide, parallel double disc and split wedge are all variations of the gate valve. Figure 4 shows a parallel slide gate valve.

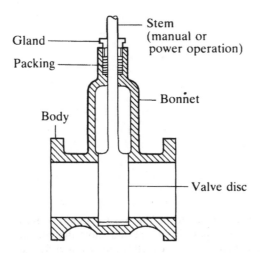

Figure 4. Parallel slide gate valve

## 3.5 PINCH VALVE

The pinch valve consists of a reinforced sleeve, manufactured from natural rubber or some synthetic elastomer, that is pinched and flattened to effect a closure. The main advantage of the pinch valve, over many other valve types, is that the operating mechanism is totally isolated from the transported medium which may be abrasive, corrosive or both. Pinch valves may be manually operated, as seen in Figure 5, however pneumatic, hydraulic and electrically driven pinch valves are used (18).

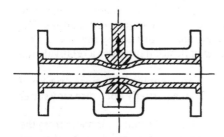

Figure 5. Manually operated pinch valve

The major limiting factor of the pinch valve is the relatively low maximum operating pressure available, however Sauermann and Webber (19), report that pinch valves have been used successfully at pressures up to 175 p.s.i., and have absorbed surge pressures up to 350 p.s.i. If pressures in excess of 175 p.s.i. are frequently encounterd it is normal practice to replace the mechanical 'jaw' device with an external pressure vessel which encloses the rubber sleeve. This pressure vessel is then pressurised to reduce effectively the differential working pressure. To completely close the valve, the pressure within the outer vessel is simply increased.

Sabbagha (20) reports that during valve tests conducted on gold slime pumping, the pinch valve proved to be the most efficient, however it is reported that development work is still required to enable higher maximum working pressures to be achieved. Another advantage of the pinch valve is the full flow capability of the fully open valve which minimises the risk of plugging.

## 3.6 PLUG VALVE

The plug valve consists of a tapered or cylindrical plug within a valve body, with, in its simplest form, a central flow passage allowing a 90° axial rotation of the plug to move the valve from the fully closed to fully open position. As with the ball valve, the plug valve is

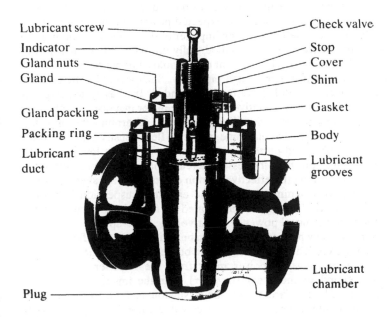

**Figure 6.** Lubricated plug valve

normally used for on/off duties, however some coarse flow control duties may be performed. The plug valve may also be adapted for multi-way flow duties by use of a multi-ported central plug.

There are two basic families of plug valve, namely lubricated and non-lubricated. It is usual practice for the lubricated version (as shown in Figure 6) to be used for slurry duty. In some non-lubricated plug valves, a 'soft' friction reducing sleeve, manufactured from PTFE, for example, is inserted between the plug and body. With the lubricated valve, an insoluble lubricant is injected under pressure between the plug and body via grooves and channels. The action of the lubricant combats corrosion and erosion of the mating surfaces.

## 4. SLURRY STORAGE

The preparation of all slurries requires the mixing of solid and liquid components. Furthermore, if settling suspensions are to be maintained in a pumpable condition, they must be agitated during storage. The design of systems to achieve these objectives requires first that the storage or feed hoppers be appropriately sized. They must then be fitted with a suitable mixing device, which might be a nozzle (jet-mixing) or an impeller. The focus in this subsection will be on mechanically stirred systems for both settling and non-Newtonian media.

Settling suspensions of mono-size particle slurries in baffled mixing vessels has been studied by Herringe (21). Figure 7a indicates the expected power consumption pattern, normalised by $\rho N^3 D^5$ the power scale for a mixing impeller of diameter $(D)$ running at a rotational speed $(N)$ within a dense liquid $(\rho)$. Under fully turbulent mixing the 'power number' $(Np)$ would be expected to be independent of Reynolds number $(\rho N D^2 \mu^{-1})$, as would the friction factor for fully turbulent rough pipe flow. Such would be the case for suspension regime 'A' in which no solids have been suspended. Increasing the impeller speed introduces regime 'B' in which the relative load on the mixer increases as solids are brought into suspension, raising the effective density of the fluid around the impeller. Beyond full suspension the solids tend to centrifuge, reducing fluid density near the rotor. In contrast to this, Herringe observed patterns more like Figure 7b. At low speeds the settled solids raised the effective bottom of the vessel, increasing initial power requirement, equivalent to a bed of solids reducing the effective section of a pipeline. A speed increase brought the power number to a minimum before the expected suspension behaviour exhibited itself. Through the work, Herringe demonstrated the difficulty of correlating the data and the inapplicability of the usual power-per-unit-volume scale-up rule.

A simple example of the way in which non-Newtonian behaviour can affect flow inside a mixing vessel is shown in Figure 8. The shear rate $(\dot{\gamma})$ can be viewed as an indicator of local mixing action, achieved at the cost of power input from the impeller. This power requirement is strongly related to the shear stress $(\tau)$ near the impeller. The radial distributions of shear rate and stress are shown for a Newtonian fluid to be mutually proportional (a), leading to acceptable mixing for an acceptable power input, given a well-designed system. Replacement of the contents by a pseudoplastic medium leads to (b),

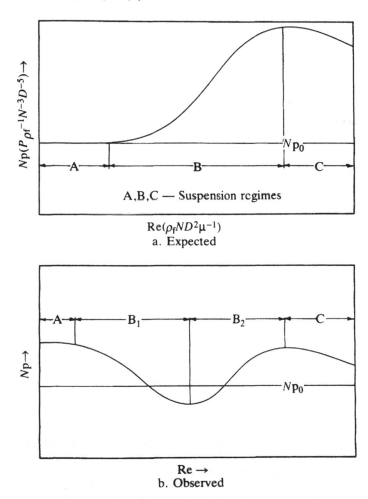

**Figure 7.** Suspension regime effect on stirring power in settling slurry (21)

where the shear rate is high near the impeller, but low everywhere else, resulting in a poor distribution of the mixing action. A dilatant mixture leads to the opposite effect (c), where the shear rate distribution is good, but the stress at the impeller very high, consuming significantly more power than the Newtonian system. These matters receive more detailed attention in the review by Silvester (22). The basic phenomena were first observed by Metzner and Taylor (23).

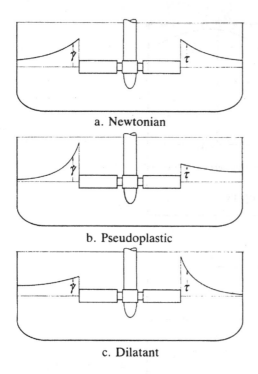

a. Newtonian

b. Pseudoplastic

c. Dilatant

**Figure 8.** Non-Newtonian effects on mixing stable solids suspensions (21)

Fortunately, the prediction or mixing power requirement for non-Newtonian fluids can be carried out by relating them to equivalent Newtonian systems. The technique by Metzner and Otto (24), was proven and developed for a wide range of dilatant and pseudoplastic fluids by Calderbank and Moo-Young (25). The concept requires first that a representative shear rate be found, a similar problem to that for pipeflow. The following proportion is usually assumed between shear rate ($\dot{\gamma}$) and impeller speed ($N$):

$$\dot{\gamma} = BN \quad \text{where } B \text{ is a function of impeller geometry} \tag{88}$$

Once shear rate is known, the corresponding apparent viscosity can be used to formulate a Reynolds number. Figure 9 indicates the general patterns of power number variation, with the familiar laminar pattern at Reynolds numbers below ten. As might be expected from the above discussion and extrapolation from pipeflow behaviour, the pseudoplastics suppress the onset of turbulence.

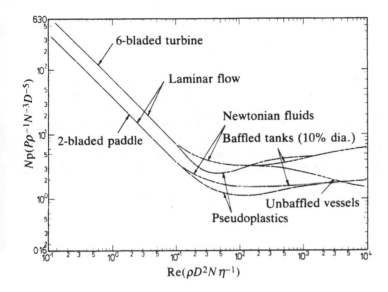

**Figure 9.** Power number characteristics for flat-bladed mixing impellers (24)

## 5. PARTICLE DEGRADATION

The ability to forecast the degradation of particles in slurry systems is an important aspect of the system evaluation. This is because the cost of dewatering and water treatment processes are closely linked with the fines content in the pipeline product; these costs increase with increases in the proportion of fines. An additional consideration is that generation of fines can also lead to undesirable changes in the hydraulic characteristics of the slurry, (26,27). Degradation of coal has received particular attention because of the commercial interest in its transportation by pipeline and it is a relatively friable substance.

### 5.1 FACTORS AFFECTING DEGRADATION

Degradation in a slurry system is a function of a number of different parameters which include:

• transportation distance,
• transport velocity,
• construction of the pump,

- construction of the pipeline,
- particle strength and shape,
- particle size distribution.

Any systematic description of particle degradation during hydraulic transport is very difficult because it is influenced by system and solids parameters; of the latter, particle strength is known to be important (27,28), and also particle size distribution (29).

Perhaps the first study where a successful attempt was made to quantify the breakage rate in a shear flow was made by Karabelas (30), using a cup and bob viscometer, with a fine bituminous coal slurry. He demonstrated that the process of degradation could be described by the grinding rate model of Austin and Luckie (31), of conventional milling processes.

The mass fraction retained on a screen j of a $\sqrt{2}$ series and passing the j–l screen varies in time in accordance with:

$$\frac{dW_j}{dt} = \sum_{i=j-1}^{1} S_i b_{j,i} W_i - S_j W_j$$

where $b$ = breakage distribution parameter
      $S$ = selection rate constant

where the first term on the right-hand side represents contributions from the breakage of larger particles. Karabelas' results showed the selection rate constants $S_i$ and $S_j$ varying nearly in proportion to particle size in the range investigated; the selection rate constants also varied with shear rate in a nearly linear fashion. However, using coarse lignite particles, Shook (32), observed that $S$ was insensitive to particle size in the range 0.7–2 mm and decreased significantly with time; the latter effect being attributed to significant 'rounding' of the particles. This work was done with material having a relatively narrow size distribution, the most satisfactory method of determing $S$ and $b$ values. This can become impractical as the cost of preparing the necessary screened solids can be excessive and in such cases it may be necessary to use a material with a wide distribution band.

Using such material to extend his earlier results Shook *et al.* (33), attempted to match the degradation of four wide size distribution coals, in a recirculating pipe loop using two different pump impellers, with the Karabelas model; he was able to arrive at a fairly satisfactory correlation. He also concluded that where substantial degradation occurs, its rate decreases approximately exponentially with time; degradation as a function of particle size for the four cases is shown in Figure 10. For the bituminous coals, the degradation process was found to produce a broad range of particle sizes but the lower rank material degraded to produce a distribution of coarse and very fine particles, Figure 11. Increasing the pump impeller size, increased the rate of degradation of the lignite.

Mikhail *et al.* (34), related coal friability to microfissuration. The microfissuration was quantified by a $\Delta P$ index obtained from a standardised procedure for measuring the rate of

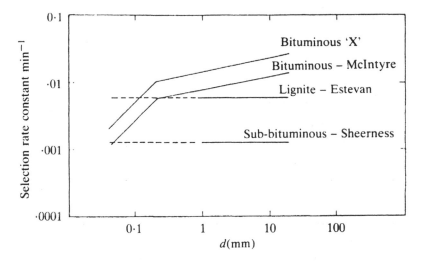

**Figure 10.** Selection rate constants as a function of particle size

methane desorption from a coal sample. They obtained a linear relationship between the amount of minus 0.6 mm coal generated in a pipe loop test to the $\Delta P$ index and suggested that this method could form the basis of a means of predicting degradation rates. They agree with Shook's results that degradation rate decreases with time and that this effect is more marked with a more friable material where the initial degradation rates would be higher.

Another measure of the mechanical properties of coal is the Hardgrove Grindability Index (HGI). Shook *et al.* (33) found that initial degradation rates increased with the value of HGI.

Shook *et al.* (33) and Traynis (35) attempted to discover the effect of concentration on degradation; the former with lignite and sub-bituminous coals at concentrations of 25–50 per cent by weight and the latter with anthracite between 0 and 33 per cent concentration by weight. No effect was detected by either.

Sold (36) studied the influence of transport velocity, particle size distribution and a change of pump design on run-of-mine coal tested in a recirculation pipe loop. An increase in velocity resulted in an increase in the degradation rate of the coarser fraction but very little significant change in the fines degradation. The degradation rate was seen to reduce with time. The coarse to fines ratio had a strong influence on degradation, and the higher the ratio the higher the rate of degradation. Of note was the result that a two-fold increase in impeller speed greatly increased the disintegration rate of the coarse granular material which is contrary to the findings of Gilles *et al.* (37). Degradation rates and characteristics in a pump are considered by Pipelin *et al.* (38) to be different from those occurring in a

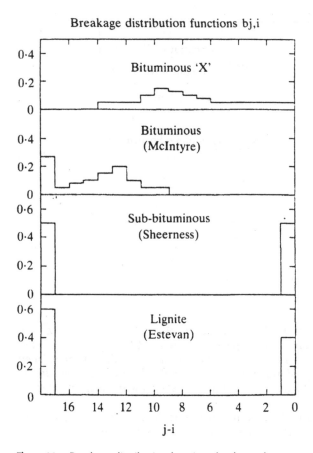

**Figure 11.**  Breakage distribution functions for the coals

pipeline; they suggest fracture to occur primarily in the pump and 'rounding' in the pipe, although Gilles comments that these conclusions are not easily extracted from the experimental results. The effect of pump impeller speeds on the breakage rate of a −50 mm coal was reported by Pipelin to vary with tangential velocity to the power 2.5 and Shook also found that lignite degradation rates appeared to increase with impeller diameter although the increase measured did not follow this power law. To determine the single pass degradation of coarse bituminous coal through a centrifugal pump, Gilles constructed a rig with a large storage tank. The study resulted in the unexpected observation that increasing the pump speed from 300 r.p.m. to 472 r.p.m. produced no significant change in degradation for 25 × 50 mm and 13 × 25 mm coals. Much higher degradation rates for coarse coal were measured than in the earlier recirculating tests (33), and it was noted that the coarser particles were particularly susceptible to breakage. It was suggested that this breakage might be due to a residual effect from stresses associated with the mining process and that it could be a cause, additional to that of rounding, of the reduction in the rate of degradation with time observed in earlier recirculating loop tests.

In a short recirculating loop driven by a conventional pumping system it would be extremely difficult to obtain data on the component of degradation which is due to the pipe only. The device, known as a 'Toroid Wheel' or 'Wheel Stand', enables slurry to flow down a pipe without the assistance of a pump and can be used to simulate degradation in a horizontal pipe. The apparatus, Figure 12, consists of a pipe bent to form the periphery of a wheel. In use the pipe is filled about one half to one third full and rotated on its axis; the slurry tends to remain stationary in the lower section of the toroid and the pipe passes over it producing hydrodynamic motion similar to flow in a horizontal pipe. Cooley and

**Figure 12.** Toroid wheel

Faddick (39), and Worcester and Denny (40). In addition to particle degradation, these wheels can be used to obtain slurry head loss and pipe wear data. Using a Toroid Wheel, Worcester and Denny studied the process of grinding of coal and developed the following expression:

$$L \propto V^3 \cdot D^{\frac{1}{2}} \qquad \text{(over the velocity range 2.5 m/s to 3.4 m/s)}$$

where $L$ is the mass of particles below 6.35 mm
$V$ is the mean slurry velocity
$D$ is the distance travelled by the slurry

The comparison made by Traynis (35), between the degradation rates and hydraulic head losses measured in a wheel and those measured in a 10 km pipeline showed good agreement; these observations support the view that the wheel simulates tolerably well the processes occurring in horizontal pipeline transport. Traynis conducted a large number of experiments on his wheel rig and obtained the following expression for degradation:

$$R_L = R_O \, e^{-ad^m(\Delta WL)^{\,b-c \,\log\, d}}$$

which relates the quantity of solids $R_L$ retained on a screen having an aperture size $d$ and the distance pumped $L$; $a$, $m$, $b$ and $c$ are empirical coefficients which depend on the coal characteristics.

## 5.2  CONCLUSIONS

It is apparent that some of the principles governing particle degradation are beginning to emerge from experimental studies. However, the variability of the characteristics of the solids and differences between pipe systems and pump designs mean that for any proposed pipelines where degradation is an important consideration (such as the design and specification of dewatering plant) it is still necessary to undertake individual experimental studies to be able to predict reliably the degradation which is likely to occur in that system.

Almost all the work reviewed concentrates on the degradation of coal; and this is the result of its friable nature and the commercial interest in pipelining this commodity.

## 6.  GENERAL COMMENTS

Although much research and development work has been expended on the prediction of pressure-drop flow-rate characteristics of slurries, the practical aspects discussed in this chapter are of considerable importance. Advances in electronics and computing have progressed so rapidly that normal system control should not present undue difficulties. Start-up and shut-down procedures, however, are perhaps still in an area requiring experimental verification. The suspension of solids in mixing tanks is a well practised art but still subject to considerable research. Particle degradation is only partly understood but there is useful practical experience of the subject.

# 7. REFERENCES

1. McElvain, R. E. and Hardy, T. H. (May 1978) 'Operational report on a 5 mile pit-to-flotation-plant transport line with a simplified approach to slurry transport', *Proc. 5th Int. Conf. on the Hydraulic Transport of Solids in Pipes*, Hanover, Federal Republic of Germany, Paper J3, BHRA.
2. Jordan, D. (April 1979) 'Hydromechanical winning and hydraulic conveying', *Coal International Supplement*, pp. 42–7.
3. Tczap, A., Pontins, J. and Rattel, D. N. (May 1978) 'Description, characteristics and operation of the Grizzly Gulch tailings transport system', *Proc. 5th Int. Conf. on the Hydraulic Transport of Solids in Pipes*, Hanover, Federal Republic of Germany, Paper J4, BHRA.
4. Buckwalter, R.K. (May 1976) 'Instruments and control systems for positive displacement pumped slurry pipelines', *Proc. 4th Int. Conf. on the Hydraulic Transport of Solids in Pipes*, Alberta, Canada, Paper H4, BHRA.
5. McCain, D. L. *et al.* 'Slurry transport system operation', (Loveridge Mine), *AIME*, Pre-print No. 81–403.
6. Burnett, M., Harvey, A. C. and Rubin, L. S. (September 1979) 'A self-controlling slurry pump developed for the US Department of Energy', *Proc. 6th Int. Conf. on the Hydraulic Transport of Solids in Pipes*, Canterbury, UK, pp. 305–14, BHRA.
7. Thorley, A. R. D. and Enever, K. J. (September 1979) 'Control and suppression of pressure surges in pipelines and tunnels', *CIRIA Report 84*, 112 pp.
8. Guzman, A., Beale, C. O. and Vernon, P. N. (25–27 August 1982) 'The design and operation of the Rössing tailings pumping system', *Proc. 8th Int. Conf. on the Hydraulic Transport of Solids in Pipes*, Johanessburg, S.A., Paper A2, pp. 17–35, BHRA Fluid Engineering.
9. Rouse, W. R. (25–27 August 1982) 'Operating experience with a residue disposal system', *Proc. 8th Int. Conf. on the Hydraulic Transport of Solids in Pipes*, Johannesburg, SA, Paper A6, pp. 77–90, BHRA Fluid Engineering.
10. McCain, D. L., Doerr, R. E. and Rohde, E. G. (1981) 'Slurry Transport system operation', *Society of Mining Engineers of AIME*, preprint 81–403.
11. White, J. F. C. and Seal, M. E. J. (25–27 August 1982) 'Seasand handling and pumping at Cockburn Cement', *Proc. 8th Int. Conf. on the Hydraulic Transport of Solids in Pipes*, Johannesburg, S.A., Paper A5, pp. 63–75, BHRA Fluid Engineering.
12. Montfort, J. G. (23–26 March 1982) 'Operating experience of the Black Mesa Pipeline', *Proc. 7th Int. Conf. on Slurry Transportation*, Lake Tahoe, Nevada, USA, Washington DC, USA, pp. 421–9, Slurry Transport Association.
13. Jenning, M. E. (26–28 March 1980) 'The SAMARCO pumping system', *Proc. 5th Int. Conf. on Slurry Transportation*, Lake Tahoe, Nevada, USA, Washington DC, USA, pp. 276–86, Slurry Transport Association.
14. Aude, T. C. and Gandhi, R. L. (2–4 March 1977) 'Research and development for slurry pipeline system design', *Proc. 2nd Int. Conf. on Slurry Transportation*, Las Vegas, Nevada, USA, Washington DC, USA, pp. 56–64, Slurry Transport Association.
15. Olada, T. *et al.* (25–27 August 1982) 'Experiments on restart of reservoir sediment slurry pipeline', *Proc. 8th Int. Conf. on the Hydraulic Transport of Solids in Pipes*, Johannesburg, S.A. Paper H3, pp. 399–414, BHRA Fluid Engineering.

16. Pitts, J. D. and Hill, P. A. (January 1978) 'Mining operations and slurry pipelines', *Proc. 10th Annual Meeting of the Canadian Mineral Processors*, Ottawa, pp. 198–240.
17. Baker, P. J. and Jacobs, B. E. A. (1979) 'A guide to slurry pipelining systems', Cranfield, Bedford: BHRA Fluid Engineering, 63 pp.
18. Kemplay, J. (ed.) (1980) 'Valve users manual', *Mechanical Engineering Publication for British Valve Manufacturers Association*, 103 pp.
19. Sauermann, H. B. and Webber, C. E. (4–6 November 1980) 'High pressure pinch valves in hydraulic transport pipelines', *Proc. 7th Int. Conf. on the Hydraulic Transport of Solids in Pipes*, Sendai, Japan, Paper A4, pp. 33–40, BHRA Fluid Engineering.
20. Sabbagha, C. M. (25–27 August 1982) 'Practical experiences in pumping slurries at ERGO', *Proc. 8th Int. Conf. on the Hydraulic Transport of Solids in Pipes*, Johannesburg, S.A. Paper A1, pp. 1–16, BHRA Fluid Engineering.
21. Herringe, R. A. (April 1979) 'The behaviour of mono-size particle slurries in a fully baffled turbulent mixer', *Third European Conference on Mixing*, York, Paper D1, pp. 199–216, BHRA, Cranfield, Bedford, UK.
22. Silvester, R. S. (May 1983) 'Mixing of non-Newtonian media – a literature review', *Confidential Report CR 2009*, 61 pp., 38 figs, BHRA, Cranfield, Bedford, UK
23. Metzner, A. B. and Taylor, J. S. (March 1960) 'Flow patterns in agitated vessels', *A.I.Ch.E. Journal*, Vol. 6, No. 1, pp. 109–14.
24. Metzner, A. B. and Otto, R. E. (March 1957) 'Agitation of non-Newtonian fluids', *A.I.Ch.E. Journal*, Vol. 3, No. 1, pp. 3–10.
25. Calderbank, P. H. and Moo-Young, M. B. (1959) 'The prediction of power consumption in the agitation of non-Newtonian fluids', *Transactions of the Institution of Chemical Engineers*, Vol. 37, pp. 26–33.
26. Haas, D. B. and Husband, W. H. W. (8–11 May 1978) 'The development of hydraulic transport of large sized coal in Canada – Phase 1', *Proc. 5th Int. Conf. on the Hydraulic Transport of Solids in Pipes*, Hanover, Federal Republic of Germany, Vol. 1, Paper H1, pp. H1-1–H1-12, BHRA Fluid Engineering.
27. Bohme, F. and Ebbecke, K. (June 1975) 'Ergebnisse und Probleme experimenteller Untersuchungen des hydraulischen Transports Korniger braunkohle', *Neue Bergbautechnik*, Vol. 5, No. 6, pp. 433–9.
28. Lammers, G. C. (1958) 'A study of the feasibility of hydraulic transport of a Texas lignite', *US Dept. Interior Bureau of Mines Report of Investigations 5404*, 42 pp.
29. Schriek, W. (15–17 May 1974) 'Experimental studies on the hydraulic transport of coal', *Proc. 3rd Int. Conf. on the Hydraulic Transport of Solids in Pipes*, Golden, Colorado, USA, Paper B1, pp. B1-1–B1-19, BHRA Fluid Engineering.
30. Karabelas, A. J. (July 1976) 'Particle attrition in shear flow of concentrated slurries', *AIChW J.*, Vol. 22, No. 4, pp. 765–71.
31. Austin, L. G. and Luckie, P. T. (1971/72) 'Methods for determination of breakage distribution parameters', *Powder Technology*, Vol. 5, pp. 215–22.
32. Shook, C. A. (August 1978) 'Breakage rates of lignite particles during hydraulic transport', *Canadian J. Chem. Engng.*, Vol. 56, No. 4, pp. 448–454.
33. Shook, C. A., Haas, D. B., Husband, W. H. W. and Small, M. (26–28 September 1979) 'Degradation of coarse coal particles during hydraulic transport', *Proc. 6th Int. Conf. on the Hydraulic Transport of Solids in Pipes*, Univ. Kent, Canterbury, UK, Vol. 1, Paper C1, pp. 125–36. BHRA Fluid Engineering.
34. Mikhail, M. W., Mikula, R. J. and Husband, W. H. W. (23–26 March 1982) 'Prediction of size degradation of coarse coal during slurry pipeline transportation', *Proc. 7th Int. Conf. on Slurry Transportation*, Lake Tahoe, Nevada, USA, Washington DC, USA, pp. 399–404. Slurry Transport Association.

35. Traynis, V. V. (1977) 'Parameters and flow regimes for hydraulic transport of coal by pipe lines', Transl. of 1970 Russian ed.; ed. by W. C. Cooley and R. R. Faddick, Rockville, USA, 265 pp. Terraspace Inc.
36. Sold, W. (June 1982) 'Particle degradation of ROM coal and washery tailings up to 63 mm size with horizontal; hydrotransport', *Bulk Solids Handling*, Vol. 2, No. 2, p. 261.
37. Gilles, R. *et al.* (25–27 August 1982) 'A system to determine single pass particle degradation by pumps', *Proc. 8th Int. Conf. on the Hydraulic Transport of Solids in Pipes*, Johannesburg, S.A., Paper J1, pp. 415–31, BHRA Fluid Engineering.
38. Pipelin, A. P., Weintraub, M. and Orning, A. A. (1966) 'Hydraulic transport of coal', *US Dept. Interior Bureau of Mines Report of Investigations 6743*, 31 pp.
39. Cooley, W. C. and Faddick, R. R. (16 December 1976) 'Review of Russian data on hydrotransport of coal' Washington, DC, US Dept. Interior Bureau of Mines, p. 18, USBM Open File, Dept. 90–70.
40. Worster, R. C. and Denny, D. F. (September 1954) 'Transport of Solids in Pipes', SP 496, 40pp. BHRA, Cranfield, Bedford, UK.

# 7. ALTERNATIVE FORMS AND APPLICATIONS

This chapter is concerned with applications, other than the conventional pipelining of comminuted minerals, where hydraulic slurries are now being used or may be used in the foreseeable future. The applications covered are:

- vertical hoisting of minerals,
- coal water mixtures,
- coarse particle conveying,
- waste disposal,
- ship loading and unloading,
- non-aqueous slurrying media,
- three-phase mixtures in coal conversion technology.

None of these can strictly be termed 'new' but improved techniques and changes in economic conditions and environmental pressures such as fluctuating oil prices and concern about surface waste disposal have served to renew or increase interest in these applications.

## 1. VERTICAL HOISTING OF MINERALS

There are three main areas of application where the vertical hoisting of minerals by hydraulic means is currently of either commercial or technical interest. These are hydraulic hoisting in land-based underground mines, deep sea mining and hydraulic borehole mining. These systems are generally used to transport coarse material and may be associated with underground and/or surface pipeline transport systems.

### 1.1 HYDRAULIC HOISTING

An alternative to conventional methods of raising material to the surface in a mine is hydraulic hoisting which is the transport of a slurry upward in a vertical or near vertical pipe. The upward motion of solids is normally achieved either by using a pump to energise the mixture or by injecting air at the base of the column of slurry, the latter technique is known as air-lifting. These techniques can be particularly attractive if the material is mined hydraulically but its technical and economical feasibility has been proved in applications where both conventional and hydraulic mining techniques are used; Table 1 shows some of the mines where hydraulic hoisting systems are operating.

At the Consolidation Coal Company's Loveridge Mine (1), in North West Virginia, coal is received from continuous miners, crushed to 100 mm and pumped 450 m horizontally in a 350 mm diameter pipe to a multiple feed sump; a travelling dredge reclaimer feeds the vertical hoisting system with slurry at a constant rate and concentration. Seven slurry pumps in series hoist the coal through a cased borehole 274 m to a surface pumping station. Buffer storage is provided at the surface station which then pumps the slurry a further 4 km overland to a coal preparation plant. The first coal was pumped in April 1980 and by October all major modifications had been completed and the system was in full three-shift operation.

Particular aspects of the hoist where development was required were concentration control, dissipators which provide back pressure on pump start-up and some support equipment such as valve actuators, gland water quality and compressed air systems. Plugging of the vertical pipe after an emergency shut-down is avoided by dumping the contents into the feed sump; plugging of the horizontal section is cleared by back flushing and water injection through ports located at intervals in the pipe. Solids pumping rates of 425 tonnes/hour have been sustained for several hours with peaks in coal production above this rate being transported by rail.

**Table 1.** Examples of hydraulic hoisting installations

| Mine | Material | Particle size (mm) | Vertical lift (m) | Tonnes hoisted per hour | Pipe size (mm) | % Concentration by volume | Feeder type |
|------|----------|-----------|-----------|-----------|-----------|-----------|-----------|
| Yoshima Mine: Japan | Coal | −50 | 260 | 100 | 165 | 20 | Hitachi Hydro-Hoist |
| Sunagwa Colliery: Japan | Coal | −30 | 520 | 100 | 190 | 25 | Hitachi Hydro-Hoist |
| Konkola div. of ZCCM: Rhodesia | Mineral ore | −6 | 2 lifts 198 & 204 | 240 | 204 | 35.5 | Centrifugal pumps |
| Lengede Mine: West Germany | Iron ore | −35 | 130 | 300 | 224 | 20 | Modified dredge pump |
| Vaal Reefs: South Africa | Gold ore | −2 | 2200 | 25 000 t/month | 152 | 16.5 | Mars pump |
| Hansa Colliery: West Germany | Coal | −60 | 850 | 250 | 250 | 20 | Three-chamber pipe feeder |
| Loveridge Colliery: USA | Coal | −100 | 275 | 100 | 300 | 40 | Centrifugal pumps |

The operators claim that the system has increased production from all sections of the mine by releasing large numbers of rail wagons and by allowing the existing haulage facilities to be used more efficiently; safety, productivity and economics of production have all been improved. Consolidated Coal has announced that it will licence technology for this system.

In a so-called hydro-mine where the mineral is removed by water jet monitors, the advantages of hydraulic hoisting and slurry transport are self-evident, i.e. a uniform transportation system from the coal face through to the preparation plant without dust problems and exposed moving machinery, and with a relatively low space requirement. The Hansa hydro-mine near Dortmund in West Germany (2,3), operated for three years, until unexpectedly difficult geological conditions forced premature closure. After hydraulic removal from the face and a pre-preparation stage, coal is transported by flume and then, after further crushing to a top size of 60 mm and screening, by pipeline to a buffer store prior to being hoisted 850 m to the surface. The hoist pumping arrangement, unlike that in the Loveridge mine, is devised so that the solids do not pass through the high pressure pumps. The system, which is called a three-chamber feeder has been described in chapter 2. Each chamber is alternately filled with slurry from one end by a low pressure pump and then discharged into the riser by injecting high pressure water from the other end of the chamber. The provision of three chambers allows continuous feeding of slurry from one chamber whilst a second is full and ready to be discharged and a third is in the process of being filled.

Work at the Hansa mine has shown that it is possible to eliminate the low pressure slurry pump and charge the chambers using only the static pressure from a vertical slurry bunker. The system achieved throughput rates of over 500 tonnes/hour compared with an expected throughput of 250 tonnes/hour; the increase resulted because it ultimately proved possible to handle slurries of up to 50 per cent concentration by volume instead of an anticipated concentration of 10–20 per cent by volume. Average energy consumption for vertical haulage was 12.9 kWh/tkm and it is claimed that in a similar application, an optimised system would operate with specific energy consumptions of about 7–9 kWh/tkm.

The efficient and reliable operation of hydraulic systems relies on their effective control; some problems were experienced which stemmed from variable coal to dirt ratios and inadequate slurry concentration measurements. Development to overcome these difficulties had not been completed before the mine was closed, nevertheless the trial showed that in appropriate circumstances hydraulic hoisting can compete with conventional haulage systems.

A system similar to that used in the Hansa mine but called the 'horizonal Hydrohoist' is employed to raise minus 0.75 mm coal at 27 per cent concentration by weight, 520 m vertically and 2000 m horizontally from the Sunagawa Mine in Japan (4). The hydrohoist is used in conjunction with hydraulically mined faces and the material raised represents 9 per cent of the 0.65 Mtpa mined hydraulically. The system has been in use since 1963 and was originally designed to raise coarse as well as fine material at a rate of 95 tonne/hour. Due to rising energy costs it has proved more economical to transport the plus 0.75 mm coal to the surface in bins; the fines can be raised at a rate of 55/tonnes/hour. Lower transport velocities and smaller diameter pipes are largely responsible for the better economics when pumping fines. Given the severe conditions in this mine, faulty, steeply inclined seams and gassy coal, hydraulic mining and hoisting has proved to be a safe and effective method of coal extraction. The Hitachi Company, which developed and built the Sunagawa hydro-hoist unit, now has several other similar units in operation pumping coal, sand and mud; these are summarised in Table 2. A development of this system is the 'vertical hydrohoist' in which the three chambers are aligned vertically and the slurry and water in these chambers is separated by a float (5).

Of the eleven hydraulic mines in the USSR (6), at least two, the Yubilegnaya and the Krasnoarmeyskaya (Red Army) mines, employ hydraulic systems for raising coal to the surface; the former uses an hydraulic hoist and the latter an air lift system. The Red Army mine is one of the few hydraulic mines in existence utilising an air lift recovery system.

To summarise, the use of hydraulic hoisting in mines does not present technical difficulties which cannot be overcome. The economics of using this system for raising material will vary with each application, such aspects as cost of power, availability of water, transportation requirements on and below the surface, material properties, mining method, particle size required, dewatering costs and preparation plant process, etc., would need to be considered before the cost effectiveness can be evaluated. Such an example is given by Sellgren (7) where the integration of a proposed hydraulic hoisting system with mine dewatering is discussed.

## 1.2  SEABED MINING

Although mineral extraction from the sea began as long ago as 2000 BC, when the Chinese evaporated sea water from salt and more recently (1907) when alluvial tin deposits were

**Table 2.**  Horizontal Hydrohoist installations

| Installation | Materials | Flow rate Pressure |
|---|---|---|
| Yoshima Colliery of Furukawa Kogyo Co., Ltd | Raw coal | 280 m³/h 53 bar |
| Sunagawa Colliery of Mitsui Sekitan Co., Ltd | Raw coal | 270 m³/h 85 bar |
| Tsuchiura Wks of Hitachi Ltd | Silica sand | 104 m³/h 11 bar |
| Rinkai Construction Ltd (Inba Lake) | Mud | 1300 m³/h 9.2 bar |
| Electric Power Development Ltd (Sakuma Dam) | Silica sand | 83 m³/h 50 bar |

dredged from the coast of Thailand (8), current commercial interest since 1964 has centered upon so called 'manganese nodules'. These nodules contain varying amounts of manganese, copper, nickel and cobalt, the former in most abundance hence their name; the latter, although in the smallest concentrations, is considered the most valuable. Commercially interesting concentrations of these nodules are located in waters up to 5000 metres in depth and therefore their gathering and raising presents some severe technical problems. Two methods of raising involving hydraulic conveying have been proposed, the air-lift and hydraulic hoisting; other competitive systems which do not use hydraulic techniques are a bucket dredge proposed by the Japanese and a capsule system (9,10). Relatively complete systems of all these configurations have been tested albeit at sizes substantially smaller than would be used commercially.

Currently, the leading contender is the air-lift. In this device solids are raised in an open ended vertical submerged pipe into which compressed air is injected at some point below the surface. The resultant three-phase mixture is less dense than the sea water surrounding the pipe causing the static pressure at the base of the pipe to force the mixture to the surface. This system, although being relatively inefficient in terms of energy consumption, has the very major advantage of having no submerged moving parts. The hydraulic hoisting technique is similar to that described in the previous sub-section but would be powered either by a centrifugal or jet pump; centrifugal pumps may be placed either at the surface, in which case there will be added losses due to the descending feed pipe or at the base of the pipe in which case maintenance of the system will be extremely difficult.

In the case of the submerged centrifugal system, the pump would need to be capable of handling particles of up to 150 mm in size which cannot be achieved without a reduction in efficiency. The jet pump arrangement overcomes the latter problem but still requires a conventional pump to generate the primary jet; again this can be placed either at the surface or submerged with the attendant disadvantages of either added losses or difficult maintenance respectively. Furthermore, a jet pump is fundamentally an inefficient device and the intrusive jet nozzle, a feature of some but by no means all designs, would be liable to damage from the nodules.

Air-lifting from great depths poses two major technical problems; the first is that of structural loads on the riser, due to the net external pressure, at the air injection point which will require additional stiffening in this region. Burns and Suh (11) claim that this can be done provided that the stresses are predicted accurately enough to allow low safety factors. The second difficulty is caused by the expansion of the gas as it rises; this results in high velocities and friction losses and the additional risk of forming flow regimes where the solid transport mechanism is pneumatic instead of hydraulic (12). The costs of developing these systems to recover economical quantities of the nodules are extremely high, requiring syndicated efforts (13). Discussing the future prospects of deep sea mining, Silvester (14) makes the point that metal prices hold the key to mining economics, which together with the unsatisfactory conditions of the 'Law of the Sea' concerning the predictability of access required by a mining company, already faced with massive technical risks, would not appear to provide the satisfactory economic, political and legal conditions required to attract investors for exploration of this mineral resource. Nevertheless, the potential mineral recovery is enormous, as are the technical problems associated with its exploration, and assuming that political and technological progress can proceed in parallel, it might be expected that the nodules could be mined within the next decade.

Other deep sea minerals have been found; these include polymetallic sulphides formed at oceanic spreading zones by a leaching action and metalliferous mud in the Red Sea. The quantities of sulphides are not yet known but the Red Sea deposit is of commercial interest and is to be exploited by the joint owners, Saudi Arabia and Sudan. A system for extracting the mud and pumping it some 2000 m to the surface is described by Hahlbrock (15).

## 1.3   HYDRAULIC BOREHOLE MINING

Hydraulic borehole mining, also called slurry mining, is a concept in which the mineral seam is accessed by a single borehole down which is introduced a unit incorporating a water jet coal cutting tool and a down-hole slurry system. The tool is operated from the surface, with high pressure water jets from the cutting tool breaking up the coal, while the slurry system removes the fragmented material and water from the resultant cavity to the surface.

Net resource recovery by this method is estimated, for coal, at approximately 45 per cent (16,17), the remainder being left behind for roof support or not recovered by the slurry system. From costs based on this estimate it is concluded that existing surface mining techniques would be preferable to hydraulic borehole mining within the current 100 m depth limit of the borehole system. However, the borehole system compares favourably with other methods of extraction from thick near vertical seams. The subject is treated in greater detail in an excellent summary of hydraulic borehole mining techniques given by Wood (18).

## 2.   COAL WATER MIXTURES (CWM)

The rise in oil prices in 1973/74 provided the incentive for research into methods by which energy costs could be reduced. Two of the ideas which arose were COM (coal-oil-mixtures), discussed later in this chapter, and CWM; the objective being to produce a fluid with the handling and burning advantages of oil which could be fired directly in existing oil or pulverised fuel burners, but which also resulted in a much lower cost per unit of energy.

CWM is known by a number of proprietary names such as Carbogel, Co-Al, Densecoal and F-Coal which appear to differ mainly in particle size distribution and the viscosity reducing and suspension stabilising additives employed. In the USA today there are six CWM preparation facilities operational with a total capacity of $1 \times 10^6$ barrels per year and four combustion plants, in operation. Development work is also being carried out in West Germany, Sweden, the Netherlands, Japan and Italy.

In general, the maximum particle size is less than 300 $\mu$m with 75–80 per cent less than about 74 $\mu$m, that is to say similar to or finer than conventional pulverised fuel. The concentration by weight of coal is usually about 60–70 per cent; the presence of water will clearly reduce the net calorific value of the fuel but this does mean that lower ash fusion temperatures can be tolerated and that nitrogen oxide emissions are reduced. The estimated cost of CWM including additives and pulverisation is quoted as about $2.8/ Gigajoule assuming coal at $30/tonne (March 1982) (19).

Although combustion efficiences are about 94–98 per cent, boilers adapted for CWM use are de-rated by 15–50 per cent; this is partly because the ash in coal causes more fouling or slagging of the heat exchanges than oil and partly because CWM has only about half the energy density of fuel oil (20). Early problems of combustion stability have been overcome by improved atomisation and combustion air pre-heating. Installing soot blowers is one way to remove fouling of the heat transfer surfaces. Increasing the slope of boiler floors from 15° to 45°, common in oil fired boilers, to 55° is sufficient to allow the ash to settle by gravity; this is considered by some to be the most expensive single modification in the conversion to CWM (20). Reducing the ash content by making CWM with beneficiated coal would eliminate major boiler modifications, however the technology of cleaning coal has not yet been fully developed and it is difficult to determine realistic costs for the process (20). One of the most critical break-throughs, according to SEC of the Netherlands, in the industrial and utility markets is the availability of coal with an ash content of less than 2 per cent. Commercial fine coal cleaning systems have been used to de-ash high quality bituminous coal but it is clear that only chemical methods will be capable of reducing ash content to 0.5 per cent (21).

At one stage it was thought that long distance pumping of CWM would not be economical because it has relatively high viscosities, some 20 to 100 times that of a Black Mesa type slurry (22). Sommer and Funk (19) describe the pumping pressure drops of their Co-Al product as being similar to those of coal-oil mixtures and higher than those of a No. 6 oil. A more realistic proposition would be transportation in conventional or stabilised slurry form to the point of use followed by partial dewatering (which may not be necessary if a stabilised mixture is used) and wet grinding. CWM is not only being considered for existing oil fired plant (22); some gasification and liquefaction processes involve feeding coal water mixtures into a pressurised reactor; these plants would require large quantities of coal and would therefore be suited to supply by slurry pipeline.

Much of the work has concentrated on the preparation and combustion aspects of CWM. It was thought that pumping of CWM should not present too many technical difficulties, since the high viscosities (compared with conventional coal slurries) could be tolerated provided short distances were anticipated. The high concentration and fine particle size will suppress or eliminate settling in storage and pumping, and reduce pumping velocities. If longer distance transport is contemplated it may be necessary to lower the concentration, to reduce the viscosity and pumping power, but at the risk of increased settlement rates. This could affect both pumping and storage. The effects of particle size distribution and additives on slurry viscosity are described by Batra *et al.* (23). Additionally much work on the combustion of CWM has been reported in the Orlando conferences organised by the US Department of Energy (24). European Conferences on Coal Liquid Mixtures have been organised by the Institution of Chemical Engineers (UK) (25). The use of CWM for improving the overall transport process from mine to user is discussed by Mahler (26).

Depending on viscosities and transport distances, either centrifugal or positive displacement pumps will be used for conveying. Helical rotor (Mono type) positive displacement pumps have proved satisfactory for this material, particularly as variable speed units would assist controllability of burner feed (19); centrifugal pumps have also been satisfactorily used. The low pumping velocities will help to minimise abrasion of pumps, pipes and fittings. To reduce the possibilities of settlement and plugging, pipeline systems should contain long radius elbows, conventional tees should be replaced with Y branches and valves need to be of the full flow type. Other aspects of CWM which require further consideration and study are temperature effects on stability and rheology, variations in slurry properties between different suppliers, feed systems and their control.

Any change from the use of oil to coal for electric power generation would be a major undertaking involving high investment. It should be remembered that the use of CWM is just one of the potential options, such as new or re-built firing of boiler plant, COM or coal liquefaction and gasification; there will be no universally accepted technology and choice will be highly site specific.

In spite of initial work indicating CWM would be unsuitable for long distance transport, developments have taken place which have allowed the design and construction of a 256 km pipeline from Belovo to Novosibiersk to carry this material.

## 3.  COARSE PARTICLE CONVEYING

There are a number of important advantages which the ability to convey coarse particles, when compared with fine particle conveyance, would confer on a potential pipeline operator. Firstly, the cost of crushing the solids prior to slurry preparation and pumping is reduced; typically, a coarse particle slurry top size may be about 50 mm, whereas the top size of a fine particle heterogeneous, e.g. Black Mesa, type of slurry, is about 1–2 mm. Secondly, the dewatering costs may, for difficult materials such as coal, be reduced significantly, perhaps by as much as 50–75 per cent; most of the solids can be separated from the water by use of wash screens leaving only the minus 0.5 mm fraction requiring centrifuging and thickening. Lastly, a coarse product may be a more valuable and saleable commodity than the same material ground to a smaller size distribution.

Unfortunately, problems arise when coarse particles are simply mixed with water as a dilute slurry, of say 20–30 per cent concentration by weight. All these problems stem from the fact that large particles settle faster in a liquid than fines and consequently higher flow velocities (between 2 m/s and 3 m/s for a 50 mm top size ROM coal in a 200 mm diameter pipe) are required to provide sufficient turbulence to maintain the large particles in suspension; high velocities lead to higher abrasion of the pump, pipe and valves and high specific power requirements. In addition, larger solids are more difficult than fines to re-suspend in the pipe after a shut-down with slurry in the pipeline. Even so, there are applications where the advantages of dilute coarse particle conveying outweigh the disadvantages; these are almost universally short distance high throughput systems (up to about 5–10 km) where most of the capital and running costs are associated with the crushing and dewatering plant. In such instances, the operator can more than offset the higher power and maintenance costs against the lower crushing and dewatering costs.

Other factors may influence the decision to adopt a coarse particle system, for example, when considering the installation of a system to lift hydraulically coal from a mine (see section 1.1), reduced dust levels, increased safety, space constraints, cost of expanding or replacing conventional transport systems and the mining method are important.

Although the ability to pump coarse mixture over long distances economically has great commercial potential, it is not with us yet. The possibility that coarse material could be pumped at low velocities with relatively low pressure drops was first raised when, in the 1960s, Elliot and Gliddon (27) were conducting tests on a heterogeneous slurry containing 13 mm coal. For low velocity pumping to be possible they concluded that:

- The concentration should be of the order of 60 per cent by weight.
- Fine coal should be about 25–30 per cent of the total solids.
- The size distribution should conform to the conditions of maximum packing density.

Using a 100 mm diameter tube viscometer and a stabilised mixture, Lawler *et al.* (28) demonstrated the stabilised flow principle, they called it 'stab-flo' and they suggested that the reduced grinding and dewatering costs would make this mode of conveying cheaper than conventional slurries up to distances of 80 km if a power station or utility used the product. Duckworth *et al.* (29) are of the opinion that low velocity coarse coal transport would be economical over longer distances, particularly where an ocean link is involved.

Much of current research is focused on methods of reducing the power consumption, abrasion and start-up difficulties of coarse particle conveying. Broadly speaking, two types of slurry have been proposed, one termed 'stabilised flow' seeks to suspend the coarse particles in a viscous carrier fluid which can either be composed of fines (usually of the same material as the coarse) and water or produced by adding a thickening agent, and the other, sometimes called 'dense phase flow', where particle support is by contact load with the submerged weight being transmitted to the pipe wall. These two concepts are difficult to distinguish in practice and it is probable that the real support mechanisms lie somewhere between the two. They do however form the basis of two popular theories of the behaviour of high concentration coarse particle flow in horizontal pipes namely the sliding bed model of Wilson (30), based on the dense phase concept and the plug flow models which are derived by assuming the existence of a stable core surrounded by a region of shear flow adjacent to the pipe wall (31,32,33). Most of the published work on plug flow models is with reference to concrete pumping although work dealing with coarse coal transport has appeared more recently.

The suspending properties of the stable carrier fluid are all important and depend on the particle size distribution and concentration. Ideally, the carrier should be capable of supporting the coarse fraction at zero velocity, i.e. it should be a fluid with a yield stress. The rheological properties of the carrier also play an important role in the resultant friction losses of the whole slurry during pumping. As suspension of the coarse fraction does not rely on slurry turbulence stabilised flows, which could be transported in the laminar flow regime, it could be pumped at velocities down to zero m/s, thus reducing particle attrition and system wear and potentially eliminating start-up problems. Another aspect of the flow of these mixtures in pipes, which is receiving attention, is instability in the slurry concentration whilst in the pipeline, this could, over very long pipe runs, eventually lead to the production of a settling slurry and the possibility of gradual deposition of the coarse fraction over long distances.

A very limited amount of work has apparently been done on mixtures termed 'enlightened' coal slurries (22). Here additives similar to those used in the froth flotation separation process are mixed with the slurry so that air bubbles will adhere to the coarser particles. Concentrations of over 70 per cent with particle size of 13–25 mm have been achieved.

For stabilised mixtures it is anticipated that separation of the coarse from the majority of fines and water at the end of the pipeline will be accomplished by simple washing and screening; the fines can then be pumped back to their source, via a return pipeline, thickened and re-used thus avoiding the necessity of providing crushing and grinding plant. The return of water and fines, and high slurry concentration, means that system water consumption is reduced, which can be a crucial factor in places where water is scarce.

No long distance coarse particle stabilised flow or dense phase system is yet in operation; the nearest candidate is The Rugby Portland Cement pipeline, opened in 1964, which can carry $1.7 \times 10^6$ tonnes of limestone per year, 92 km from Kensworth to Rugby, as a fine particle stabilised mixture; one object of using this type of slurry was to minimise the water content at delivery to the cement kilns. A dense phase vertical hoisting system at the Doornfontein Gold Mine in South Africa is described by Streat (34), where mud and sundry undersized material is pumped at low pressure to the surface from a depth of 1830 m. Other uses of coarse particle stabilised mixtures include concrete pumping and mine pump-packing systems; both these applications are over short distances with relatively high friction losses which would make them uneconomical over longer distances. However, there is interest in reducing these losses in packing systems to allow transport over greater distances using existing pump delivery pressures.

There is continued industrial interest in coarse particle transport, particularly coal, with the result that steady experimental and theoretical progress is being made in several countries including the UK, USA, Canada, Australia and South Africa.

## 4. WASTE TRANSPORT AND DISPOSAL

The current annual production of major mineral tailings, e.g. coal, P.F.A. (pulverised fuel ash), china clay, iron and steel slag, in the UK is estimated at $90 \times 10^6$ tonnes adding to existing waste stocks estimated at $4 \times 10^9$ tonnes (35). Approximately similar annual tonnages of coal spoil are apparently generated in the USA and USSR (36); Down and Stocks (37) estimate the world annual production of non-ferrous metallic ore tailings alone to be $1500 \times 10^6$ tonnes to which must be added very large quantities of coal tailings and P.F.A. Only a small percentage of this material can be absorbed usefully as building aggregate or in the manufacture of building blocks for instance.

Disposal of this waste is a problem which will increase as, presumably, mineral, coal and energy demands grow. By far the largest proportion of this material is disposed of by surface dumping and it is estimated that in the UK alone 10 000 hectares are covered with colliery spoil and about 1000 hectares with china clay waste.

For economic reasons the dumping site is usually located near the mineral body; selection of a suitable site is fundamental to safe design and construction of waste containments and hence the tendency to choose a site close to the mine is being eroded by the need to take account of environmental and safety factors. Environmentally, the point of most concern with ores is usually the residual metal levels; the presence of sulphide minerals is also of importance because a combination of sulphides and metals frequently results in severe pollution effects. Amongst other possible pollution sources are alkaline or acidic run-off and the presence of residual agents which have been used in the ore processing phase. Examination of the stability of stockpiles, impoundment walls and foundations with a view to increasing safety, after several major disasters such as Aberfan, has resulted in the intensive study of the engineering characteristics of tailings and in the case of colliery tips, slopes have been reduced which further increases land usage.

Surface stockpiling near to mines however, continues in the face of increasing pressure to develop alternative means of disposal, i.e. on the surface but at a site which is safe and environmentally secure but not necessarily near the waste source, in water or in opencast or underground exhausted workings. Thus it is evident that, because of increasing concern about the environment and a diminishing number of suitable local disposal sites, there is likely to be increasing interest in cost effective methods of transporting waste. The economics of any transportation system are highly site specific but in many existing tailings disposal applications hydraulic conveying has been selected as the most appropriate method, (see Table 1, chapter 8 for existing installations) and there seems to be little doubt, as pipeline technology advances, that this trend will persist and possibly increase in the future.

The engineering problems encountered in such systems present a considerable challenge. However, techniques, materials and equipment are available to overcome these problems which result largely from high flow rates, variable particle size distributions and concentrations, and the highly abrasive nature of some of the material.

In the UK, full scale experimental work in this field has been undertaken by British Coal, who have constructed a pilot pipeline to dispose of 200 tonne/hour of spoil from the Horden Colliery on the Durham Coast; the pipe is 1.2 km in length and has a diameter of 250 mm (38,39). Future prospects are that such spoil may be used for land reclamation or for filling mineral workings. Several proposals for very large reclamation schemes have been examined including Maplin $400 \times 10^6$ tonnes, Spurn Head $1000 \times 10^6$ tonnes and Pyewipe Flats Grimsby $80 \times 10^6$ tonnes (40), although not all of these have necessarily proposed the use of pipeline conveying. A scheme has also been proposed to use colliery and shale-oil waste to reclaim 615 hectares of land at Grangemouth. It has been concluded that the cheapest form of transport would be by a 10 km long hydraulic pipeline (41). A number of schemes are in progress for pumping P.F.A.; at Longannet, ash from two power stations is piped from lagoons up to 13 km to create new grazing land off the Firth of Forth and at

Peterborough, ash from three power stations is being used to fill 250 hectares of old clay pits (42). Schemes involving pipelining have been proposed for the disposal of china clay waste which is located entirely in South West England (43). None of these proposals has been taken up but should surface disposal be prohibited in the future the cheapest solution could be pumping to the coasts for subsequent disposal at sea by hopper barge, although it is unlikely that this option will remain open for long because of increasing pressures to protect the environment.

Well established in the mining industry, pipelining is used to convey a variety of tailings to disposal sites and a world review of tailings disposal practice has been conducted by Down and Stocks (37,44). Several large pipelines handling gold tailings are reported in South Africa; one, 35 km long, conveys $1 \times 10^6$ tonne/year and other shorter examples convey several million tonne/year. The longest reported tailings pipeline, conveying copper tailings over 70 km, is in Japan (45).

Back-filling of underground workings is a partial alternative to surface disposal (generally only about 50 per cent of the tailings can be returned underground because, for example, the packing density of return spoil is reduced and fines can be unsuitable for back-fill material) and although tailings have been used for several decades as a fill medium it is only relatively recently that attempts have been made to develop a rational scientific approach to its preparation, placement and performance (44).

In Europe underground disposal of mine coal tailings was undertaken extensively in the past, in recent years however the practice has declined drastically, mainly because the stowing process has been unable to keep pace with the rapid advance rate of highly mechanised faces. This reduction in underground stowing has aggravated the already serious problem, in Europe, of mine waste. Large volumes have been, and are being, dumped on the surface as spoil banks or in lagoons. With the possibility of increased coal production in the future, waste disposal will become an even greater problem. Underground stowing provides one disposal method that should be considered to help overcome the problem. This practice, although not widespread, is already undertaken in several countries, e.g. West Germany, 3–4 $Mm^3$ annually; France, 6–7 $Mm^3$ annually; Poland, 4–5 $Mm^3$ annually; and Czechoslovakia, 2–3 $Mm^3$ annually (46). Difficulties associated with surface disposal have prompted a UK study to consider alternatives (Commission on Energy and the Environment 1981).

There are three basic methods by which waste material can be stowed underground, namely hydraulically, pneumatically or mechanically. Hydraulic stowing has certain advantages which include relatively small space requirement (pneumatic stowing shares this advantage) and low potential for increasing dust concentrations (compared with mechanical or pneumatic methods). Additionally, hydraulic fill has a low potential for spontaneous combustion, methane emission can be reduced and the degree of packing is usually denser than with pneumatic systems. These considerations need to be weighed against the disadvantages of hydraulic stowing which are increased flood potential, possible deterioration of roof and floor conditions caused by water seepage, high humidity, large water requirements, acid drainage and the need for protection against freezing above ground.

Within the coal mining industry hydraulic stowing methods have been used fairly extensively where it is often a necessary part of the mining process helping to improve roof control, reduce face subsidence and surface subsidence. The most extensively used fill material is sand, mainly because of its low cost and good draining qualities. The use of mine waste as fill material is, at the moment, much less widespread for the reason that it may not necessarily have the required characteristics; the most important of which are listed below:

- The variable characteristics of the waste, particularly size distribution and potential for cementitious reactions, may be a problem if the fill has to provide a support function.
- The material may require preparation before pumping, e.g. crushing, screening or dewatering if taken from a lagoon.
- Fines may need to be removed – too high a fines content will impair the drainage characteristics of the fill.
- Chemical properties, e.g. toxicity, acidity.
- Combustible content – not such a hazard if the waste is hydraulically placed.

For further information on underground stowing of colliery waste the reader is directed to an extensive summary of the subject given by Wood (46).

The disposal of coal fired power station ash is becoming an increasing problem in South Africa. Currently $1.3 \times 10^6$ tonnes of ash is generated each month. Disposal methods include surface disposal on tips and in lagoons and to a lesser extent mine back-fill. Transport to the disposal site can be by mechanical, pneumatic or hydraulic means, however, the back-fill is placed hydraulically. The practice has been to convey P.F.A. at 15–25 per cent concentration by weight but this has resulted in excess water problems during back-filling. To reduce the backfill water content tests have been conducted at the University of Witwatersrand (47), on pumping P.F.A. at higher concentrations 65–75 per cent by weight. The increased concentration has, not unexpectedly, resulted in high pressure gradients but interestingly the addition of coarse bottom ash to the fines has been demonstrated to significantly reduce the gradient; the effect is thought to be associated with the reduced wetted area which the coarse particles present compared with that of the fines they replace. An additional benefit which the coarse particles are said to confer is that of removing scale on the inner pipe wall which is observed to form when only fine solids are included in the slurry. This reduction in gradient has important implications for planned waste back-fill operations and could result in extended pumping distances which in some applications is a critical factor.

Development of a back-filling system by the Anglo-American Corporation of South Africa has been continuing for a number of years. Here, gold tailings in the form of very fine slimes are concentrated to about 78 per cent by weight and pumped into used stopes (69,70). The programme is being shared with other gold mining organisations through the Chamber of Mines who are also involved with waste back-filling programmes using coarser solids obtained from non-gold-bearing rock.

Extensive testing of hydraulic back-fill material has been undertaken by Mount Isa Mines Ltd (48). The work, instigated by the need to dispose of some 500 000 tonnes/year of lead/zinc/silver ore heavy media separation (HMS) reject, showed that by adding this reject to production hydraulic fill, significant increases in concentration were obtained whilst still retaining pumpability.

Production hydraulic fill is typically pumped at about 30 per cent by weight concentration; the optimum addition of HMS reject was found to be 30 per cent by weight of the dry solids at a slurry concentration of 70 per cent by weight. Increasing or decreasing the proportion of HMS reject increased the required pumping power. The addition of fines in the form of ground copper reverbatory furnace slag was tried but this did not produce any significant effect. The work continued (49), with the object of quantifying the relevant mixing and placement parameters and of providing data necessary for planning and design.

The method of placing fill into the voids of abandoned workings through boreholes either pneumatically or hydraulically is generally termed 'flushing'. Flushing has generally been conducted in abandoned underground workings underlying urban areas in an attempt to reduce subsidence, but could also be used to stow waste in mined out areas of active mines. Flushing operations to date have been carried out at depths of less than 90 m; deeper mines could be filled by this method but borehole costs would be considerably increased. Two types of hydraulic flushing, blind and controlled, were practised prior to 1970 whereby water and solids were poured down the borehole using the force of gravity. A more recent development using pumps instead of gravity is termed pumped slurry injection. Several pumped slurry injection programmes have been undertaken in the USA, mainly for the purpose of providing roof support to reduce subsidence (50,51,52); both sand and mine wastes have been used and both wet and dry mines injected.

As separation processes improve and mineral values increase, it is becoming economical to re-process existing waste stockpiles in order to extract the residual ore. Due to the low concentrations of ore in the waste it is usually necessary to process large throughputs and this makes hydraulic conveying particularly suitable for this application. One such example is at ERGO in South Africa (53), where approximately $20 \times 10^6$ tonnes of waste is re-treated annually. In this case each tonne of waste contains only 0.5 grammes of gold. The waste is located at surface sites and pipelines are used to convey material to the central processing plant.

## 5. SHIP LOADING AND UNLOADING

The price of oil has fluctuated considerably in recent years. After the previous large increase in the price of oil compared with that of coal there was a period in which the latter became more important as an energy source for electrical power generation. The growing interest and activity, particularly in the USA, centred on hydraulic conveying of coal and the desire to export has naturally lead to the idea of loading and unloading bulk carriers with coal in slurry form. An additional incentive to coal exporting countries that have a preponderance of shallow coal terminals, which large bulk carriers cannot use, is that slurry loading could well provide a cheaper and more efficient means of loading and unloading these vessels than dredging and maintaining deep-water channels and using existing shore-side loading facilities. Wang (54) indicates that only four coal terminals in the world at that time, were capable of loading bulk carriers of 100 000 dwt plus. It is desirable to use large vessels because the higher the capacity the lower the transportation costs. Furthermore, a proportion of the world's coal terminals are overloaded or at full capacity; South Africa, Canada and the USA have substantial expansions planned or underway and Australian capacity was expected to be exceeded by the mid 1980s.

In the last few years several substantial coal shipping studies have been conducted where the possibility of slurry loading systems have been investigated. For example, the Boeing Pacific Bulk Commodities Transportation System encompasses slurry pipelining of coal from Utah to California and slurry loading of ships 5 km off-shore by twin 1.0 m diameter pipeline with a 0.9 m diameter black water return to give a loading rate of 6000 tph (55,56). Engineering and environmental studies have been undertaken for twin 0.75 m diameter pipelines to carry coal slurry across Staten Island to a coal export terminal (57). This, however, was abandoned a number of years ago.

Several pipelining projects involving coal teminals have been proposed for Europe; the most interesting from the slurry loading aspect is the project based on the Port of Rotterdam, where it was intended that bulk carriers of 250 000 dwt would be handled. Initial cost comparisons showed that investments for a slurry unloading dewatering and inland transport system would be about 25 per cent less than those for a similar sized conventional terminal (58).

Whilst current interest is focused on coal loading and unloading, it should be noted that successful loading of iron sand slurry at Waipipi on the West coast of North Island, New Zealand, has been carried out since the mid 1970s (59). This system employs an S.P.M. (Single Point Mooring) as the discharge point of an off-shore pipeline, which enables bulk carriers to be loaded in deep water 3 km off-shore.

Conventional ship loading terminals generally use one of two basic types of equipment, the travelling gantry loader and the slewing loader (54); both use conveyor belts to transport dry coal to the hold. The main disadvantage of both types is that they require the ship to be located alongside the loader shore position. The cost of moving this equipment off-shore at the head of a jetty, for instance, would be high because of the substantial support structure required. The weight of a 6000 tph travelling gantry may be between 800–1500 tonnes. Conventional unloading systems include bucket wheels, grabs, bucket and closed screw elevators: these all suffer from the same disadvantages as conventional loading equipment.

Future loading and unloading systems lend themselves to slurry based systems; there are three basic arrangements, all of which allow ships to load and discharge off-shore.

The Single Point Mooring buoy (S.P.M.) concept is similar in principle to the system presently used for loading oil tankers. The pipeline extends underwater from a shore base pump station until it reaches the buoy; the pipe can then either run vertically up the buoy support system to the surface if the buoy is of the single anchor leg type, or can be freely suspended. During the loading phase, slurry is dewatered on board and the black water is returned via a second pipe laid alongside the slurry line. To unload, coal would be slurrified using water pumped to the ship down the black water line and then pumped ashore via the slurry line. Using this concept, each collier is required to carry slurrifying equipment on board.

The second arrangement seeks to avoid the latter drawback by carrying slurrifying equipment, such as the 'Marconaflo' system, on a crane barge. The barge moves alongside the collier, unloads the cargo and then pumps the slurry to shore via the S.P.M. The main disadvantage here is that discharging cannot take place in such heavy weather as the S.P.M. only system.

Finally, because of their relative lightness, these slurry systems could be adapted for use on existing off-shore mooring facilities, such as jetties, which are not substantial or large enough to carry conventional coal handling equipment. In this instance slurrifying units and slurry transfer pumps would be mounted on cranes on the jetty head and the slurry and water pipe run along the jetty. Such an arrangement has been suggested by Moore (60), where conversion from oil to coal of a power station was under consideration; here the oil was transported to the station by sea and pumped ashore via a jetty and it was desirable to retain the use of as much of the existing oil unloading facilities as possible for unloading coal which would be brought in by bulk carrier.

An important factor which will determine the design of any coal unloading system is the hydraulic characteristic of the slurry to be handled. These characteristics are largely determined by the slurry concentration and the solids particle size distribution. Three slurry types are of interest; fine particle heterogeneous slurries, low concentration coarse particle slurries and coarse particle stabilised mixtures.

The technology required to convey fine particle heterogeneous (settling) slurry, similar to that pumped in the Black Mesa line, is well-developed and therefore from the conveying aspect this would be the simplest material to tranship initially. The fine slurry is normally pumped at about 45–50 per cent concentration by weight and the low concentration coarse mixture can be conveyed at concentrations of up to about 30 per cent by weight. However these are far too low for economic transportation by ship; to optimise use of the space available in the holds and hence minimise transportation costs, it is essential to reduce the water content as much as possible. In addition concentrations in this range could result in cargo instability during the voyage. It would be practical to thicken the fine homogeneous slurry on-shore and pump it on-board at a concentration of about 60 per cent by weight. Final dewatering on-board would be by decanting and/or gravity drainage which are the simplest and least costly techniques. Coal slurries containing a significant proportion of fine particles are difficult to dewater and drainage tests on small (4 kg) samples of this type of slurry showed that after two hours concentrations of 68 per cent by weight were achieved (61). Although the container used in no way represented a ship's hold, this result indicates the moisture levels which may be reached by this dewatering method. In practice the available dewatering time would be greater than this because holds would have to be filled sequentially to avoid over-stressing the hull. At an average loading rate of 1000 tph of coal a 100 000 tonne vessel can be filled initially in about 50 hours. The sequence of loading would then be repeated after water had been drained or decanted off. The iron sands loaded at Waipipi, because of the good drainage characteristics and high density, contain only about 8–10 per cent moisture when aboard and dewatered. Coarse coal slurries, i.e. with a top size of approximately 50 mm at a concentration by weight of about 30 per cent, dewater more readily than fine material, the little work published to date suggests that moisture levels of 10–15 per cent should be achievable by gravity dewatering alone (61,62,63). Fine particles are easier than coarse to restart after an unplanned shut-down. The risks of blockage can be minimised by sensible system design, controlled start-up procedures, provision of flushing water injection points and stand-by pipes. Experience on the Black Mesa line (64,65) has provided solutions to most of these problems, however, one area of concern is blockage formation at the base of the riser. Land based systems with large vertical lifts can afford the luxury of dumping the riser contents in the event of a shut-down, the method used at the Loveridge mine, however this solution would not be allowed on a submerged pipe. Fine particles can, according to Faddick (61) be fluidised by low start-up flows and then loosened as pump speeds increase; unfortunately published information of the particle size distributions which are amenable to vertical hoisting after shut-down is very scarce.

Stabilised flow mixtures should not present problems after shut-down because they are designed to prevent settlement of the larger solids when stationary. These mixtures have a relatively low moisture content, approximately 30 per cent by weight (28), which means that less dewatering is necessary than with heterogeneous slurries. Because these slurries are intended to be non-settling they cannot effectively be dewatered by gravity or decanting only; in some circumstances their relatively high concentration may make dewatering before shipment unnecessary. It is not known if there would be any particle segregation of this slurry at pumping concentrations due to the movements of the ship during transportation. Another possibility would be to separate the coarse fraction, that is to say the material above about 0.5 mm, by wash screening from the fines either on-board or in a specially equipped barge moored alongside the vessel. The diluted fines would then be pumped back ashore for thickening and re-use in the transportation process. The remaining coarse material would be pumped into the holds as a dilute coarse mixture which could then easily be gravity dewatered. It should be stressed that very little, if any, work to date has been done on large scale dewatering of this type of slurry and the methods described above have yet to be proven.

In most applications it will be necessary to have slurry storage facilities ashore to provide a buffer between ship loading and unloading requirements and the supply (loading terminal) or demand (unloading terminal) of the shore transportation system which would probably, although not necessarily, be a pipeline. The three types of slurries can all be stored in ponds, lagoons or closed tanks. During storage, solids in the two heterogeneous slurries will settle and the supernatent liquid is pumped off, thus a method of re-slurrification is needed; the stabilised mixture has the advantage that it should not settle, this has been demonstrated in small containers but has yet to be extended to very large storage vessels. Similarly, it will be necessary to re-slurrify the solids in the hold before unloading unless they are carried in a stabilised form. Fine coals are relatively easy to slurrify using a water jet system such as the Marconaflo technique. A great deal of experience of slurrifying these mixtures has been accumulated at the Mohave Power Station (supplied with coal from the Black Mesa pipeline) where the slurry is stored, in a partially dewatered state, in lagoons before being recovered by Marconaflow units for final dewatering and then burning (66). This technique is not limited to coal, it is successfully used to slurrify the Waipipi Iron Sands and to date approximately 10 000 000 tonnes have been loaded. Other granular materials, such as copper, lead and zinc tailings and various mineral bearing sands, can be slurrified by using water jets (67).

The published work on slurrifying coarse materials using jets of water seems to be limited to medium scale tests using a few hundred tonnes of material. These tests (63) were conducted by Marconaflo Inc. on a 50 mm × 0 mm coal using a virtually standard Marconaflo unit and resulted in the conclusion that the material could be recovered by this type of equipment. Concentrations in the delivery pipe from the unit of up to 45–50 per cent solids by weight were achieved and material from a 20 m radius (the limit of the test area) was recovered.

Other methods of re-slurrification, e.g. the travelling dredge type reclaimer as used in the Loveridge mine hydraulic lift system, or one of the various feeding arrangements described by Sauerman (68), are available and could be used on-shore to reclaim stored solids. However it would appear that they would not offer such a practical solution for the ship unloading phase as the Marconaflo type of unit because they tend to be bulky and require mechanical drives. The most suitable would appear to be either reciprocating pocket or rotary type feeders, which can feed both slurries or dry solids into a pipeline. One disadvantage is that they can require a supplementary system to supply them with solids or slurry.

One area of concern is that of degradation of coarse particle slurries during loading and unloading. Faddick (61) comments that in general this issue has been side stepped and supports the view that the short runs, use of large pumps and the wide size distribution of the coal will suppress degradation to such an extent that it will not be a major problem.

There are four main items of equipment associated with slurry loading and unloading not so far discussed in detail, where development will be required namely S.P.M., submarine pipe, flexible hoses and ship modifications.

The S.P.M. which was first used in 1971 for oil transfer duties has subsequently been adapted for iron sand slurry and on the Waipipi system it has performed well after initial problems caused by bearing seizure. The experience there suggests that this type of hardware would be able to handle coarse coal slurries effectively. On some proposed coal installations it will be necessary to build larger versions of the S.P.M. to accommodate higher loading rates and lower density of the solids.

The technology associated with the submarine slurry pipeline will be very similar to that for conventional oil and gas pipes. No additional problems are posed except the pipes will be thicker walled and carry a more dense and abrasive medium.

Flexible hoses are used to connect the loading vessel to the buoy or barge. These will be of the order of 0.5 m in diameter and will have to suffer considerable abuse from contact with the buoy and ship as they do in an oil loading application. The additional stresses caused by pumping slurry will include internal abrasion and the possibility of higher dynamic forces on the bends when the density of slurry is higher than that of crude oil.

Most existing marine carriers are specialised vessels (container ships, bulkers, tankers) and therefore restricted to the type of cargo they can transport; however one of the fastest growing class of merchant vessel is the combination tanker, a vessel equipped to carry dry and liquid cargoes, e.g. ore/oil and ore/bulk/oil. This ability to carry different types of cargoes offers a great increase in flexibility and profitability in trading; these advantages would also apply to ships which could carry slurry as one of their cargoes (provided that slurry transhipment becomes widespread). Currently there is a severe excess of merchant tonnage which has resulted in some interest in conversion of single cargo vessels such as tankers into combination carriers with the ability to carry slurries. An oil tanker, however, is not suitable for carrying slurry and extensive modifications must be carried out; the most major of these are as follows:

● It would need a slurry loading/unloading system (if it was to be used in conjunction with a S.P.M.) which could also be integrated with an oil loading system.
● Either a decant system to pump off surface water as the solids settled (this would also have to be able to compensate for the different levels in the hold during the filling phase) and/or a screen and drain system in the bottom of the hold to remove water which percolates through the solids would be required.
● It would be necessary to strengthen the hull to accept a denser cargo. and slurry stability requirements during transit may dictate changes from conventional hold configuration.

## 6. NON-AQUEOUS SLURRYING MEDIA

The objective of this section is to examine the feasibility of fluids other than water as slurrying media. Also excluded are gaseous vehicles, as pneumatic transport is a well established area in its own right. The incentive to consider non-aqueous liquids stems from three basic considerations:

● An adequate water supply may not exist local to the mine or primary plant.
● Dewatering the product for its end use may be costly and inefficient.
● There may be difficulties facing environmentally acceptable effluent disposal.

For a variety of reasons, coal is the only payload for which non-water slurry transport has been seriously proposed. Its end use, for instance, is frequently as a fuel, where it is seen as a preferable alternative to petroleum. Any water present in a coal furnace or boiler must be vapourised, imposing an energy penalty. Another feature of coal is that it can be processed to produce substances such as methanol, synthetic fuel and carbon dioxide, any of which might then be used to transport the coal and subsequently used or sold in their own right. The total payload would then be the entire contents of the slurry line, with obvious economic benefits.

Most of the literature on non-water coal slurries is American. United States operators have had encouraging experience with the 272 mile Black Mesa coal-water line from Colorado to Nevada. The country is a major coal producer, conscious of its need to reduce its dependence on imported oil. Therefore a number of slurry lines have been proposed, each intended to provide a large and reliable coal supply to the end-user for the design life of the pipeline (71). Many of the inland coal producing regions, especially those in the belt from Wyoming to New Mexico, experience a low annual rainfall, inhibiting large-scale installation of water based slurry lines. Santhanam *et al.* (72,73) have claimed that opposition to water rights acquisition or transfer has been the chief obstacle to water slurry line approval for the Western States.

Figure 1 illustrates some of the basic process options available to the user of one of the three major coal slurrying media. Both carbon dioxide and methanol can be produced from coal, and this source is assumed on the flow diagrams. The carbon dioxide must be separated from the coal at the end of the pipeline (Figure 1a); the ease of such separation is one reason for proposing $CO_2$. At this point it may be marketed (open loop) or returned for reslurrification (closed loop). Similar alternatives exist for a methanol coal system (Figure 1b) with the additional choice of burning the methanol with the coal. Coal-oil slurries (Figure 1c) are the most difficult of the three to separate, thus finding their strongest application as direct fired fuels.

The text of this section has been subdivided into sections on oil, methanol and carbon dioxide media, followed by a discussion of the relative economics of the various processes. The largest proportion of the literature on non-water media concerns coal-oil slurries, conceived either as means of transporting coal on existing crude oil lines, or as mixtures available for direct firing into boilers. Some attention will be paid to the modifications required to convert an existing pipeline to slurry use. Coal transport in methanol is also considered in two categories, the simple mixture of ground coal and methanol, as in water slurries, and 'methacoal', a specially prepared thixotropic product involving some surface chemical reaction between the coal and methanol. There have been a variety of rheological models proposed for methacoal. They seem to fit existing experimental data equally well, yet predict significantly different design parameters when extrapolated to full scale. Carbon dioxide slurrying, despite some promising advantages, is the least researched of the three technologies and rheological information is correspondingly scarce. It will be discussed in relation to a system proposed for the production and transportation of naturally occurring carbon dioxide.

Before proceeding to the detail of the above three slurry media, it is worth mentioning a concept proposed by Bradley (74) for the transport of coarse coal slurries, with a maximum particle size of order 25 mm (1 inch). The transport medium is a high concentration water-in-oil emulsion. The purpose of a high concentration of internal aqueous phase is to impart structure to the emulsion, generating a sufficiently high yield stress to support large coal pieces. The choice of a suitable oil and surface active agent allows the attainment of up to 95 per cent volumetric concentration of internal phase. A yield stress of 10 Pa was obtained for an internal phase concentration of 82 per cent. This value of yield stress was the minimum recommended by Bradley for coarse coal transport. Although the grinding step and much of the dewatering could thus be eliminated, the emulsion would have to be returned to the coal source. There would be the dual risk of emulsifier absorption by coal particles and of inversion to a water-in-oil emulsion, with dramatic losses in yield stress. The concentration

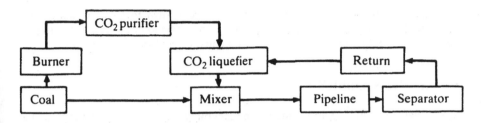

a. Carbon dioxide carrier

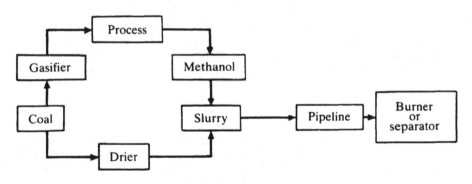

b. Methanol carrier

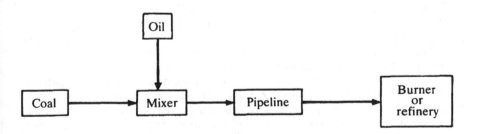

c. Hydrocarbon carrier

**Figure 1.** Sample flowsheets for non-water coal slurrying media

of fines would also have to be controlled as these are known to have surface active properties. Bradley suggested that if emulsion losses could be kept below 5 per cent per cycle, the process would compete favourably with conventional water slurrying.

## 6.1 COAL–OIL SLURRIES

These can be naturally divided into crude oil and fuel oil based mixtures. The purpose of a crude oil slurry is to maintain the capacity of an existing crude oil pipeline during depletion of the field that it serves. At the end, the oil and coal could either be separated or refined together in a specially adapted refinery. At present, such a refinery would require a considerable advance in chemical engineering, as coal and oil are not compatible feed-stocks. Coal and fuel-oil mixtures, on the other hand, are designed for direct combustion as a step towards the replacement of oil by coal as a fuel.

Sandhu and Weston (75) have conducted a feasibility study into the conversion of existing crude oil and natural gas lines to either water or oil slurry use. The thickest walled pipe sections lie just downstream of the pumping stations. Closer spacing of the pumping station, or a reduction in the flow rate, contribute to lower maximum pressures, hence wall thickness requirement. For a hypothetical line of 610 mm diameter, 1250 km long, and carrying 18 Mta (million tonnes per annum) of crude oil, the study recommended the replacement of 55 km with thicker walled pipe. For the coal-water case the resulting throughput of coal was estimated to be 8.6 Mta at $12/tonne (1980 prices) compared to $49 for a new coal-water pipeline. A similar coal-oil system came to 5.9 Mta at $16/tonne coal. The cost per tonne product would be reduced if the payload value of the oil were considered, partly offset by the greater difficulty of de-oiling then watering coal slurries. Furthermore, the value of the separated oil would depend strongly upon its freedom from coal. Another factor not included in the quantitative study, however, was the reduced impact on the pipeline of an oil-based slurry. The oil would not corrode the pipe, as might water, and its higher viscosity would provide better lubrication against wear. Sandhu and Weston concluded that both C/W and C/O systems were technically feasible and looked economically attractive in comparison with a new system.

Clearly, efficient separation of the coal and oil could increase the end-use options for a crude oil slurry. Smith *et al.* (76) considered both separation and preparation problems. For reasons of safety, they recommended wet grinding of the coal, followed by thermal drying and mixing with the oil. If the grinding mill could be sealed against the loss of light crude fractions, the coal could be ground with the oil. This factor also required the separation plant to be sealed against evaporation. Attempts at bench-scale filtration resulted in membrane blockage by crude oil waxes, a problem that could not be cured by heating. Centrifugation at 3000 r.p.m. was more successful, with an evaporation loss of 2.4 per cent, an oil content in the cake of 5–30 per cent and an effluent solids concentration of 0.2–5.9 per cent. The coking properties of the metallurgical coal samples tested were not affected by retained oil, and the quality of combustion coal was improved thereby. The presence of coal in oil, however, reduced its value considerably, principally by raising the Conradson

Carbon Residue (CCR), which tends to smother the cracking catalyst. Either the fines concentration would have to be controlled at source, an improved effluent separator developed, or a means be found to cope with a high CCR, before crude oil/coal slurries can be considered a commercial option.

Coal-oil-mixtures (COM) are based on heavy fuel oils, loaded to a weight concentration of 40–50 per cent pulverised coal, about the same as a crude oil or water based slurry. They are designed for direct firing into boilers, hence require no separation. Fuel oils, having had the light ends removed by fractionation, impose a far less stringent vapour sealing requirement than is typical of crude oils. Pan *et al.*(77) report the results of an American Department of Energy (DoE) COM development programme. Provided suitably wear resistant inserts were used in the burning nozzles, the mixture maintained at 70 °C, and provisions made to prevent plugging, 50 per cent COM could be burned with 97 per cent carbon conversion efficiency.

The Japanese COM programme (78) is directed towards increasing the coal import level from 58 Mta in 1977 to 101 Mta in 1985 and 144 Mta in 1990. The Electric Power Development Company (EPDC)/industry (e.g. Mitsubishi) programme has focused on COM preparation and transportation, considering the combustion technology to be well advanced. Drawing from both the American and Japanese research results, an economic study by Eberle and Hickman (79) recommended COM for boilers above 100 MW thermal.

Another end use option for COM, proposed by Kreusing and Franke (80) is direct injection into a blast furnace. A simplified flow diagram appears in Figure 2, indicating a heating requirement for both oil and slurry storage, as well as agitation to produce and maintain an homogeneous suspension. The main reason for agitated storage was the coarse coal (upper size limit 5 mm) to be fed to the furnace. The economics of such a system is often determined by the balance between particle size, suspension stability and pumping viscosity. Ideally, the system should exhibit some thixotropy, thick enough at rest to prevent settling yet a low apparent viscosity after the application of shear.

The rheology of coal-oil mixtures is not yet fully established, depending as it might on coal type, oil type, additives, temperature, particle size distribution (PSD) solids concentration and storage history. Munro *et al.* (81) looked at the effect of temperature, concentration and average particle size on the rheology of essentially monodisperse slurries of subbituminous coal. Their results indicate Newtonian behaviour and little dependence of relative viscosity upon temperature for concentrations up to 40 per cent. Particle size, prepared for testing in batches of 45–75 $\mu$m, 75–90 $\mu$m and 90–104 $\mu$m respectively was found to have no significant rheological effect, a result probably due to the narrow range tested. The Newtonian features, and the steep viscosity increase of 40–55 per cent solids, are in good agreement with an analysis by Ackermann and Hung (82) for suspensions of rigid spheres in a viscous liquid at moderate shear rates, neglecting chemical and electrical interactions. At high shear rates they predicted a dilatant (shear thickening) behaviour resulting from particle collisions, unlikely to apply to coal-oil slurries. More importantly, they predicted a viscosity minimum if coarse particles were mixed with fines (at least ten times finer) in a ratio of about 3:2.

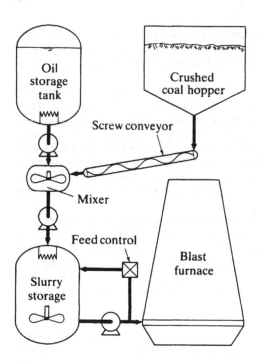

**Figure 2.**  Flow diagram for injection of COM into a blast furnace (80)

The particle size distribution was one of the parameters examined by Kreusing and Franke (80), a selection of whose results have been plotted on Figure 3. The concentration range of 40–44 per cent demanded high resolution at low shear rates and a large range of measurable shear stress, revealing a non-Newtonian (pseudo-plastic) behaviour for the thicker mixtures. Significantly lower viscosities would appear to be a feature of wider PSD (curves B and C), supporting the idea that the fines reduce interaction between coarse particles, thinning the slurry if the fines themselves are not too concentrated. The inserted bar graph supports the idea that a large ratio of maximum to minimum particle size tends to produce thin slurries. Otherwise, it might be concluded that coal type (anthracite or lignite) was the determining factor. This possibility should not be ignored, as lignite is both less dense and more porous than anthracite, and thus would occupy a larger volumetric proportion at the same concentration by weight. The absence of any indication of yield stress might have prompted the authors (80) to recommend agitated storage. The temperature for the data on Figure 3 was 50 °C, and the viscosity of order 10 Pas (10 000 cP), clearly difficult to pump. The recommended operating temperature was 80 °C, high enough to effect a fourfold drop in viscosity, yet not sufficient to vapourise the moisture content of the coal.

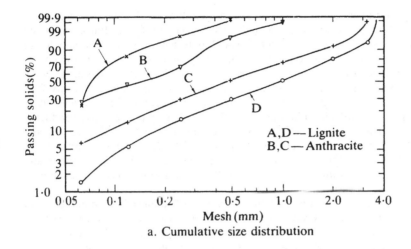

a. Cumulative size distribution

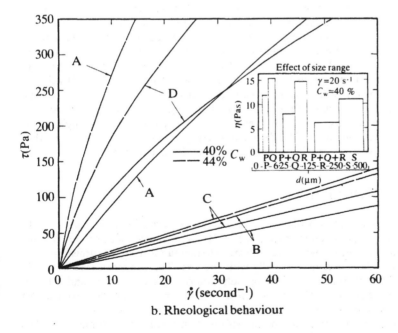

b. Rheological behaviour

**Figure 3.** Effect of particle size distribution on coal slurry rheology

Other rheological studies of coal-oil-mixture have been done by Nakabayashi *et al.* (83), Yagamata *et al.* (84), Ekmann and Klinzing (85), Bhattacharya (86) and Battacharya and Barro (87). Nakabayashi *et al.* (83) conducted pipe flow tests indicating essentially linear variation of pressure drop with velocity, from which they inferred a Newtonian viscosity. Transition to turbulent friction behaviour was found at a Reynolds number of 3000, supporting the Newtonian theory. No settling behaviour was observed under any flow condition, despite a maximum particle size of 5 mm and an average of 1 mm, a 30–95 °C temperature range and a velocity minimum of 0.23 m/s. Yamagata *et al.* (84) conducted rotational viscometer as well as pipeflow tests, and found both time dependence and Bingham plastic behaviour. However, the fuel oil itself was found to exhibit these characteristics, emphasising the importance of testing the slurry medium in its pure state. Minimum pressure loss was found at a fines (under 20 $\mu$m) concentration of 20 per cent, in line with the foregoing discussion.

Ekmann and Klinzing (85) examined settling behaviour of stabilised COM in the 50 per cent concentration range. They recommended considerable care in the choice of surface active agents, as these could lead to agglomeration of coal particles and settling of the slurry. They conducted the sedimentation tests with ultrasonic probes, avoiding the difficulties of flow visualisation. Bhattacharya (86) investigated the effect of storage, for up to ten days, on coal-oil suspension rheology. Over this period the yield stress increased from zero to 250 Pa for a 36 per cent lignite suspension. Half this increase was observed in the first fifty hours (one fifth total time). Bhattacharya and Barro (87) investigated the phenomenon further with a capillary rheometer and found storage history to have a marked influence. For example, a sample of suspension stored undisturbed for 24 hours exhibited over three times the viscosity shown by a sample disturbed twice for testing. One apparent contradiction worth resolving is that between the yield stresses quoted by Bhattacharya (86) and the pure Newtonian behaviour indicated on his shear-stress strain-rate plots. Nevertheless, it can be concluded that undisturbed storage can increase the viscosity, whether or not actual settling occurs. Therefore some means of agitation ought to be provided in all COM storage tanks. The degree of stirring required is a matter for further research, as is a method for preparing a mixture that needs no agitation.

## 6.2  THE METHANOL VEHICLE

Like crude oil, methanol owes its attractiveness as a slurry vehicle to its value as payload. In this capacity it has been recommended by Othmer (88) as a substitute transport medium for natural gas from Alaska. The gas would be converted to methanol at source, utilising the 22 per cent carbon dioxide occurring naturally with the gas. The proposed system would eliminate the need for compression, cooling, dehydration, gas separation, and a large, insulated, thick-walled pipeline. Othmer claimed that the methanol system would cost under half that of a gas pipeline per unit product.

Gödde (89) examined the use of methanol with a 25 m test loop of 80 mm pipe. His interest in the methanol alternative was derived from its low freezing point ($-98$ °C), avoiding the need to insulate against frost, as well as its payload value. He tested three different size distributions ranging from a coarse mix with over 11 per cent above 3 mm to a fine mix with over 49 per cent under 45 $\mu$m. The range of carrying media included water, methanol and mixtures of water in methanol (40:60 and 10:90). The methanol slurries gave substantially higher pressure losses if prepared with the coarse coal mix. Although the results did not indicate a critical settling velocity, it is likely that settling effects were partly responsible. Methanol's low viscosity and specific gravity (0.55 cP and 0.79 at 23 °C) would promote both settling and inter-particle collisions, both contributing to the total pressure loss. For a 300 $\mu$m average particle size the settling velocity for a 50 per cent slurry is so high (2.7 m/s) that the head loss becomes prohibitive. The problem can be solved by cutting the average size to 170 $\mu$m, at some cost in grinding and end use separation.

Retaining the concept of a 'conventional' methanol-coal slurry, Aude and Chapman (90,91) considered not only the need to maintain flow velocities above the critical settling value but also to ensure turbulent flow. The concept is illustrated in Figure 4a, for which laminar flow is assumed to lead to settling, usually the case for an unstabilised slurry. Within the turbulent flow regime, settling can still occur if there is insufficient lift to maintain the top size in suspension. Various criteria besides the minimum pressure drop can be used, and Aude and Chapman selected the specification of a maximum top size concentration ratio of 0.75 from the top of the pipe to the centre. They subjected methanol-coal slurries to rheological tests, assumed Bingham plastic behaviour, and predicted the transition velocities from a theory by Hanks (92). Settling velocities appear to have been inferred from static sedimentation tests, though this is not clear from the paper (90). Figure 4b shows results for two grinds with respective average size (50 per cent retention) of 200 and 120 $\mu$m respectively. It is plain that the finer grind can be transported at a much lower minimum velocity but its concentration is limited by the transition to laminar flow. Using grind 'B' as a basis, the authors estimate a pumping energy requirement 32 per cent greater for a methanol than for a water slurry, mainly due to a 15 per cent higher velocity requirement. They offset this against a 222 per cent estimated increase in payload, based on the weight of methanol carried. No mention was made of the production and separation costs of methanol in relation to its market price. Nevertheless, 200 per cent is an attractive increment.

A process that seeks to bypass the above settling and transition difficulties involves the mixing of fine dried coal with methanol, leading to 'methacoal', claimed to be a stable, yet readily pumpable, coal suspension by its principal inventor, Keller (71). The stability is said to be such as to avoid any need for pipeline slope and velocity restrictions. Restarting would be easier, obviating the normal storage requirements at both ends of the line. The methanol would be produced from the coal with minimal refining, leading to the fuel grade known as 'methyl fuel'. The choice of a low-grade, low-cost carrier means that methacoal can be burned directly without the cost of separation or the loss of a valuable product. This is a useful option considering that some chemical reaction between the coal and methanol is involved in the methacoal process, which could make efficient separation difficult. Keller also suggests that the current (1979) half a million tonnes annual methanol production overcapacity in the United States could be diverted to methacoal, particularly as many

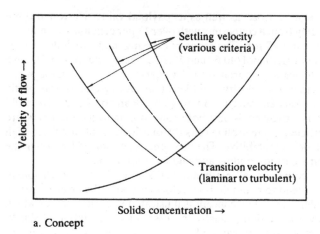

a. Concept

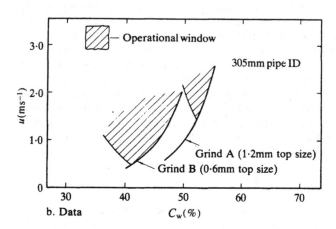

b. Data

**Figure 4.** Velocity and concentration limits for MeOH-coal flow (90,91)

older plants are not suited to high grade production. The essential process sequence for methacoal is sketched in Figure 1b, where it is distinguished from conventional slurrying principally by the drying step.

Methacoal rheology is interesting for a number of reasons. Firstly, if it is stable in laminar flow the power requirement is directly proportional to the apparent viscosity. Secondly, if the effective yield stress is sufficiently high it may be possible to load the slurry with a proportion of coarse coal without the risk of settling. Thirdly, the transition velocity may be of interest so that the high head losses associated with turbulent flow can be avoided. On the other hand, if slurry stability is in question turbulent flow may be required as a safety measure. Therefore, prediction of apparent viscosities over a wide range of shear rates, concentrations and size distribution is required.

Figure 5 illustrates the forms of the various models assumed for methacoal rheology. Each has special advantages quite unrelated to its modelling accuracy. For instance, the power law parameters, consistency ($m$) and index ($n$), can be readily extracted from a log-log plot of shear stress against strain rate. Such a plot is also convenient for representing wide ranging data. The Bingham plastic model, on the other hand, incorporates a yield stress ($\tau_y$), allowing some assessment of slurry stability independently of actual settling tests. It also allows a relatively simple regression procedure. The Casson model, however, seems to be suited to substances with a yield stress, yet whose data traces an obvious curve on a linear plot. It is possible to add a third degree of freedom, adding a yield stress to the power law. This model, known as the Herschel-Bulkley model, was claimed by Darby (93) to be unsuccessful for methacoal. As the Bingham plastic and power law models were found to be acceptable (93), it is worth noting that these are special cases of Herschel-Bulkley with an index of unity and a yield stress of zero respectively. There may be statistical difficulties regressing three degrees of freedom around multiparametric data. Before leaving Figure 5,

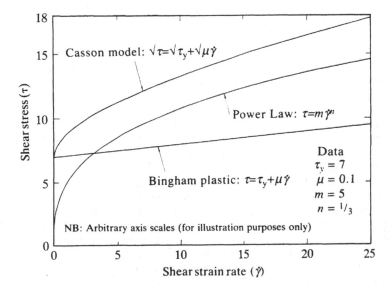

**Figure 5.**   Rheological models used and proposed for methacoal (93)

it is instructive to note the parallelism of the Casson and power law models between strain rates of 5 and 15. Assuming data, with very little scatter, to be available within these limits, the models might be expected to fit them equally well. Nevertheless, the practice of extrapolating either model up or down could lead to serious errors. As most rheological data possess a great deal of scatter, the importance of a wide range of shear rate cannot be over-stressed.

One of Darby's (93) main interests was to correlate apparent relative viscosity ($\eta_r$) against the volume fraction of solids under 30 $\mu$m ($\phi_{30}$). He hypothesised that surface chemical reaction between fine coal particles and methanol created an adsorbed layer of liquid around each particle, neutralising its buoyancy and separating it from other particles. The resultant absence of settling or collisions would serve to lower the apparent viscosity whilst maintaining stability. The results for lignites dried to 5 per cent and 15 per cent appear in Figure 6. The range of experimental shear rates was 2–120 inverse seconds. Although the Casson model was stated to fit the data well, only the Bingham plastic and power law parameters were quoted over a fines concentration range ($\phi_{30}$) of 28–45 per cent. Over this range the power law gave a rising consistency ($m$) and a falling index ($n$), whilst the Bingham plastic indicated a rising yield stress ($\tau_y$) and minimum plastic viscosity ($\mu$) at an intermediate $\phi_{30}$. The curves of Figure 6 represent an extrapolation of these findings over the 0–50% range of fines concentration. The curves for lignite with 5 per cent moisture show a constant viscosity minimum between concentrations of 20–30 per cent fines, with sharply rising trends either side of the minimum. The 15 per cent curves show less sensitivity to fines concentration, as might be expected from Darby's hypothesis, but the optimal $\phi_{30}$ value depends strongly upon the assumed rheological model. For instance, the power law correlation at 10 s$^{-1}$ seems to recommend no fines at all, returning one to a conventional slurry!

Some of the above difficulties may be overcome by the use of the Casson model, as proposed by Hanks (94) in his development of a theory for the transition velocity for such fluids. He indicated that the model has been shown to be a special case of a more general theory based on bond formation and breaking for agglomerating suspensions. The apparently surface active features of methacoal might put it in this class, along with paint, blood and printer's ink. Laminar flow is to be desired to minimise pressure loss, however turbulent flow may be required to ensure homogeneity in the event of agglomeration. There appears to be no data on the settling behaviour of methacoal, or other property changes resulting from undisturbed storage.

Ethanol has also been considered as a carrier, with the advantage over methanol of a lower toxicity, offset by greater difficulties preparing it from coal. The rheology of 'ethacoal', a mixture produced by grinding the coal and ethanol together in a ball mill, has been studied by Davis *et al.* (95). They concluded that the mixture was pseudoplastic (power law) in contrast to methacoal which was Bingham plastic! They tested coal concentrations of 52–67 per cent by weight and particle sizes of 150–420 $\mu$m. They found property changes during the wet grinding in ethanol, the mixture changing from a paste to a fluid and back to a paste as grinding progressed. They attributed this to absorption of the alcohol to the particle surface. They did not mention having predried the coal, nor did they

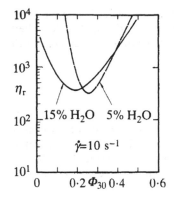

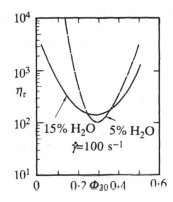

a. Bingham plastic

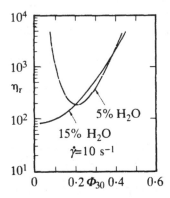

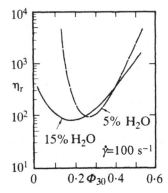

b. Power law

**Figure 6.** Viscosity minima for various volumetric fines concentrations (93)

report any symptoms of a chemical reaction. All suspensions were reported to be stable within the concentration range 52–67 per cent, addition of more than 48 per cent ethanol destabilised the system. The apparent viscosities (at 500 s$^{-1}$ shear rate) at 52 per cent and 67 per cent were 1000 cP and 15 000 cP, a fifteenfold increase. On the basis of their power law correlation the authors made a prediction of pressure loss for various large pipe diameters and recommended a minimum to 760 mm for a concentration of 57 per cent (optimal) and a throughput of 2 Mta of coal. Nevertheless, the strain rate at the wall under these conditions was estimated to be of order 2 s$^{-1}$, less than half the minimum applied in the rheological study. Bearing in mind the high apparent viscosities at low strain rates further work in this regime is obviously required. The rising sphere rheometer, described by Bradley (74), is designed to measure yield stress directly (zero strain rate) and may well provide the answer.


## 6.3   CARBON DIOXIDE

In contrast to COM and some end uses of methacoal, carbon dioxide owes its attractiveness as a slurrying vehicle to its ease of separation from particulate coal. The technology is yet in its infancy and being developed in the USA by a collaboration between Arthur D Little Inc. (ADL) and WE Grace and Company. Santhanam (96) outlined the prospective advantages of $CO_2$ as follows:

- It can be recovered from burning coal without consuming water, unlike methanol or ethanol. Stack gas and natural sources also exist.
- Liquid carbon dioxide is at least one fifteenth as viscous as water, allowing it to be loaded to higher coal concentrations (80 per cent vs 50 per cent).
- Very little $CO_2$ is absorbed by the coal, making it easy to separate, and avoiding agglomeration or swelling problems.
- Separation can be achieved by simple evaporation of the carrier liquid, yielding dry ground coal and purified $CO_2$.
- Carbon dioxide, in large quantities, could become a marketable product for such end uses as enhanced oil recovery (EOR).


The areas most in need of further work were listed as wear rate determination for such a low viscosity carrier fluid, and emergency response in the event of a line rupture. The results to date from ADL research have included a preliminary rheological study of highly concentrated (>80 per cent) slurries, indicating that they can be pumped, and a look at separation procedures, which appear to have been simpler than expected.

Carbon dioxide is not a medium commonly handled on the same bulk scale as have been oil (crude or refined), natural gas (liquefied or compressed) and methanol (fuel grade or chemical). It must be pressurised to maintain it in a liquid state, cooling alone being insufficient. It is acid in the presence of water, traces of water in a $CO_2$ system presenting a more severe corrosion problem than traces of $CO_2$ in a water system. With this in mind, it is worth looking at the design of a production unit for naturally occurring $CO_2$, described for the Sheep Mountain reservoir in Colorado by Renfro (97). Assuming the well to be a good producer, the high pressure in the reservoir would be blown down to produce a two-phase mixture at the well head. This must then be heated to enable the separation of connate water from a now superheated carbon dioxide. The superheat must be sufficient to avoid hydrate formation (13°C) as well as $CO_2$ condensation at any stage of the treatment process. Final stages are compression to the operating pressure followed by cooling to ambient temperature, yielding some heat for recovery and a fluid density sufficiently high for economical mass flow rates. The pressure is well above critical, so that pipeline losses and elevation differences (over 2000 metres in this case) can be handled without internal vapourisation. The Sheep Mountain Unit is within one of the coal producing areas of Colorado and may well provide an attractive starting point for a prototype slurry plant.

If the carbon dioxide is to be produced from the coal itself, purification processes are required in addition to dehydration. These are sketched on the upper half of Figure 7, a diagram based on the flowsheets and recommendations of Santhanam (72,73,96). Sulphur dioxide and water vapour would have to be removed early, as these would be absorbed by any alkaline recovery agent for $CO_2$. An alternative would be to use a selective solvent for carbon dioxide, such as propylene carbonate, though this is a relatively new technology.

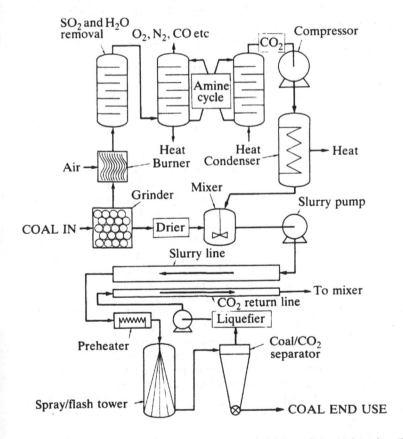

**Figure 7.** Flowsheet for preparation and separation of $CO_2$-coal slurries based on (96)

Carbon dioxide absorption has to date been focused on removing it as a contaminant (e.g. natural gas sweetening) rather than purifying it as a product. One of the major barriers, for example, to the use of flue gas as a $CO_2$ source for EOR are the logistics of treatment and transport on the scale required. If carbon dioxide slurry systems are to make an impact on the EOR market, the scale of operation must be sufficient to justify the effort. Such a market is essential to the 'open loop' (96) system, avoiding the need for $CO_2$ recirculation. A further requirement is for a suitable oilfield to be nearer the coal delivery station than the length of the slurry line. Otherwise it would be cheaper to recycle the $CO_2$ for reslurrying.

The requirements at the end of the pipeline are illustrated on the lower half of Figure 7. The slurry must be preheated so that the carrier can be evaporated without the formation of dry ice. The resulting pneumatic suspension would be passed through a conventional filter or cyclone separator to remove the coal and reliquefy the $CO_2$ for closed loop recycling or open loop marketing. Santhanam suggested locating the $CO_2$ production facilities at the downstream end for closed loop operation, saving the cost of burners and liquefaction plant at the source point. This would, however, make the entire slurry line dependent upon reliable production at the downstream end, so that downtime at the coal consumption plant could jeopardise the supply of slurry vehicle. Before commissioning the system, coal would have to be transported to the end point by some other means, a process that may have to be repeated on a large scale in the event of a failure. Finally, producing $CO_2$ at the end point would be more costly to convert to an open loop should a market be found for $CO_2$ some time during its operation. Either the return line would have to be retained, or new production plant built at the upstream end.

Santhanam quoted 500 p.s.i. (35 bar) as a minimum pressure for a $CO_2$ slurry line. The saturation temperature for $CO_2$ at this pressure is about $0\,°C$, requiring some form of insulation or supercooling for the line. This does not include any allowance for elevation or hydrodynamically induced cavitation. The quoted pressure may yet be feasible, as Coulter (98) has indicated for a system with rather more stringent requirements. He conducted a feasibility study for a slurry system with frozen crude oil suspended in LNG. It would be located in North West Canada over a distance of 1000 miles (1610 km). The upper temperature and lower pressure limits were respectively $-84\,°C$ and 48 bar. He concluded that a 750 mm pipe with 13 mm of insulation would be suited to the design LNG flow of 85 million standard cubic metres per day. A turn down ratio of six was chosen to give the system some flexibility. Pumping station spacing is determined by the maximum flow rate. Cooling station spacing, however, is determined by the minimum flow rate, for which environmental heat input predominates. Although the cooling power requirements may be low, spacing is influenced by the rise in slurry temperature over a given length of line. Relevant to both LNG and sub-ambient $CO_2$ lines is the question of cooling in the event of a pumping station failure. The study suggested the idea of pumping concentrated slurries, though the rheological properties of frozen oil in LNG have not been established. The feasibility study was based upon water-granite slurry behaviour, the solid to liquid density ratios being similar at 2.54 and 2.65. It may be feasible to subsitute coal for frozen oil, as Gödde (102) has suggested.

Arthur D Little and W E Grace decided to pursue the commercialisation of carbon dioxide slurrying on a consortium basis. For this reason Santhanam *et al.* (72,73,96) published no quantitative data, other than to state that 75 per cent coal-$CO_2$ slurries produce a similar hydraulic resistance to 50 per cent coal-water slurries. From this they argue that a given pipeline could carry 100–300 per cent more coal in a $CO_2$ slurry than in water. Despite its attractions, the coal-$CO_2$ slurry concept cannot be evaluated further without more specific information on such matters as grinding/drying requirements for the coal and the resulting transition/settling velocities for the slurry.

## 6.4 ECONOMICS

Comparative economic studies of non-aqueous slurry media have been limited to oil and methanol. Souder (99) used a 500 mile (805 km), 24 inch (610 mm) pipeline carrying 10 Mta of coal in 50 per cent water slurry. The transport cost per energy unit of coal was assessed at 16.8 per cent more for a methanol line than for water, whilst the increment for No. 2 fuel oil was only 13.5 per cent. If the energy content of the carrier were added to that of the coal the resulting transport costs were less for methanol and oil than for water, by respective decrements of 38.7 per cent and 58 per cent. However, adding the cost of carrier to that of transport meant that a methanol system would cost 257.5 per cent more than water and fuel oil an extra 206.8 per cent. Souder concluded that the use of a non-water medium would only be justified if the medium were separated from the coal at the end use point. It is significant that Souder did not take dewatering costs or coal losses into account. For instance, it is widely held (77) that conversion of an oil-fired power plant to COM is much cheaper than complete conversion to coal. In such a case COM would be the most economical transport medium, provided the coal and oil were produced in the same place. Thus oil refineries might be optimally located near coal mines. Concerning methacoal transport, it may be that the unit cost of methanol might be dramatically reduced by locating a fuel grade plant at the minesite. Nevertheless, Souder's increments are large enough to suggest further research into the separability of methacoal.

Ghandhi *et al.* (100) compared five candidate systems, namely, conventional water slurries, high concentration water slurries, coal-oil (No. 2 fuel grade), water slurries of upgraded lignite, and water slurrying with water recycle. They found conventional slurrying to be the best for most of the cases considered. These included a 1600 km and an 80 km line, each with coal throughputs of 2–18 Mta and water available at $0.40 m$^{-3}$ and oil at $130 m$^{-3}$. The coal-oil transport system became competitive when the coal throughput was reduced to below 7 Mta for the long pipeline. For the short pipeline it was by far the most expensive. At the top throughput (18 Mta), oil was the most economic medium for the 1600 km line provided the cost of water was raised beyond $4.50 m$^{-3}$. For the purpose of the study the coal was assumed to retain 10 per cent oil after separation, though it is not clear how the oil was evaluated as a payload. If it were assessed on the basis of its energy value alone, it would make more sense to fire the mixture directly than to separate it, improving its economics relative to water slurries.

Banks (101) compared direct-fired coal-oil and various modes of methanol slurry operation, both set against the Black Mesa line as a reference. Of the methanol systems, direct-fired, separated-returned and separated-sold, only the separated-sold process emerged as economic. As he found direct-fired coal-oil to be uneconomic, Banks concluded that coal-oil transport could not be justified until the separation technology was further developed. From the favourable result for separated methanol, he outlines some of the end uses to which it could be marketed. These included turbine fuel, automotive fuel and chemical feed stock. Of these applications, the first could probably use 'methyl fuel' crudely separated from methacoal, the degree of required separation as yet an unknown quantity. The use of pure or blended alcohols in vehicles is not yet widespread and would require a more complete removal of particulate coal. The highest level of purification is needed for chemical grade methanol and this channel would seem to defeat the purpose of a low-cost methyl fuel plant at the minesite. Banks referred to a counter example claimed by Keller to show direct-fired methacoal to be economic. No published reference was given.

More general discussions of the place of non-water vehicles in the context of coal slurrying are provided by Gödde (102) and Lee (103). Lee (104) also includes an economic analysis of coal-water lines and a general treatment of non-water media without bringing the two together. In (103) Lee makes the point that coal size can be reduced but not increased, worth remembering if the transport medium requires a finer grind than does the end use. He also mentions coal-water-mixtures (CWM), consisting of high concentrations of ultrafine coal for direct-firing in boilers. The idea arose from the oil crisis years of 1973/74. Additives are required to maintain a stable suspension and the system viscosity is at least 20 times that of a Black Mesa slurry, which at that time appeared to make it unsuitable for long distance transport. Presumably, a Black Mesa type slurry could be ground to a CWM at the end-use point, which may be a coal refinery requiring water as an ingredient. A typical example of such a refinery is a methanol plant, which could be feeding another slurry line. Lee (103) concludes that none of the individual systems should be considered in isolation, rather that each should be integrated into a total pattern of economic and environmentally acceptable coal transport.

## 6.5   CONCLUSIONS ON NON-AQUEOUS MEDIA

Non-water slurry media are available for coal transport, though none has yet been tested on a pilot scale. The major candidates are crude oil, fuel oil, methanol/ethanol, and carbon dioxide. Other possibilities are water-in-oil emulsions, for coarse coal, and LNG, originally proposed for frozen oil transport. Basic conclusions concerning oil, alcohol and $CO_2$ media are listed below:

● More work is required on the separation of coal from oil before crude oil can be considered an economic transport medium. At present, too much of the value of the crude is likely to be lost through contamination.
● The combustion of coal-oil-mixtures (COM) is a well-established technology. The combination of high oil viscosities (fuel oils) and fine coal leads to high mixture viscosities at ambient temperature, the major limitation concerning long distance transport.

- Methanol-coal slurries can be prepared and pumped in a manner similar to coal-water slurries but require a finer grind of coal if head losses and transition velocities are to be kept down.
- 'Methacoal', prepared from fuel grade methanol and dried coal, offers the possibilities of stabilised transport in the laminar regime. Its rheology is not fully established, especially under pipeflow conditions.
- Liquid carbon dioxide offers a readily separable slurry medium, largely unreactive with the coal itself. Its low viscosity allows higher concentrations than water for the same head loss. Major unknown areas are the resultant wear rate and required emergency response for a $CO_2$ system.

## 7. THREE-PHASE MIXTURES IN COAL CONVERSION TECHNOLOGY

Direct combustion of coal, although highly developed, is not easily applied to transportation systems such as aircraft or road vehicles. There is considerable interest therefore in the conversion of coal into clean burning gaseous and liquid fuels. The degree of interest is a function of oil prices, which have fluctuated considerably, but in the long term will tend to rise as supplies become scarcer. The production of a fuel with as low a content of pollutants as possible is at least as important as the production of a liquid substitute for oil (105).

Liquefaction processes are either direct or indirect. In the direct process the coal is heated in the presence of a solvent, which is usually derived from the coal, and under pressure in a hydrogen atmosphere. In the indirect process the coal is first gasified and then turned into a variety of other hydrocarbons. The direct processes are of greater relevance to this book since they make greater use of slurry handling. The Solvent Refined Coal process (SRC), the Exxon Donor Solvent process (EDS) and the British Coal Solvent Extraction and Hydrocarbon process are examples of direct systems.

Heating of the coal and solvent, prior to reaction, may be required and the rheological properties of hot coal slurries need to be known so that the handling equipment can be properly designed, see Figures 8 and 9. Sanghvi and Tolan (106) have conducted experiments on two-phase coal slurries under process conditions (temperature 449–507 °K) using tube viscometers of 3 and 5 mm bore. The nominal particle size was approximately 1 mm. In addition, they wrote computer programs with the intention of predicting and then making comparisons with experiment of the pressure drop caused by pipe flow.

For non-settling slurries the computer program correlated the data within 5 per cent. For a settling slurry the data was correlated to within 17 per cent. Wall slip was considered to be important along with the effects of stratification and the migration of particles away from the pipe wall.

A paper by Segev and Golan (107) describes the development of a computer program to understand the flow behaviour in the tubular preheater of the Exxon coal liquefaction pilot plant, see Figure 10. In this plant, coal, mixed with recycled hydrocarbon solvent, is fed with hydrogen to the preheat furnace. Here the three-phase mixture is heated to approximately 425 °C at 135–170 bar before passing to the liquefaction reactor. It was found that

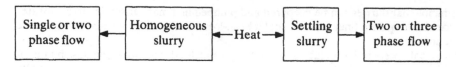

**Figure 8.** Effect of heat on slurry composition

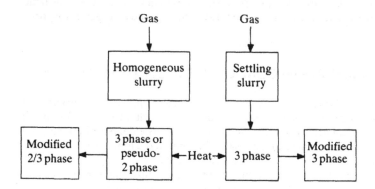

**Figure 9.** Addition of heat and gas to slurry

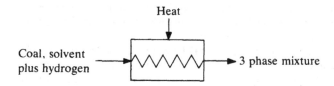

**Figure 10.** Coal conversion preheater

the slurry could be regarded as a homogeneous fluid thus simplifying the three-phase mixture to one having only two components. Conventional two-phase flow analysis (liquid-gas) could then be applied. The effect of temperature on viscosity was also found to be important.

The flow will depend on the heat transfer rate since not only will it affect the viscosity of the slurry but will also affect the ratio of liquid to gas, which in turn will change velocity and friction factors. The heat transfer can be described as forced convection boiling to a three-phase mixture. The liquid and gas phases are multi-component and the solid phase undergoes physical and chemical changes. The paper claims that this complicated process has not been described previously in the literature. For the case of convective boiling heat transfer the liquid and gas phases have different compositions and these change as the more volatile components are boiled away thus altering the heat transfer coefficients. A special correlating function was proposed based on 500 experimental results of convective heat transfer to two-phase flows and included terms to take account of voidage and solids content. The correlation was:

$$\frac{Nu}{Nu_s} = \frac{Re_v^{0.36}}{(1 - \alpha_c)^{2.1} X_{tt}^{0.34}}$$

where Nu = Nusselt number, (Surface heat transfer coefficient) × D/thermal conductivity of three-phase mixture
$D$ = pipe diameter
$Nu_s$ = Nusselt number of slurry
$Re_v$ = Reynolds number for vapour conditions
$\alpha_c$ = voidage fraction in liquid due to presence of coal particles
$X_{tt}$ = Lockhart Martinelli 2-phase flow function

The boiling effect is encompassed in the vapour Reynolds number, together with the Lockhart Martinelli parameter. The solids content is expressed by $1 - \alpha_c$ in addition to changes in fluid properties caused by the presence of solids and accounts for the coal dissolution. At the upper end of the range of experimentally measured heat transfer coefficients the values are predicted within a range of ±6 per cent. At the lower end of the range, however, the accuracy falls to about ±20 per cent.

Use of the program showed 25 per cent change in heat flux produced only a 0.3 per cent change in pressure drop. Conversely a change of tube diameter from approximately 81 to 50 mm changed the pressure drop by a factor of eight and reduced the outlet temperature by approximately 95 °C. Thus the program demonstrated the considerable sensitivity of the operating parameters to the physical size of the components. Such marked changes in parameters would have a most substantial effect on the operating performance of the equipment.

The majority of the coal dissolution takes place in the dissolver, Figure 11, even though a considerable degree of liquefaction occurs in the preheater. For effective operation the solids must remain suspended at the required concentration and the plant must be able to operate at reduced throughput without the solids settling. Knowledge of scale effects is required to enable commercial plant to be designed from pilot plant tests. Ying *et al.* (108) and Sivasubraminian *et al.* (109) discuss the requirements for a coal dissolver as outlined above and present experimental data. An important factor in dissolver design is voidage. This depends on liquid surface tension and viscosity as well as gas superficial velocity. Irwin *et al.* (110) reviewed the literature but found none of the correlating equations was entirely satisfactory. The predictions required knowledge of the axial dispersion coefficient and Ying *et al.* (108) found that only at higher velocities was there agreement between the various equations they compared. The information was not good enough to provide scale-

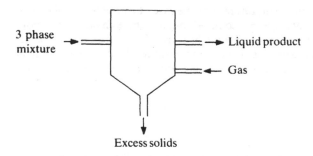

**Figure 11.** Simplified coal dissolver

up. At a later stage Mojaes (111) obtained data from a 1.83 m diameter dissolver, to enable the design of a 3.35 m diameter reactor. Concern was expressed relating to the effectiveness of this data under different conditions of temperature and pressure.

## 8.  GENERAL COMMENTS

Part of this chapter has considered the flow characteristics of slurries composed of solids and non-aqueous carrying media and as such can be considered to be a purely technical problem. The reader may well ask, however, which of the various forms of slurry usage are likely to increase in the future? In the author's opinion the use of hydraulic transport for movement of waste material may well be a candidate. If the oil price rises CWM will become more widely used. Three-phase mixtures are also likely to be used in coal conversion technology. Other forms of slurry technology, such as borehole mining, will be employed as and when the economics dictate. Long distance non-aqueous lines will require special circumstances to become viable.

## 9.  REFERENCES

1.  Alexander, D. W. (15–18 March 1983) 'Loveridge coarse coal slurry transport system performance and applications', *Proc. 8th Int. Conf. on Slurry Transportation*, San Francisco, USA, Slurry Transport Assoc.
2.  Harzer, J. and Kuhn, M. (25–27 August 1982) 'Hydraulic transportation of coarse solids as a continuous system from underground production face to the end product in the preparation plant', *Proc. 8th Int. Conf. on the Hydraulic Transport of Solids in Pipes*, Johannesburg, S.A. Paper J4, pp. 461–70, BHRA Fluid Engineering.
3.  Kortenbusch, W. (25–27 August 1983) 'Latest experience with hydraulic shaft transportation at the Hansa hydromine', *Proc. 8th Int. Conf. on the Hydraulic Transport of Solids in Pipes*, Johannesburg, S.A., Paper J5, pp. 471–83, BHRA Fluid Engineering.

4. Wakabayashi, J. (June 1979) 'Slurry pumped 3000 feet vertically', *Coal Age*, Vol. 84, pp. 84–7.
5. Sakamoto, M. *et al.* (26–28 September 1979) 'Vertical type hydrohoist for hydraulic transportation of fine slurry', *Proc. 6th Int. Conf. on the Hydraulic Transport of Solids in Pipes*, Canterbury, UK, Paper F1, pp. 257–68, BHRA Fluid Engineering.
6. Gontov, A. E. (28 November – 2 December 1977) 'Hydraulic coal mining in the USSR', Proc. 3rd IIASA Conf. on Energy Resource (Future Coal Supply for the World Energy Balance), Moscow, USSR, pp. 205–12, Publ. by Pergamon Press, Oxford, UK.
7. Sellgren, A. (23–26 March 1982) 'Integration of mine dewatering installations with a hydraulic hoisting system', Proc. 7th Int. Tech. Conf. on Slurry Transportation, Lake Tahoe, Nevada, USA, pp. 75–80, Slurry Transport Assoc.
8. Mero, J. L. (1978) 'Ocean mining: an historical perspective', *Marine Mining*, Vol. 1, No. 3, pp. 243–55.
9. Victory, J. J., Siapno, W. D. and Meadows, R. (August 1977) 'Mining manganese nodules from the ocean floor', *Mechanical Engineering*, Vol. 99, No. 8, pp. 20–5.
10. Polderman, H. G. (1982) 'Design rules for hydraulic capsule transport systems', *J. Pipelines*, pp. 123–136.
11. Burns, J. Q. and Suh, S. L. (30 April – 3 May 1979) 'Design and analysis of hydraulic lift systems for deep ocean mining', *Proc. 11th Offshore Technology Conf.*, Houston, USA, Vol. 1, Paper OTC 3366, pp. 73–84.
12. Weber, M. (1982) 'Vertical hydraulic conveying of solids by air-lift'. *J. Pipelines*, No. 3, pp. 137–52 (1982).
13. Flipse, J. E. (January 1983) 'Deep ocean mining economics', *Sea Technology*, Vol. 24, No. 1, pp. 41–3.
14. Silvester, R. S. (July 1983) 'Areas of application for research into three-phase flow', BHRA Internal Report.
15. Hahlbrock, U. (September 1979) 'The development of a vibrating suction head for mining deep sea metalliferous mud', *Proc. 6th Int. Conf. on the Hydraulic Transport of Solids in Pipes*, Canterbury, UK, Vol. 1, Paper G4, pp. 329–40, BHRA Fluid Engineering.
16. Boyce, T. A. (December 1978) 'Review of the borehole hydraulic coal mining system', *Engineering Societies Commission on Energy Inc.*, Final Report FE-2468-37, 53 pp. Washington DC, USA.
17. Boyce, T. A. (4 December 1978) 'Coal mining R&D review: borehole mining', *ESCOE Echo*, vol. 2, No. 23, pp. 2–3.
18. Wood, P. A. (September 1980) 'Less-conventional underground coal mining', *International Energy Agency Coal Research Report ICTIS/TR12*.
19. Sommer, T. M. and Funk, J. E. (1981) 'Development of a high-solids, coal-water mixture for application as a boiler fuel', *ASME*, Paper 81-JPGC-FU-4, 4 pp. American Society for Mechanical Engineers.
20. Johnson, E. P. (January 1982) 'Pay-off time is near for coal/water slurries', *Chemical Eng.*, Vol. 89, No. 1.
21. (March 1983) 'Slowly but surely. . . .', *Bulk Systems Int.*, Vol. 4, No. 11, pp. 53, 55 and 57.
22. Lee, H. M. (25–27 August, 1982) 'An overview of proposed coal slurry technologies and their cost-saving applications', *Proc. 7th Int. Tech. Conf. on Slurry Transportation*, Lake Tahoe, Nevada, USA. pp. 217–23, Slurry Transport Assoc.
23. Batra, S. K., Kenney, R. B. and Batra, R. K. (15–18 March 1983) 'Stability and rheological characteristics of coal water slurry fuels.' *Proc. 8th Int. Conf. on Slurry Transportation*, San Francisco, USA, Slurry Transport Assoc.
24. US Department of Energy (1983) *Fifth International Symposium on Coal Slurry Combustion and Technology.*
25. Institution of Chemical Engineers (UK) (5–6 October 1983) *First European Conference on Coal Liquid Mixtures.*

26.   Mahler, H. W. (15–18 March 1983) 'Dense coal pipelining of highly concentrated stabilized fine coal slurries from the mine to the customer', *Proc. 8th Int. Conf. on Slurry Transportation*, San Francisco, USA, Slurry Transport Assoc.

27.   Elliott, D.E. and Gliddon, B. J. (1–4 September 1970) 'Hydraulic transport of coal at high concentrations', *Proc. 1st Int. Conf. on the Hydraulic Transport of Solids in Pipes*, Warwick, UK, Paper G2, pp. G2-25–G2-56, BHRA Fluid Engineering.

28.   Lawler, H. L. *et al*. 'Application of stabilized slurry concepts of pipeline transportation of large particle coal', *Proc. 3rd Int. Techn. Conf. on Slurry Transportation*, Las Vegas, USA, pp. 64–178, Slurry Transport Assoc.

29.   Duckworth, R. A., Pullum, L. and Lockyear, C. F. (unpublished) 'Hydraulic transport of coal at high concentration', C.S.I.R.O. Division of Mineral Engineering, Australia.

30.   Wilson, K. C. (January 1970) 'Slip point of beds in solid-liquid pipeline flow', *Proc. American Society Civil Engineers J. Hydraulics Div.*, Vol. 96, No. HY1, pp. 1–12.

31.   Valore, R. C. (unpublished) 'Pumpability aids for concrete', Valore Research Association, Ridgwood N.J. 07450.

32.   Ede, A. N. (November 1957) 'The resistance of concrete pumped through pipelines', *Magazine of Concrete Research*, Vol. 9, No. 27, pp. 129–40.

33.   'Testing of pump mixes for blockage sensitivity', (Unidentified report).

34.   Streat, M. (23–24 August 1982) 'Dense phase flow', In: *Hydraulic Transport of Solids in Pipelines Short Course*, Johannesburg, S.A., Chapter 11, 48 pp. SAICE, *et al*.

35.   James, J. G. (March 1980) 'Pipelines considered as a mode of freight transport: a review of current and possible future uses', *Minerals and the Environment*, Vol. 2, No. 1.

36.   Grishechko, N. D. *et al*. (1980) 'Complex utilization of washery wastes and environmental protection', *Koks i Khimiya*, No. 12, pp. 12–14 (In Russian) and *Coke and Chemistry*, No. 12, pp. 19–25 (In English).

37.   Down, C. G. and Stocks, J. (1976) 'The environmental problems of tailing disposal at metal mines', *G.B. Dept. Environment Research Report 17*.

38.   Onley, J. K. (26–28 March) 'The hydraulic disposal of colliery waste materials', *Proc. 5th Int. Tech. Conf. on Slurry Transportation*, Lake Tahoe, Nevada, USA, pp. 120–6, Slurry Transport Assoc.

39.   Paterson, A. C.and Watson, N. (26–28 September 1979) 'The National Coal Board's pilot plant for solids pumping at Horden colliery', *Proc. 6th Int. Conf. on the Hydraulic Transport of Solids in Pipes*, Canterbury, UK, Vol. 1., Paper H1, pp. 353–66, BHRA Fluid Engineering.

40.   (1979) 'Reclamation of Pyewipe using waste from the Yorkshire coalfield', *Strategic Conf. of County Councils of Yorkshire and Humberside (SCOCC)*, Wakefield.

41.   Brown, D. W. (1976) 'The possible use of industrial waste and its transportation for reclamation purposes', *Proc. Land Reclamation Conf.*, Thurrock Borough Council, Grays, Essex.

42.   Wright, D. and Brown, J. (26–28 September 1979) 'Hydraulic disposal of P.F. ash from C.E.G.B. Midland Region power stations', *Proc. 6th Int. Conf. on the Hydraulic Transport of Solids in Pipes*, Canterbury, UK. Vol. 1, Paper H3, pp. 379–88, BHRA Fluid Engineering.

43.   Corner, D. C. and Stafford, D. C. (1972) 'Chinaclay sand: liability or asset?', Devon County Council, Exeter.

44.   Down, C. G. and Stocks, J. (May, July, September 1977) 'Methods of tailings disposal', *Mining Magazine*, Vol. 136, pp. 345–59, Vol. 137, pp. 25–33, Vol. 137, pp. 213–23.

45.   Couratin, P. (May 1969) 'Tailing disposal', *World Mining*.

46.   Wood, P. A. (April 1983) 'Underground stowing of mine waste', *International Energy Agency Coal Research Report ICTIS/TR23*, 67 pp.

47.   Verkerk, C. G. (25–27 August 1982) 'Transport of fly ash slurries', *Proc. 8th Int. Conf. on the Hydraulic Transport of Solids in Pipes*, Johannesburg, S.A., Paper F4, pp. 307–16, BHRA Fluid Engineering.

48. Thomas, E. (April 1977) 'H.M.S. float fill development. Stage 1: pumping', *Technical Report No. Res. Min. 48*, Mount Isa Mines Ltd.
49. Davidson, C. W. (November 1978) 'Cemented aggregate fill: underground placement', *Technical Report No. Res. Min.53*, Mount Isa Mines Ltd.
50. Paterson, R. M. (22–24 October 1974) 'Closed system hydraulic backfilling of underground voids', *Proc. 1st Symp. on Mine and Preparation Plant Refuse Disposal*, Louisville, KY, USA, pp. 161–4, National Coal Assoc.
51. Jankousky, C. K. (1 May 1977) 'Disposal of Coal Refuse Slurry Underground', *American Mining Congress 1977 Coal Convention*, Pittsburgh, PA, USA. Paper 24, 12 pp. CONF-7705111, National Technical Inf. Service.
52. Colaizzi, G. J., Whaite, R. H. and Donner, D. L. (July 1981) 'Pumped-slurry backfilling of abandoned coal mine workings for subsidence control at Rock Springs, Wyoming'. *US Bureau of Mines; National Technical Inf. Service PB82-120593*, 107 pp.
53. Sabbagha, C. M. (25–27 August 1982) 'Practical experiences in pumping slurries at ERGO', *Proc. 8th Int. Conf. on the Hydraulic Transport of Solids in Pipes*, Johannesburg, S.A. Paper A1, pp. 1–16, BHRA Fluid Engineering.
54. Wang, G. S. C. (23–26 March 1982) 'Coal slurry shiploading world trade applications', *Proc. 7th Int. Tech. Conf. on Slurry Transportation*, Lake Tahoe, Nevada, USA, pp. 81–9, Slurry Transport Assoc.
55. Kopeikin, L. I. (26–28 March 1980) 'Pacific bulk transportation system', *Proc. 5th Int. Tech. Conf. on Slurry Transportation*, pp. 12–15, Slurry Transport Assoc.
56. Sterry, W. M. (23–26 March 1982) 'Pacific bulk transportation system', *Proc. 7th Int. Techn. Conf. on Slurry Transportation*, Lake Tahoe, Nevada, USA, pp. 17–19, Slurry Transport Assoc.
57. Beteille, G. L. *et al.* (1981) 'Proposed coal export terminal at Stapleton, Staten Island, New York, using coal slurry technology', *Dept. of Ports and Terminals Report*, 151 pp. City of New York.
58. Hoogendoorn, D. (26–28 March 1980) 'A coal slurry terminal for Rotterdam?', *Proc. 5th Int. Techn. Conf. on Slurry Transportation*, Lake Tahoe, Nevada, USA, pp. 46–52, Slurry Transport Assoc.
59. Pitts, J. D. and Aude, T. C. (June 1977) 'Iron concentrate slurry pipelines, experience and applications', *Trans. Soc. Min. Engrs. of AIME*, Vol. 262, pp. 125–33.
60. Moore, T. J. M. (February 1983) 'Preliminary economic and feasibility study on a coal slurry ship unloading system at the Cloghan Point fuel terminal', *BHRA Report RR 1966*, 32pp.
61. Faddick, R. R. (23–26 March 1982) 'Technical aspects of shiploading coal slurries', *Proc. 7th Int. Tech. Conf. on Slurry Transportation*, Lake Tahoe, Nevada, USA, pp. 45–9 Slurry Transport Assoc.
62. (June 1981) 'Coarse coal pipelining: applications for coal exports', Kamyr, Inc., USA, 21 pp.
63. (1982) 'Coarse coal handling by Marconaflo slurry systems', *Internal Report*, 44 pp. Marconaflo Div., McNally Group, USA.
64. Montfort, J. G. (26–28 March 1980) 'Operating experience of the Black Mesa pipeline', *Proc. 5th Int. Tech. Conf. on Slurry Transportation*, Lake Tahoe, Nevada, USA, pp. 16–23, Slurry Transport Assoc.
65. Montfort, J. G. (23–26 March 1982) 'Operating experience of the Black Mesa pipeline', *Proc. 7th Int. Tech. Conf. on Slurry Transportation*, Lake Nevada, USA, pp. 421–29, Slurry Transport Assoc.
66. Cobb, D. B. *et al.* (29–31 March 1978) 'Coal slurry storage and reclaim facility for Mohave Generating Station', *Proc. 3rd Int. Tech. Conf. on Slurry Transportation*, pp. 58–68.
67. Andersen, A. K. 'Marconaflo systems state of the art'. Marconaflo Inc.
68. Sauermann, H. B. (23–24 August 1982) 'Hydraulic hoisting and backfilling', In: *Hydraulic*

272                                                      *Design of slurry transport systems*

*Transport of Solids in Pipelines Short Course*, Johannesburg, S.A., SAICE, *et al.* Chapter 8, 42 pp.
69. Patchet, S. J. (23–25 May 1978) 'Fill support systems for deep level gold mines', *12th Canadian Rock Mechanics Symposium*, pp. 48–54, Canadian Inst. of Mining and Metallurgy.
70. Patchet, S. J. and Currie, G. E. R. (July 1981) 'Operation and performance of a slimes fill system in a deep mine', *Special Project 11/81*, Anglo-American Corporation, S. Africa Ltd. Rock Mechanics Dept.
71. Keller, L. J. (January 1979) 'Methacoal slurry transportation technology', *Proc. Interpipe '79, 7th International Pipeline Technology Convention*, Houston, Texas, pp. 149–57, Slurry Transport Association, USA.
72. Santhanam, C. K., Dale, S. E. and Nadkarni, R. M. (March 1980) 'Non-water slurry lines – potential techniques', *Proc. 5th International Technical Conference on Slurry Transportation*, Lake Tahoe, Nevada, pp. 93–105, Slurry Transport Association, USA.
73. Santhanam, C. K., Dale, S. E. and Nadkarni, R. M. (June 1980) 'Non-water slurry lines', *Oil and Gas Journal*, Vol. 78, No. 26, pp. 128–40.
74. Bradley, G. M. (March 1982) 'Novel slurry transport medium' *Proc. 7th International Technical Conference on Slurry Transportation*, Lake Tahoe, Nevada, pp. 411–18, Slurry Transport Assoc.
75. Sandhu, A. S. and Weston, M. D. (March 1980) 'Feasibility of conversion of existing oil and gas pipelines to coal slurry transport systems', *Proc. 5th International Technical Conference on Slurry Transportation*, Lake Tahoe, Nevada, pp. 85–92, Slurry Transport Association, USA.
76. Smith, L. G., Haas, D. B., Richardson, A. D. and Husband, W. H. W. (May 1976) 'Preparation and separation of coal-oil slurries for long-distance pipeline transportation', *Proc. Hydrotransport 4*, Banff Springs Hotel, Alberta, Canada, Paper E5, pp. 63–78. BHRA Fluid Engineering, UK.
77. Pan, Y. S., Bellas, G. T., Joubert, J. I. and Lunifeld, D. (January 1981) 'Coal-oil mixtures: a near term approach to conserving petroleum', *Proc. ASME*, Paper No. 81-Pet-22.
78. Ushio, S. (January 1980) 'Blends of coal and oil are readied for Japanese fuel tests', *Chemical Engineering*, Vol. 87, No. 2, pp. 62E-62F.
79. Eberle, J. W. and Hickman, R. H. (1979) 'The feasibility of firing a coal-oil mixture', *Coal Processing Technology*, Vol. 5, pp. 60–3.
80. Kreusing, H. and Franke, F. E. (September 1979) 'Investigations on the flow and pumping behaviour of coal-oil-mixtures with particular reference to the injection of a coal-oil slurry into the blast furnace', *Proc. Hydrotransport 6*, Canterbury, UK, Paper C2, pp. 137–52, BHRA Fluid Engineering, UK.
81. Munro, J. M., Lewellyn, M. M., Crackel, P. R. and Bauer, L. G. (March 1979) 'A characterisation of the rheological properties of coal-fuel oil slurries', *AIChE Journal*, Vol. 25, No. 2, pp. 355–8.
82. Ackermann, N. L. and Hung, T. S. (March 1979) 'Rheological characteristics of solid-liquid mixtures', *AIChE Journal*, Vol. 25, No. 2, pp. 327–32.
83. Nakabayashi, Y., Matsuura, Y., Nagata, K., Nagamori, S. and Yano, T. (November 1980) 'Flow behaviour and pressure loss in pipes on coarse COM', *Proc. Hydrotransport 7*, Sendai, Japan, Paper G2, pp. 277–90, BHRA Fluid Engineering, UK.
84. Yagamata, Y., Kokubo, T., Suzuki, S. and Moro, T. (November 1980) 'Rheological study of viscosities and pipeline flow of coal-oil mixtures', *Proc. Hydrotransport 7*, Sendai, Japan, Paper G1, pp. 259–75.
85. Ekmann, J. M. and Klinzing, G. E. (August 1981) 'Analysis of coal-oil mixture sedimentation by modified continuity wave theory', *The Canadian Journal of Chemical Engineering*, Vol. 59, pp. 417–22.
86. Bhattacharya, S. N. (August 1980). 'Some observations of the rheology of time dependent

suspension on long-term storage', *8th Australian Chemical Engineering Conference (I.Chem.E., I.E.Aust.)*, Melbourne, Victoria, p. 95.

87.  Bhattacharya, S. N. and Barro, L. (May 1980), 'Flow characteristics of coal-oil slurries', *The Australian Institute of Mining and Metallurgy, New Zealand Conference*, pp. 275–82.
88.  Othmer, D. G. (November 1982) 'Methanol is the best way to bring Alaska gas to market', *Oil and Gas Journal*, Vol. 80, No. 44, pp. 84–5.
89.  Godde, E. (May 1981) 'Possibilities for transporting coal through pipelines using methanol as a carrier liquid' (in German), *Technische Mitteilungen Krupp Forschungberichte*, Vol. 39, No. 1, pp. 23–8.
90.  Aude, T. C. and Chapman, J. P. (March 1980) 'Coal/methanol slurry pipelines', *Proc. 5th International Technical Conference on Slurry Transportation*, Lake Tahoe, Nevada, pp. 106–14, Slurry Transport Association, UK.
91.  Aude, T. C. and Chapman, J. P. (July 1981) 'Coal/methanol slurry lines show promise', *Oil and Gas Journal*, Vol. 79, No. 29, pp. 135–40.
92.  Hanks, R. W. (May 1963) 'The laminar-turbulent transition for fluids with a yield stress', *AIChE Journal*, Vol. 9, No. 3, pp. 306–9.
93.  Darby, R. (March 1979) 'Rheology of methacoal suspensions', *Proc. 4th International Technical Conference on Slurry Transportation*, Las Vegas, Nevada, pp. 183–94, Slurry Transport Association, USA.
94.  Hanks, R. W. (December 1981), 'Laminar-turbulent transition in pipeflow of Casson model fluids', *Trans. ASME*, Vol. 103, pp. 318–21.
95.  Davis, P. K., Muchmore, C. B., Missavage, R. J. and Coleman, G. N. (August 1981) 'Rheological and pumping characteristics of ethacoal', *Journal of Pipelines*, Vol. 1, No. 2, pp. 139–44.
96.  Santhanam, C. K. (March 1982) 'Development of coal/liquid $CO_2$ slurry transportation – current status', *Proc. 7th International Technical Conference on Slurry Transportation*, Lake Tahoe, Nevada, pp. 203–9, Slurry Transport Association, USA
97.  Renfro, J. J. (November 1979) 'Sheep Mountain $CO_2$ production facilities – a conceptual design', *Journal of Petroleum Technology*, Vol. 31, No. 11, pp. 1462–7.
98.  Coulter, D. M. (December 1976) 'LNG-crude oil slurry cryogenic pipelines', *ASME Publication 76-WA/PID-8*, Process Industries Division.
99.  Souder, P. S. (February 1980) 'Coal slurry pipeline economics for alternative fluidising media', *Proc. ASME*, Paper 80-Pet-46.
100. Gandhi, R. L., Weston, M. D. and Snoek, P. E. (September 1979) 'Economics of alternative coal slurry systems', *Proc. Hydrotransport 6*, Canterbury, UK, Paper C3 pp. 153–60, BHRA Fluid Engineering, UK.
101. Banks, W. F. (March 1979) 'Economics of non-water coal slurry systems', *Proc. 4th International Technical Conference on Slurry Transportation*, Las Vegas, Nevada, pp. 165–70, Slurry Transport Association.
102. Gödde, E. (February 1981) 'Alternative carrying media and agglomeration processes for transportation of coal', *Bulk Solids Handling*, Vol. 1, No. 1, pp. 65–70.
103. Lee, H. M. (March 1982) 'An overview of proposed coal slurry technologies and their cost saving applications', *Proc. 7th International Technical Conference on Slurry Transportation*, Lake Tahoe, Nevada, pp. 217–23, Slurry Transport Association.
104. Lee, H. M. (August 1978) 'The economics of coal slurry pipelines', *I.E.A. Coal Research, Economic Assessment Service, Working Paper*, No. 32, Study D.
105. Reviews of Modern Physics: (October 1981) 'Primer on coal utilisation technologies', 53 (4), 553–72.
106. Sanghvi, S. M. and Tolan, J. S. (1982) 'Prediction of pressure drop of two phase coal slurries in pipelines', DE83 003382 ORNL/MIT-349, Springfield, VA, USA, National Technical

Information Service, 82 pp.

107. Segev, A. and Golan, L. P. (1983) 'Multi Phase Flow and heat transfer in the EDS slurry preheat furnace', *American Institute of Chemical Engineers Symposium*, Series 79 (225); pp. 360–72.

108. Ying, D. H. S., Sivasubramanian, R., Moujaes, S. F. and Givens, E. N. (1982) 'Gas slurry flow in coal-liquefaction processes (fluid dynamics in a three phase flow column)' DE83 001643, DOE/ET/14801-30, Springfield, VA, USA, 365 pp, National Technical Information Services.

109. Sivasubramanian, R., Ying, D. H. S. and Given, E. N. (9 August 1981) 'Experimental simulation of solids distribution in coal liquefaction dissolves', *IECEC Conference, Atlanta, CA, USA, Conference 810812, IEEE*, 2, pp. 1092–100.

110. Irwin, C. F., Sincali, A. J. and Wong, E. W. (1981) 'Hydrodynamics of coal liquefaction dissolver configurations'. *ORNL/MIT-326*, Oak Ridge, USA Oak Ridge National Laboratory.

111. Moujaes, S., Sivasubramanian, R. and Ying, D. H. S. (8 August 1982) 'Solids dispersion modelling in coal liquefaction dissolvers', In: *17th Intersociety Energy Engineering Conference*, Los Angeles, CA, USA, Conf-829144, IEEE, pp. 850–3.

# 8. PRACTICAL APPLICATIONS AND ECONOMICS

## 1. INTRODUCTION

The following chapter reviews a number of solids handling pipeline installations that are operational, as well as one system that has been proposed but not yet constructed. Examples of slurry systems have been selected on a worldwide basis and are representative of the general interest displayed in this mode of transport.

Although pipelines are used for the transportation of many different solids materials, it has been decided to concentrate on the larger, long distance mineral slurry pipelines, as information tends to be more readily available than for the smaller 'in-house' applications. The author wishes to stress, however, that many successful examples of slurry pipeline systems exist that are not reported in the available literature. Many 'in-house' applications which are relatively short (less than 5 km) are in constant use, particularly in mixing and material processing applications.

An attempt has been made to give some indication of the related economics, both with respect to capital expenditure and operating costs. This information, however, was not readily available in all cases. A brief discussion concerned with the major alternative methods of solid transportation has been included, as is an introductory view of pipeline economics. It must be emphasised that due to the complexity of pipeline economics, no 'in depth' analysis has been attempted.

An American Standard, ANSI/ASME 'Slurry Transportation Piping Systems' B31.11–1986, has been written which deals with the mechanical aspects of pipe design.

## 2. EMINENT DOMAIN

One of the major potential growth areas, within the solids handling pipeline industry, would appear to be that of coal transportation, particularly in the USA and to a lesser extent in Europe. The potential is closely associated with oil prices and hence interest in this mode of transport tends to be cyclic. However, in the long run, the potential is likely to be fulfilled.

Cross-country transport by pipelines presupposes that it is possible, legally, to lay the pipeline where required. At one stage there were some 10 major proposals for coal transport pipelines in progress in the USA, but many of these systems were relying on the right of eminent domain being made law. Ref. 1 states that it is unlikely that any major, long-distance slurry pipelines will be constructed and commissioned without the right of eminent domain being made law.

Effectively, eminent domain is a right which makes it possible for private property to be purchased at a fair price, as determined by the courts, for projects which are for public benefit. It is interesting to note that Federal eminent domain in the USA was granted to oil pipelines in 1940, and to natural gas pipelines in 1947. Much of the land, over which pipelines would have to pass, appears to be owned by operating railroad companies, who consider the transportation of coal by pipeline to be contrary to their interests. Much is made, therefore, of the respective economics concerned with pipeline and rail transport methods, and this subject is covered in greater detail in section 4.

The right of eminent domain has been passed in a number of the individual States in America (3), however, several references state that it is necessary for the law to be passed by the federal government, enabling a nationwide 'procedure' to be followed. Conversely, Desteese (3) states that the eminent domain problem appears to be of comparatively small concern when related to the overall issue.

## 3. WATER ACQUISITION AND REQUIREMENTS

Hydraulic transport systems, utilising a solids/water mix, require large amounts of water although it is normal to make use of any water at the processing plant or power station located at the pipeline terminal. Typically, a pipeline may contribute 10–15 per cent of the water requirements for the power station cooling system (9). Normally, water is secured from local (to the start of the pipeline) sources, such as rivers and purpose built deep wells. In some of the pipeline proposals, however, water has to be transported over considerable distances by separate pipeline systems. Some of the American States, however, operate a 'no-export' policy that is applicable to water reserves which adversely affect some proposals which have been made.

Under some situations, such as within areas of very limited water supply, or to overcome prohibitive legislation, it may become necessary to operate a water recirculation system. This type of operation requires only very small additional amounts of 'make-up' water due to loss by evaporation in settling pools and surface moisture on the dewatered coal cake. The re-circulation type of installation will however increase the capital expenditure as two pipelines are required. Some doubts have been raised, particularly in the USA with respect to pollutants from the coal in the waste water. Adam (5) states, however, that the Mohave power station has been using the slurry water, as received from the Black Mesa pipeline for more than 10 years, with no environmental ill-effects.

## 4. ECONOMICS

### 4.1 BACKGROUND

There have been many studies that examine the respective economics of slurry transport by pipeline and alternative modes of transport including rail, barge, road and combinations thereof. Due to the complexity of information available, and coupled with the fact that individual applications all contain almost unique governing factors, it is intended to give only a brief overview of the economics of pipeline slurry transport.

Although, as described in the introduction, many different materials are transported by pipeline, the majority of relevant information associated with economics tends to originate from coal transport systems. Nevertheless, it is suggested that many apsects that relate to the transport of coal, will also be applicable to the transport of other minerals.

## 4.1.1 Estimating techniques for feasibility studies

Costs associated with major capital items required for slurry transport plant can be obtained from capacity-cost relationships available from cost engineering literature. Where necessary these can be augmented by manufacturers' data. These costs can then be processed by the factored cost estimating method (10) whereby the capital item costs are multiplied by certain factors for additional equipment, construction, installation, indirect costs, etc. Providing the appropriate cost information is available, the accuracy of the estimates is claimed to be about 30 per cent.

## 4.2 SLURRY PIPELINES

Pipeline systems are generally very capital intensive, but as a result, tend to be less prone to rising costs due to inflation than most other transport systems.

It has been suggested (11) that this widely accepted view may only be superficially true and is possibly indicative of a small rate of return on initial investment only slightly above the existing rate of inflation. It is also stated that in the longer term, the investor will require adequate protection against inflation.

Ref. 1 suggests that capital investment, i.e. fixed costs, represents 84 per cent of the total expenditure. Variable costs, which includes power, labour and general supplies, account for the remaining 16 per cent. Obviously, these figures are dependent upon many factors relating to individual applications. Greco (7), for example, states that the latter proportion of costs is nearer 30 per cent, which agrees with the figures quoted by Pitts and Hill (4). In general, pipeline tariffs tend to escalate slowly, usually at about 20–30 per cent of the overall inflation rate.

It is generally accepted that any economic advantages associated with slurry pipeline systems improves with increasing throughput and distance. Ref. 1 states that relatively short-distance pipelines tend not to be as economically attractive as unit trains, for example, but when the distance goes above 800 km and/or the annual throughput goes above 10 million tonnes, the pipeline system will start to show an economic saving over the alternative systems. Very vew economic analyses of short 'in-house' solids handling systems have been located in the literature. Unfortunately, it appears that any published cost figures tend not to isolate specific sections of overall treatment or handling plant, such as pipelines. Coal usage requires that many millions of tonnes per year are required to be transported. Slurry pipelines will be economic for other minerals at much lower through-puts, particularly for the in-house types.

Although the availability of water has presented problems for some of the proposed pipeline systems, the problems are normally associated with political or environmental aspects and not economics. Lee (6) states that it is unlikely that water costs would be a constraint on any pipeline scheme even if the overall system requires the construction of a water supply pipeline of significant length.

Before it is possible to gain any overview of the economics related to slurry pipeline systems, it must be realised that the different types of slurry system available each have specific economic structures. Five different slurry systems are briefly described, however non-water slurries have been covered in greater detail in Chapter 7.

## 4.2.1   Conventional slurry transport

In conventional coal slurry systems coal is transported in nearly homogeneous or mixed-flow conditions which are usually designed to operate at slurry velocities close to the deposition velocity to reduce overall power consumption. Gandhi *et al.* (8) suggest that a typical conventional system may utilise coal of less than 14 mesh at a solids concentration of 40–50 per cent by weight. Water requirements would be of the order of 0.85 m$^3$ per tonne of delivered coal.

There is also a requirement for relatively low concentration coarse particle slurry transport. Concentrations vary up to a maximum of approximately 30 per cent by weight. Slurry particles tend to settle up to this concentration and therefore velocities have to be kept reasonably high as indicated by the high velocity quoted in the (now abandoned) Staten Island project. Settling slurry systems were also used at the Hansa Hydromine, West Germany.

## 4.2.2   Dense phase or stabilised flow

More recently, interest has been shown in dense phase or stabilised flow conditions which enable a much higher concentration slurry to be transported. Also, larger lumps of coal may be transported which, in theory, would increase the scope of the market. In dense phase transport, coal is carried in laminar flow conditions, which, together with the high concentration, exhibit non-settling characteristics. Gandhi *et al.* (8) state that if a batching system were used, dilution at the 'tail' of the solids may take place with the following water, thus forming a settling slurry, but, if a continuous 'plug' of solids were transported, this possible limitation would be overcome. Typically, solids concentration may be as high as 60 per cent by weight, or even higher, therefore reducing the water requirements to approximately 0.5 m$^3$ per tonne of coal delivered.

### 4.2.3 Upgraded coal

Sub-bituminous coal has an inherently high moisture content which drastically reduces the heating efficiency. In some cases, a typical bituminous coal may deliver up to 40 per cent more 'heating value' than the equivalent amount of sub-bituminous coal, and therefore, economic savings would be realised if the moisture content could be reduced. By capturing the released moisture, the fluidity of the coal may be improved, therefore allowing a higher concentration slurry to be used and decreasing the overall water requirements, in some cases, by as much as 43 per cent.

### 4.2.4 Recirculated water systems

In conventional slurry systems, water that is used as the carrier fluid is normally utilised at the pipeline terminal as, for example, make-up water at a power station. In areas of low water availability, which may be due to either economics or environmental constraints, it may become necessary to recycle the carrier water back to the head of the pipeline for re-use. The only additional water then required would be that to compensate for loss during dewatering and evaporation in settling ponds. Obviously, this system necessitates the construction of a second, usually smaller diameter, pipeline. In some situations, it may be possible to obtain water from a source that is some distance from the mine but requires a shorter additional pipeline.

### 4.2.5 Non-aqueous slurry media

Recently, interest has been shown in non-water based slurries, particularly in connection with coal transport, and in areas known to be scarce of water. Various options have been identified, namely crude and fuel oil, methanol, ethanol and carbon dioxide. Non-aqueous slurry media systems have been extensively reviewed in chapter 7.

### 4.2.6 Comparative costs of various slurry systems

An economic study was performed by Gandhi *et al.* (9) enabling a direct comparison of costs between the five different slurry systems indicated above. In each case, comparable equipment such as pipe material and pumps were selected. Operating costs were estimated on the existing electricity tariff, plus labour and component usage and the total annual cost calculated using a 20 per cent capital charge in addition to direct operating costs. As the systems were specifically aimed at coal transport, a comparative throughput in terms of heating value, was used. The predicted costs for throughputs of 15 and 30 Mta over distances of 800–2400 km, showed the conventional slurries to be the cheapest. For the higher throughput, and a distance of 1600 km, conventional slurries cost half as much as coal-oil mixture.

A more recent study by Brookes and Snoek (12) compared costs of stabilised mixtures, 'Stabflow', with conventional slurries, fine coal-water mixtures and rail, in which approximately 50 per cent of the rail was existing. This analysis, based on further development work for the stabilised slurries, showed these to be the most economic, see Figure 1. Due to the continuing effects of inflation the costs are shown in a proportionate manner. It is expected that changes in economic structures will occur less rapidly than inflation itself, thus the relative positions of the curves should remain approximately constant for a significant period. The cross over point for the rail and conventional fines pipeline, occurring at 450 km, coincided with a cost of approximately 3.8 US cents/tonne km, at 1986 prices. The curves are applicable for flow rates of 5–10 Mt/a.

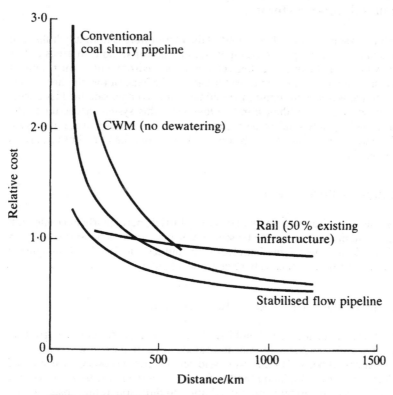

Throughput 5-10 Mt/a
Cost structure based on 1986 data
Costs include dewatering

**Figure 1.** Comparison of unit tariff rates (12)

## 4.3 ALTERNATIVE TRANSPORT SYSTEMS

### 4.3.1 Railway

A 'unit train' is a complete train consisting purely of one cargo, in this instance coal, that is loaded at one point and delivered to another. The trains are loaded and unloaded at purpose built installations without uncoupling the cars and therefore a relatively quick turn-round time may be achieved.

Lee (6) states that less than 50 per cent of all coal transported by rail is by unit train, however this proportion is steadily increasing, particularly in view of the vast quantities that are likely to be moved in the future.

According to Skedgall (2), it is estimated that 1400 million tonnès of coal are expected to be shipped in the USA in 1990. Of this total, 840 million tonnes or 60 per cent are expected to be transported by rail. For comparison, 750 million tonnes were 'shipped' in 1979, of which 485 million tonnes were moved by rail. It is interesting to note that if all the proposed major American coal slurry pipelines were to have become operational by 1990, the total maximum capacity would only amount to 170 million tonnes/year.

The cost rate for a particular route, applied to unit trains, is generally developed by the carrier and takes into account track conditions, terrain, motive power, tonnage and distance. In some countries, USA for example, the coal waggons may be supplied by the 'carrier', 'user' or 'shipper'.

It is generally accepted that rail transport is usually less capital intensive than pipelines. Due to the relatively high dependences upon labour and fuel, railways tend to be more susceptible to rising costs related to inflation than, say, pipelines. Indeed, a study conducted by Ebasco (7) concluded that for rail transport, approximately 95 per cent of annual cost is subject to escalation, as opposed to only 30 per cent for slurry pipelines.

### 4.3.2 Barge/inland waterway

According to Lee (6), inland waterways can provide the cheapest form of coal transportation. However, there are only two major systems of inland waterways in the Western world. These are the Mississippi River Basin in the USA and the system based around the Rhine river in North West Europe. As there are very few cost figures available, and barge/towboat transportation only applies with any significance to the two aforementioned areas, it is not proposed to pursue this area of transport further. However, Figure 2 includes proportionate cost figures for coal transportation as applied to barge and inland waterway systems.

### 4.3.3 Comparative costs of various transport modes

Greco and Sherlock (7) conducted a study where four different 'energy' transport systems

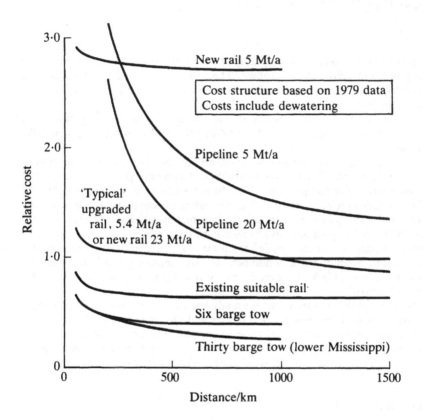

**Figure 2.** Coal transportation costs (6)

were assessed from an economic standpoint. The four systems were slurry pipeline, rail and barge, rail only and transmission line. The study assumes a movement of various amounts of energy, 1600, 3200, 6400 and 9000 MW over three different distances of 800, 1500 and 2400 kilometres.

Various economic assumptions were made to enable comparative costs to be calculated. Fixed charges were taken at 13 per cent, based on 80 per cent debt and 20 per cent non-utility type financing. Energy costs were based on actual utility rates and included pumping, preparation and dewatering costs. Material and labour costs were escalated to 1986 figures for initial start-up. It was shown that the coal slurry pipeline system presents the cheapest transport system from the alternatives studied. This view is shared, at least in part, by Ercollani and Ferrini (13), who looked at coal and coal-derived energy transportation systems. Systems involving the delivery of 26 TWh/year over 800 km were studied, and included coal slurry coal-derived electricity and gas or liquid fuels produced at the coal mine. They concluded that the slurry and transmission lines were the most economical.

A major economic study conducted by Lee (6), looked at various coal transport systems, including slurry pipelines. Lee states that the vast majority of data relating to coal transport originates from the USA, although some cost figures have been obtained which stem from West Germany. Figure 2 shows the estimated costs for different methods of coal transport. The curves for slurry pipeline, barge and rail apply to the USA only, and Lee states that these values may vary significantly for other countries.

As with Figure 1 the costs are shown in proportionate terms to suppress the effects of inflation. It is interesting to note that pipeline transport is shown as being less economic than rail transport when existing rail is utilised, but becomes more attractive when new or upgraded rail has to be used. There are some differences between Figures 1 and 2 as shown by the cross over points for conventional pipelines and upgraded rail. These reflect differences in the assumptions, both engineering and economic. It must be stressed that in reality, the curves represent a range of values and, therefore, in some situations, pipeline transport may be cheaper, irrespective of type of rail system used. Although river transport is the most economic proposition, it is only applicable where suitable navigable waterway systems exist, such as the Rhine or Mississippi regions.

Lee concludes by saying that rail transport will continue to be the most widely used form of coal transport but slurry pipelines can be cheaper in certain circumstancs particularly when new or upgraded rail has to be used.

## 5. EXISTING SLURRY PIPELINE INSTALLATIONS

The appendix lists some slurry pipeline systems that are operational or have been operational and have since closed. The following section gives further details not included in the tables, of existing systems including economics where available. The economic data generally refers to the date at which the pipeline commenced operation, or soon afterwards. No attempt has been made to allow for inflation or changing exchange rates.

Although the majority of installations listed refer to major overland schemes, a few 'in-plant' systems have been covered and particular note should be made, for instance, of the comparatively short pipelines in South Africa handling gold slime.

## 5.1 SLURRY PIPELINES IN COAL MINES

As previously mentioned, many smaller solids handling pipeline systems operate in and around material handling and processing plant, particularly coal mines. According to Moore (14), the two best known examples are the Hansa Hydromine, near Dortmund,

West Germany (15), and the Loveridge Mine, West Virginia, USA (16). The Hansa mine operated successfully conveying up to 2000 tonnes of coal per day through a 0.25 m diameter pipeline of 5.95 km total length, including an 850 m vertical rise.

It is known that installations of this type exist in the USSR, the most widely known being the Yubileyena mine where coal is transported at the rate of 908 m³/hour, 300 m vertically and 10 km horizontally, in a 0.3 m (12 inches) diameter pipe. In another system, at Krasnoarmeyskaya (17), coarse coal is transported over 1 km in a 0.42 m (16.5 inches) diameter pipe at the rate of 1.5 Mta.

Other systems reported in the literature include a Japanese miné at Sunagawa (18), where coal is transported 2.52 km, including 520 m vertical lift. Although initial operation was on coarse coal, the system now conveys only sub 0.75 mm particles as it was found to be more economic to use 'coal-bins' for the large particle transport.

Pipelining systems exist in UK mines, the most widely known being located at Horden Colliery.

## 5.2   ARGENTINA, SIERRA GRANDE – IRON ORE PIPELINE

The Sierra Grande pipeline system is similar to the Pena Colorada installation in Mexico, in that it utilises an elevation drop enabling only one pumping station to be necessary. The pump station houses three vertical triplex plunger pumps, each of 450 kW. The pipeline is 32 km in length and has a diameter of 0.219 m (8.6 inches). Capacity of the system is 2 million tonnes/year.

Pitts and Aude (19) state that the primary use of the ore concentrate was again domestic steel production.

## 5.3   BRAZIL, SAMARCO – IRON ORE PIPELINE

**Background**

In 1971, Marcona made an agreement with Samitri (S.A. Mineracao Trindade) to study the possible developments of iron ore reserves in the Iron Quadrangle, State of Minas Gerais, Brazil, already owned by Samitri.

A study concluded that deposits of Germano were of prime interest, however an efficient mode of transport was required to deliver ore to the coast for export. Samitri already operated a mine at Alegria, 12 km north of Germano. In this case the ore did not require beneficiation and hence the quantities to be transported were restricted. Thus rail transport, although expensive, was viable.

The subsequent development resulted in an efficient pipeline transportation system 396 km long which was initiated in May 1977 (20).

## Location

The pipeline operates between the Germano mine, Minas Gerais, approximately 400 km north of Rio de Janeiro, and the pellet plant and port facility near Ponta Uba in the state of Espirito Santo (21). The highest point of the pipeline is approximately 1200 metres above sea level.

### Pipeline/technical specification

Most of the 396 km long pipeline is made from 0.508 mm (20 inch) diameter pipe (API 5LX60), however, for a 40 km section, 0.457 mm (18 inch) diameter is used, to assist with the dissipation of potential energy immediately prior to valve station number 2.

Two pumping stations are used and both use seven Wilson-Snyder triplex plunger pumps (6 operating plus one spare), each capable of pumping 0.065 $m^3$/s at 140 bar (2000 p.s.i.). The first and second pumping stations are at 1000 m and 810 m above sea level respectively. To overcome the problem of slack-flow on downhill sections, two valve stations were installed, thus isolating the upstream static head. These stations also allowed a reduction in pipe wall thickness thus reducing the pipeline costs. Wall thickness varies between 8.74 mm and 21.4 mm (0.344 and 0.844 inches).

The slurry is pumped at a concentration of 67 per cent (by weight) and the complete system is rated at a maximum throughput of 12 million tonnes/year. According to Hill *et al.* (20), the slurry is transported in batches, each averaging 23 000 tonnes of solids. During the first six months of operation, 70 batches were transported with 48 shut-downs. Shut-downs varied in time from a few hours to three days, however, it is reported that all restarts were successful.

### Economics

Jennings (21), states that the total construction cost of the system was $600 million with the pipeline itself costing $115 million. No operating costs have been located, however, Hill *et al.* (20,22) state that costs are undoubtedly lower than for existing rail or truck transportation.

## VALEP – PHOSPHATE SLURRY PIPELINE

### Background

In 1968, various mineral reserves including phosphate, titanium and niobium, were investigated for possible development near Tapira, Brazil. In 1974, however, the price of phosphate had increased substantially and in an attempt to reduce imports, it was decided to develop the phosphate reserves as opposed to the titanium ore, as was originally intended (37).

In 1975, a new company called Mineracao Vale do Paranaiba S/A was formed to initiate the Valep project which called for the production of phosphate concentrate. A slurry pipeline was chosen as being the most economic method of material transport. Design work was conducted by Bechtel and construction by Construtora de Oleodutos E Servicos Tecnicos LTDA (COEST).

### Location

The Valep pipeline runs from the mine and concentrator near Tapira to the fertilizer complex near Uberaba, in the state of Minas Geras, north-west of Rio de Janeiro.

### Pipeline technical specification

The pipeline has a diameter of 0.244 m (9.6 inches) with the wall thickness varying between 7.9 and 10.3 mm (0.3 and 0.4 inches), and is 120 km long.

Three internal diameters of 0.2, 0.23 and 0.25 m (8, 9 and 10 inches) were found to be technically feasible, however, due to the problem associated with high frictional losses in the smaller diameter and slack flow problem, associated with batch operation in the larger diameter pipe, the non-standard 0.23 m (9 inch) pipe was chosen.

The pipeline has one pumping station using three pumps, one of which is a standby. These are Continental-Emsco triplex plunger pumps with a capacity of 132 m³/hour at 149 bar (2160 p.s.i.). Each pump is driven by an electric motor rated at 670 kW. The system is designed for an annual throughput of 2.0 million dry tonnes/year at slurry concentration of 61 per cent by weight. No published details relating to economics have been located.

## 5.4   INDIA, KUDREMUKH – IRON ORE PIPELINE

### Background

The Kudremukh to Mangalore iron ore slurry pipeline represents the only major operational slurry pipeline in India. It was commissioned in 1981 and according to Basu and Saxena (23), was constructed for three major reasons:

- There was no rail link between the areas of Kudremukh and the coast near Mangalore.
- The terrain was extremely rugged.
- Finance was readily available.

In 1976, the contract was awarded by the Kudremukh Iron Ore Company Limited (KIOCL) to MET-CHEM Consultants Limited of Montreal, Canada, who completed the basic design and conceptual studies. However, further detailed work was performed by Bechtel Limited and Engineers India Limited mainly in Bangalore (24).

### Location

Kudremukh lies in the mountains of the Western Ghats, India, and the pipeline stretches from the iron ore concentrator at Malleswara, near Kudremukh to a new port site at Panambur, near Mangalore, a total distance of 67 km.

**Pipeline/technical specification**

The pipeline consists of 13.1 km of 0.457 m (18 inch) diameter and 54.6 km of 0.406 m (16 inch) diameter pipe, with an annual throughput of 7.5 million tonnes. The slurry concentration is 65 per cent (by weight), although the operational limits of the system allow a deviation of ±5 per cent $Cw$ to be handled successfully.

The pipeline system utilises four operating plus four standby centrifugal horizontal fixed speed pumps with a throughput of 1115 m³/hr, and one operating plus one standby variable speed pump. The expected life of the system is 30 years.

**Economics**

In 1976, the estimate for the total construction cost including spare parts and initial consumables was approximately $27 million (24,25). It has not been possible to confirm these figures. The cost of the total project is reported to have been approximately $700 million of which $640 million has been loaned from Iran.

No operating costs of the project have been found in the literature, however, Ref. 26 states that concentrates could realise $22/tonne as against the current realisation of $8–9 on exports of Indian iron ore.

## 5.5  INDONESIA, IRIAN JAYA – COPPER CONCENTRATE PIPELINE

**Background**

The Ertsberg copper project of Freeport Indonesia Inc. includes a 112 km (69 mile) long slurry pipeline which transports copper concentration from the mill and concentrator at Ertsberg, to the coastal port area on the Aratura Sea.

The pipeline design work was conducted by Bechtel, who also carried out the construction work in conjunction with Santa Fe Pomeroy for Freeport Indonesia Inc. (34). According to Ref. 35, the system was completed and started-up in February 1973.

**Pipeline/technical specification**

The pipeline is 112 km (69 miles) long with an O.D. of 0.108 m (4.25 inches) and has a design flow rate of 0.011 m³/s (138 gallons/minute) of slurry at a concentration 65 per cent by weight.

The two pumps (one standby) that are utilised, are Ingersoll-Rand Aldrich vertical plunger types with 224 kW electric motor drives. The pumps both have a capacity of 0.011 m³/s at 138 bar (2000 p.s.i.).

**Operational experience**

A series of pipeline failures were experienced, the first failure being experienced 11 months after start-up. After subsequent failures due to severe erosion, it was decided to bypass a complete section of pipeline where the slack flow occurred, a total length of 4.02 km. It was noticed that in each case, erosion occurred immediately after a weld, and where excessive metal was intruding into the pipe.

Although the by-pass line was introduced, it was only used when repair to the 'original' line became necessary. Subsequent discussion resulted in a 'choke' section of reduced diameter pipeline being installed just downstream of the bypass section to induce packed line flow.

A further line failure occurred in February 1974 and was repaired by replacing 215 m of pipeline. However, it is reported in Ref. 34 that no further failure occurred during the following two years.

## 5.6   MEXICO, PENA COLORADA – IRON ORE PIPELINE

Prior to the construction of this pipeline, iron ore was transported to the West coast of Mexico by road and rail, for both pelletising and steel manufacture. Bechtel were responsible for the design of the system with the construction being undertaken by Normex S.A. of Monterrey, which was completed in late 1974. The pipeline is 48 km in length, 0.219 m (8.6 inches) in diameter and can deliver up to 1.8 million tonnes/year. Slurry is mixed to a concentration of 50–60 per cent by weight.

An interesting feature of the pipeline is the gravity flow method by which slurry is transported through the pipeline. Due to the fact that the concentrator and beginning of the pipeline is at high elevation, 900 m above sea level, and the overall length of pipeline is relatively short, only one pump is required. According to Ref. 27, this pump, a 56 kW centrifugal type, is positioned in the test loop which 'monitors' the slurry concentration prior to pumping.

## LAS TRUCHAS – IRON CONCENTRATE PIPELINE

The Las Truchas system is similar to the Pena Colorada pipeline in that it utilised the elevation drop enabling only one pump station to be necessary. The pump station houses three vertical triplex plunger pumps, each of 525 kW. The pipeline is 27 km in length and has a diameter of 0.273 m (10.8 inches). Overall capacity of the system is 1.5 million tonnes/year of ore concentrate, to be used for domestic steel production.

## 5.7   NEW ZEALAND, WAIPIPI – IRON SANDS PIPELINE

This pipeline was commissioned during 1974 and involves the pumping of dredge-mined iron sands to an offshore loading buoy for export.

The pipeline consists of two sections, the first being 6.4 km and 0.203 m (8 inches) in diameter, and the second being 3.2 km long and 0.3 m (12 inches) diameter. The system must be considered, however, as two separate pipelines as the first delivers slurry to a storage pile which is then pumped to the loading buoy as shipping schedules dictate (19).

The first section is operated by five 188 kW rubber lined centrifugal pumps in series which operate at a discharge pressure of 28 bar (400 p.s.i.). The second pipeline is served by six 600 kW Ni-Hard clad centrifugal pumps operating at a discharge pressure of 56 bar (800 p.s.i.), giving a throughput of 1480 m$^3$/hour. A total capacity of approximately 1 million tonnes/year is handled by this installation.

## 5.8 NORTH KOREA, IRON ORE PIPELINE

In 1975, North Korea announced that a pipeline 98 km in length had been designed and constructed. Pitts and Aude (19) state that reliable sources have indicated that approximately 4.5 million tonnes/year of concentrates are being transported by this system for domestic steel production.

## 5.9 PAPUA NEW GUINEA, BOUGAINVILLE – COPPER CONCENTRATE PIPELINE

### Background and location

The mine and processing plant of Bougainville Copper Limited are situated on Bougainville Island, Papua New Guinea. The mine itself is at 670 m elevation and a filter and drying plant is on the coast at Anewa Bay. A slurry pipeline connects the two, transporting copper concentrate for preparation and storage prior to shipment overseas.

### Pipeline/technical specification

The pipeline is 27.4 km in length and has an O.D. of 0.15 m (6 inches) with a wall thickness of 8.7 mm. The pipe specification conforms to API 5LX-52.

The pipeline system was designed to operate in the throughput range of 59–156 tonnes/hour depending upon solids contents, which normally lies within the 55–77 per cent by weight range. This equates to an annual throughput of approximately 1 M tonnes.

The type of pump used is a reciprocating positive displacement plunger type manufactured by Ingersoll-Rand, coupled to a 522 kW electric motor drive. The pumps have a capacity of 136 tonnes/hour at 64 per cent concentration slurry. The life expectancy of the pipeline is approximately 33 years.

### Operating experience

The pipeline commenced operation in April 1972. Piercy and Cowper (36) state that no major failure had been experienced. The most severe incident involved the partial draining of the line due to a leaking block valve at the port terminal. A plug was formed in the pipeline, however this was cleared by increasing the pump pressure to 19 300 kPa.

## 5.10   TASMANIA, SAVAGE RIVER – IRON ORE PIPELINE

The Savage River installation was the first long distance iron ore (concentrate) slurry pipeline and was commissioned in October 1964. The mine and concentrator are located in North West Tasmania and the pipeline stretches for 85 km from the mine to Port Latta on the North Coast. The pipeline varies in diameter between 0.244 m and 0.219 m O.D. (9.6 inches and 8.6 inches) and is buried for the majority of its length (4,28).

The pipeline has one pumping station situated at the concentrator site which comprises four 450 kW triplex, single-acting horizontal positive displacement pumps (one of which is on standby), with normal operating pressures of 105 bar (1520 p.s.i.) at 0.0855 m$^3$/s. The system can deliver 2.25 million tonnes/year of concentrate, formed as a slurry of 60 per cent concentration by weight.

Although the original life expectancy of the pipeline was 20 years, it would appear that the system life will probably be longer due to the lower than expected corrosion rates (19).

## 5.11   UK, KENSWORTH TO RUGBY – LIMESTONE SLURRY PIPELINE

The Kensworth to Rugby limestone slurry pipeline began operations in 1965 and was built under the Authorisation of the Pipeline Act of 1962. Some of the work was undertaken jointly by the Gas Council with the design work being conducted by the Rugby Portland Cement Manufacturing Company. BHRA undertook the laboratory work on the flow characteristics of the slurry using a capillary tube viscometer and test pipe loop.

The pipeline transports material from the chalk quarry at Kensworth near Dunstable, Bedfordshire, to the cement plant at Rugby, a total distance of 92.2 km.

### Pipeline/technical specifications

The pipeline consists of 0.254 m (10 inches) diameter pipe built to API standards, and has a throughput of 1.7 million tonnes per year. The slurry has a concentration of 50–60 per cent by weight.

At Kensworth, three high pressure pumps (one standby) of 646 kW rating are driven at rather less than their full rated speed by 570 kW synchronous motors. They are driven through V-belts to provide a little flexibility between the motor and the pump, mainly to provide a safety device and stop the crank shafts from being strained in the event of a mistake. In addition they facilitate changing pump speed in the future. Each high pressure pump has a 'pulsation damper', a rubber bag accumulator charged with nitrogen, in this case to 35 bar (500 p.s.i.). Each pump is protected by a shear-pin relief valve set to 170 bar (2500 p.s.i.). These valves release much more than the pump delivery, for there is considerable elasticity in the pipeline and the pipework must be suitable for flow at full pressure maintained for several seconds. The motors were designed to start slowly so as not to produce high over-pressures, while accelerating the 92.2 km column of slurry in the pipeline.

## 5.12   USA, CALAVERAS – LIMESTONE PIPELINE

The Calaveras slurry pipeline operates between the Cataract quarry in the Californian Sierra Nevada to the Calaveras Cement plant in San Andreas. The pipeline is 28 km in length and has a diameter of 0.194 m (7.6 inches) (4).

The system uses two double acting duplex positive displacement piston pumps each rated at 175 m³/hour at 95 bar (1350 p.s.i.). One pump is used as a standby. Slurry is pumped at 70 per cent concentration by weight.

## CADIZ, OHIO PIPELINE – COAL PIPELINE (MOTHBALLED)

### Background

The Ohio 'Consolidation' coal slurry pipeline was the first long distance system to be put into operation. Although it was 'mothballed' after six years in operation, it enabled a deeper understanding of the pipelining technology to be gained.

In the early 1950s approximately 50 per cent of the cost of coal was due to transport, and the Consolidation Coal Company investigated ways of reducing the overall transportation costs. After a number of studies and test loop trials had been conducted on coal slurry systems, a full size slurry pipeline was built and put into operation in 1957.

After about six years of operation, the railroad companies reduced the transportation costs and an agreement was signed between Cleveland Electric Illuminating Company and Consol to cease operating the pipeline. The railroads agreed to let CEI have coal for all of their generating plants for $1.88 per tonne which was a significant reduction from a high of $3.47 per tonne. As the pipeline only delivered 36 per cent of the total coal requirements of the power company, the shutdown of the pipeline was economically attractive although the power company paid all outstanding debts of the Consolidation Coal Company (29).

### Location

The pipeline operated between Cadiz, Cleveland and Ohio.

### Pipeline/technical specification

The pipeline was 174 km in length and had a diameter of 0.254 m. Annual throughput was 1.3 million tonnes using a slurry solids concentration of 50 per cent by weight. Three pumping stations operated using two high pressure piston type pumps plus one standby. The capital cost of the system, including the terminals, was $13 million.

## BLACK MESA – COAL PIPELINE

### Background

The Black Mesa pipeline began operations in November 1970. The system is operated by Black Mesa Pipeline Inc., a subsidiary of the Southern Pacific Transportation Corporation.

## Location

The pipeline operates between a mine near Kayenta, Arizona, owned by Peabody Coal Company, and the Mohave Power Station in southern Nevada. The location and route may be seen in Figure 3.

## Pipeline/technical specification

Total length of the system is 439 km and consists of 0.457 m (18 inches) diameter line for the first 418 km with 0.305 m (12 inches) diameter being used for the remainder. The pipeline begins at an elevation of 1830 m and terminates at the power plant at 213 m.

Four pumping stations are required. Stations 1, 3 and 4 each have three duplex piston Wilson Snyder pumps operating in the 35–55 bar (500–800 p.s.i.) range. Station 2 has four slightly higher capacity pumps operating at 86–115 bar (1250 p.s.i. to 1650 p.s.i.). One pump at each station is on standby. Total power of the pumps is 9585 kW (30,31).

The pipeline is capable of transporting 4.5 million tonnes/annum but according to Montford (32), actual throughputs have been slightly lower due to lower requirements at the power station. Flow rates are kept within the 1.5 to 1.7 m/s range and with solids concentrations ranging from 46–48 per cent, results in a throughput of 508–599 tonnes/hour.

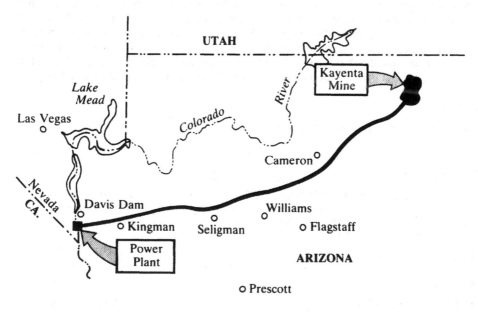

**Figure 3.**  Black Mesa pipeline

## Economics

Relatively few 'in-depth' details are available which concern the operating and capital costs of the Black Mesa system, however, certain proprietary cost figures are available which resulted from correspondence received from the Black Mesa Pipeline Company.

It is reported that preparation and transportation costs are approximately 0.87 cents per tonne km but this figure is apparently a hybrid of 1969–70 capital costs and 1983 operational and maintenance costs. Dewatering is performed by the Mohave Generating Station and costs 'several dollars per tonne'.

No specific information is available on capital costs, however, it is reported that overall costs were approximately $13 million. It is interesting to note that an equivalent rail system would have required 240 km of new track and would have covered an extra 217 km.

## E.T.S.I. PIPELINE – COAL PIPELINE

### Background

Although development of this line has been halted, prior to the date of writing, details are included to illustrate the large size of some of the proposed American pipelines. Originally called Energy Transportation Systems Inc., the E.T.S.I. Pipeline Project was formed in 1973 to develop, build and operate a coal slurry pipeline to deliver low-sulphur coal to power stations in the South-Central States.

Partners in the E.T.S.I. Pipeline Project were the following companies or subsidiaries: Atlantic Richfield, Bechtel, Kansas-Nebraska Natural Gas, Lehman Bros. Kuln Loeb and Texas Eastern.

The project progressed to the stage where basic development work had been concluded and also, of fundamental importance, E.T.S.I. had won 65 court cases it filed to validate its right of way to cross railroad tracks at key locations along the pipeline route (38).

Eventually the project was abandoned. One of the reasons was the slow progress made concerning the acquisition of rights of way.

### Location

The pipeline was proposed to operate from Wyoming to Oklahoma and Arkansas. A barge terminal would have been constructed on the Mississippi River at Cypress Bend, to ship coal to Louisiana.

### Pipeline/technical specification

The pipeline would be buried underground and would cover a total distance of 2250 km, passing through the States of Wyoming 515 km, Colorado 210 km, Kansas 550 km, Oklahoma 420 km and Arkansas 560 km. The pipeline(s) would vary in diameter between 0.406 and 1.0 m (16 and 40 inches) and would transmit slurry on a 50 per cent $Cw$ basis. Total throughput was expected to be 30 milion tonnes/year.

A total of 128 reciprocating pumps would be used, based at 19 pump stations at 130–160 km intervals along the pipeline route.

**Water acquisition and disposal**

Water for the E.T.S.I. pipeline would come from the Oahe Reservoir on the Missouri River in South Dakota and would necessitate the construction of a 435 km pipeline known as the 'West River Aqueduct'. The water rights had been agreed with the State of South Dakota and allowed E.T.S.I. to use up to a maximum of 6170 hectare metres per year. A payment of $10 million to South Dakota would be made prior to the pipeline operation beginning, and thereafter $9 million annually (indexed to inflation) over the 50 year expected life of the project.

**Economics**

According to Marcy (39) the total cost of the E.T.S.I. pipeline project would be approximately $3 billion with the 'Development' and 'Definitive' phases costing $24 million and $81 million respectively.

Few operating costs have been published, however, Wasp (40) draws comparisons between slurry pipeline and rail transportation and states that over a 30 year period, and taking into account inflation of 5 per cent, the savings could amount to $32 billion.

It is interesting to note that according to Ref. 38, the value of the materials and equipment that the E.T.S.I. project would need was expected to total in excess of $1.5 billion.

## 5.13   USSR, BAIDAJEWSKAJA-SEVERNAJA – COAL PIPELINE

Ref. 33 reports that at some Russian hydro-pits, slurry is transported by pipeline on the surface to coal preparation plants. The longest reported example is at the Baidajewskaja-Severnaja mine where run-of-mine coal is pumped a maximum of 11 km to a coal preparation plant. The pumps used are Type 12 UW-6 with eight pipelines of 0.345 m (13.6 inches) diameter in use.

## BELOVO-NOVOSIBIRSK CWM PIPELINE

In spite of initial considerations that the high viscosity of CWM would preclude its use for long-distance transport developments have allowed the design and construction of a 256 km pipeline in the Soviet Union. This system between Belvo and Novosibirsk is planned to carry 3 Mta through a 0.5 m (20 inches) diameter pipeline. The concentration is up to 65 per cent by weight with a viscosity of 750 cP. The pressure loss is expected to be of the order of 1800 m of carrier fluid. This pipeline indicates the advances that have been made in optimising particle-size distribution and the use of additives to provide both stability and fluidity. The pipeline was due to be commissioned at the time of writing this book.

# 6. REFERENCES

1. Anon. (May 1975) 'Slurry Pipelines: the most efficient way to carry coal', *Pipe Line Ind.*, Vol. 42, No. 5, pp. 29–32.
2. Skedgall, D. A. (November 1980) 'Overview of the North American coal slurry pipeline industry', *Coal Technol. 1980 Conf.*, In: *Coal Technology 1980.*, *Proc. 3rd Int. Coal Utilisation Exhib. & Cong.*, Houston, USA, Vol. 1 and 2, pp. 295–307.
3. De Steese, J. G. (14–15 June, 1978 and 1979) 'Identification and prioritization of concerns in coal transportation now through 2000'. *Pacific Northwest Labs.*, *Nat. Acad. Sci.*, In: *Proc. Symp. on Critical Issues in Coal Transportation Systems*, Washington, D.C., USA (1978), Washington, D.C., USA, Nat. Acad. Sci, pp. 321–244 (1979).
4. Pitts, J. D. and Hill, R. A. (21–23 March 1978) 'Slurry pipeline technology – an update and prospectus', *Bechtel Inc.*, *Interpipe 1978, Proc. 6th Int. Pipeline Technol. Convention*, Houston, USA, *Interpipe 1978*, pp. 166–207.
5. Adam, B. O. (January 1982) 'Railroad vs. Slurries: Facts or Friction', *Coal Mining and Processing Mag.*
6. Lee, H. M. (July 1980) 'The future economics of coal transport', *EAS Report No. D2/79*. London, UK, Economic Assessment Service, Int. Energy Agy.
7. Greco, J. and Sherlock, H. B. (1982) 'Energy Transportation – What are the alternatives?' Q1.
8. Gandhi, R. L., Weston, M. D. and Snoek, P. E. (September 1979) 'Economics of alternative coal slurry systems', Bechtel, Inc., In: *Proc. Hydrotransport 6. Sixth Int. Conf. on the Hydraulic Transport of Solids in Pipes*, Canterbury, UK, Vol. 1, Cranfield, UK, BHRA Fluid Engineering, Paper C3, pp. 153–60.
9. Gandhi, R. L., Weston, M. D. and Snoek, P. E. (27 August 1979) 'Alternatives are studied for moving coal by pipeline', *Oil and Gas Journal*, pp. 95–100.
10. Mular, A. L. (Ed.) (1978) 'Mineral processing equipment costs and preliminary capital cost estimations. *Special volume 18*, The Canadian Institute of Mining and Metallurgy.
11. Baker, P. J. and Jacobs, B. E. A. (1979) Consultant Ed. Bonnington, S. T. 'A guide to slurry pipeline systems', p. 58. Cranfield, Bedford, UK, BHRA.
12. Brookes, D. A. and Snoek, P. E. (19–21 October 1988) 'The potential for Stabflow coal slurry pipelines – an economic study', *Hydrotransport 11*, Stratford-upon-Avon, UK, pp. 33–48, Cranfield, Bedford, UK. BHRA.
13. Anon. (September 1980) 'Coal slurry transport: economics and technology', *Mining Mag.*, Vol. 143, no. 3, pP. 243, 245, 247, 249.
14. Moore, T. J. M. (February 1983) 'Preliminary economic and feasibility study on a coal slurry ship unloading system at the Cloghan Point Fuel Terminal', BHRA Report No. RR 1966.
15. Jordan, D. and Wagner, R. (May 1978) 'Supervision and Control of the hydraulic conveyance of raw coal by modern measuring equipment at the Hansa Hydromine', *Proc. 5th Int. Conf. Hydr. Trans. Solids in Pipes*, Hanover, W. Germany, Vol. 1, Paper G2, pp. G-21–G-26, BHRA Fluid Engineering.
16. McCain, D. L., Doerr, R. E. and Rohde, E. G. 'Slurry Transport System Operation', *Soc. Min. Engrs. of AIME*, Reprint No. 81–403.
17. Gontov, A. E. (July 1977) 'Hydraulic Mining: A refined and technological art', *World Coal*, Vol. 3, No. 7, pp. 55–8.

18. Wakabayashi, J. (June 1979) 'Slurry pumped 3000 ft vertically', *Coal Age*, Vol. 84, pp. 84–7.
19. Pitts, J. D. and Aude, T. C. (June 1977) 'Iron Concentrate Slurry pipelines, experience and applications', Trans: Soc. Min. Engrs. of AIME., Vol. 262, pp. 125–33.
20. Hill, R. A., Derammelaere, R. H. and Jennings, M. E. (August 1978), 'Samarco's 246 mile slurry pipeline is on line, successful, and delivering abrasive hematite', *Eng. & Min. J.*, pp. 74–9.
21. Jennings, M. E. (February 1981), 'A major step in iron transportation. Samarco's 396 km pipeline'. *Mining Engng.*, pp. 178–82.
22. Hill, R. A., Derammelaere, R. H. and Jennings, M. E. (26–28 March 1980) 'Samarco Iron Ore Slurry Pipeline', *Proc. 5th Int. Tech. Conf. on Slurry Transport*, Sahara Tahoe, Lake Tahoe, Nevada, USA, Washington DC, USA, Slurry Transport Association.
23. Basu, S. R. and Saxena, A. K. (4–6 October 1982) 'Pipeline transportation – Potential and Planning under Indian Context with special reference to coal', *Metallurgical and Engrg. Consultants Ltd.*, Paper presented at Fourth Freight Pipeline Symp., Atlantic City, New Jersey, USA.
24. Vajda, J. *et al.* (2–7 December 1979) 'The Kudremukh iron concentrate slurry pipeline North American technology in developing countries', *MET-CHFM MFT-CHEM*, Bechtel Inc., Am. Soc. Mech. Engrs., New York, Am. Soc. Mech. Engrs., 10p. (ASME Paper No. 79-WA/7-6).
25. Wilk, A. S. *et al.* (28 March 1980) 'The Kudremukh Iron Ore Concentrate Slurry Pipeline', *Proc. 5th Int. Tech. Conf. on Slurry Transport*, Sahara Tahoe, Lake Tahoe, Nevada, USA, Washington DC, USA, Slurry Transport Association.
26. Anon. (January 1978) 'The Kudremukh Iron Ore', *Mining Magazine*, pp. 26–31.
27. Anon. (November 1974) 'Pena Colorada development includes 30 mile slurry pipeline', *Engrg. & Mining J.*, p. 159.
28. Anon. (June 1979) 'Systems for slurry transportation', *Engng. & Mining J.*, Vol. 180 No. 6, pp. 150–6, 159 and 161–3.
29. Halvorsen, W. J. (1981) 'Contributions to technology from the first long-distance coal pipeline', *Consolidation Coal Co., J. Pipelines*, Vol. 1, No. 3, p. 255–9.
30. Hale, D. W. (May 1976) 'Slurry pipelines in North America, Past, Present and Future', In: *Proc. Hydrotransport 4 Int. Conf. on the Hydraulic transport of solids in pipes*, Canada, Paper G2, pp. x84–x90, BHRA, Cranfield, UK.
31. Rigby, G. R. (1982) 'Slurry Pipelines for the Transportation of Solids', *Inst. of Engrs.*, Australia, pp. 181–9.
32. Montfort, J. G. (1980) 'Operating experience with the Black Mesa Pipeline', *Interpipe 1980 Conf.*, In: *Interpipe 1980, Proc. 8th Int. Pipeline Technol. Exhibition & Conf.*, Houston, USA, Vol. 1, p. 169–81.
33. Anon. (July 1970) 'Hydraulic Mining in the Soviet Union', Extracts from: *Visit of a German mining delegation to the Soviet Union.*
34. McNamara, E. J. (May 1976) 'Operational problems with a 69 mile copper Concentrate Slurry Pipeline', In: *Proc. Hydrotransport 4th Int. Conf. on the Hydraulic Transport of Solids in Pipes*, Banff, Canada, Paper F3, Cranfield, UK.
35. Anon. (September 1975) 'Copper Slurry Pipeline: 69 mile long contributor to success of Ertsberg Project', *Engng. & Mining J.*, pp. 127–9.
36. Piercy, P. and Cowper, N. T. (August 1981) 'Bougainville Copper Limited concentrate pipeline – 9 years of successful operation', *Slurry Systems Pty. Ltd., J. Pipelines*, Vol. 1, No. 2, pp. 127–38.
37. Fister, L. C., Hill, R. A. and Finerty, B. C. (28–30 March 1979) 'VALEP – the world's first long distance phosphate concentrate slurry pipeline', Bechtel Inc., Bechtel do Brasil. In: *Proc. 4th Int. Tech. Conference on Slurry Transportation*, Las Vegas, USA, Washington DC, USA, Session IIA, pp. 60–3, Slurry Transport Association.

38. Anon. (January 1983) Private Communication *ETSI Pipeline Project*.
39. Marcy, S. (22 February 1982) 'At least One Big Coal Slurry Pipeline Moving Along', *Coal Industry News*.
40. Arlidge, J. W. (6–9 March 1980) 'Allen-Warner Valley Energy System; Alton Slurry Pipeline Project Status', In: *Proc. 5th Int. Tech. Conf. on slurry transport*, Lake Tahoe, Nevada, USA, pp. 6–9, Washington DC, USA, Slurry Transport Association.

# APPENDICES

## APPENDIX TO CHAPTER 1

| Symbol | Meaning |
|--------|---------|
| $a$ | speed of sound |
| $C$ | concentration |
| $C_D$ | drag coefficient |
| $C'_D$ | weighted mean drag coefficient of particles above a given size limit |
| $C_V$ | concentration by volume |
| $C'_V$ | concentration by volume of particles above a given size limit |
| $C_f$ | concentration by volume of fluid |
| $D$ | pipe diameter |
| $d$ | mean particle size |
| $E$ | bulk modulus, or molecular stiffness |
| $F$ | thrust |
| $F_L$ | lift force, Durand coefficient |
| Fr | Froude number |
| $F_1$ | first normal stress difference |
| $F_2$ | second normal stress difference |
| $f(f')$ | friction factor, Darcy (Fanning) |
| $g$ | acceleration due to gravity |
| He | Hedstrom number |
| $J$ | head gradient, height of water/length of pipe |
| K | loss coefficient |
| $k$ | pipe roughness |
| $L$ | length of straight pipe |
| $m$ | consistency coefficient |
| $N$ | rotational speed |
| $n$ | consistency index |
| $p$ | pressure |
| Re | Reynolds number |
| $\left.\begin{array}{l}\text{Re}_p \\ \text{Re}_d\end{array}\right\}$ | Reynolds number based on particle size |
| $r$ | radius |
| $S$ | specific gravity, or stress |
| $t$ | pipe wall thickness or time |
| $u$ | velocity, local velocity |
| $u_s$ | shear velocity |
| $V$ | average velocity in pipeline |
| $V_e$ | effective vertical velocity fluctuation |
| $V_c$ | critical transport velocity |
| $V_t$ | terminal settling velocity |
| Wi | Weissenburg number |

|        | a constant |
|--------|------------|
|        | angle between cone and plate |
|        | wall shear stress |
|        | difference |
|        | del, operator indicating gradient (pressure gradient) |
|        | porosity/dissipation rate of turbulent kinetic energy |
| $\kappa$ | kinetic energy of turbulent fluctuations |
| $\mu, \eta$ | viscosity |
| $\mu_p$ | plastic viscosity |
| $\mu_o$ | viscosity at zero shear rate |
| $\mu_\infty$ | viscosity at infinite shear rate |
| $\rho$ | density |
| $\sigma_1$ | first normal stress coefficient |
| $\sigma_2$ | second normal stress coefficient |
| $\tau$ | shear stress |
| $\phi$ | see equation 29 |
| $\psi$ | see equation 31 |
| $\Omega$ | rotational speed |

# APPENDIX TO CHAPTER 2

| $H$ | head |
|-----|------|
| $H_R$ | head ratio |
| $n$ | revolution/s |
| $Q$ | flow rate |
| $\eta_R$ | efficiency ratio |

# APPENDIX TO CHAPTER 3

| $p$ | number of particles/unit surface area |
|-----|---------------------------------------|
| $\beta$ | an exponent |

# APPENDIX TO CHAPTER 4

| $F$ | size at which 80 per cent of feed passes ($\mu$m) |
|-----|---------------------------------------------------|
| $p$ | size at which 80 per cent of product passes ($\mu$m) |
| $W$ | grinding power kWh/tonne |
| $W_i$ | work index |

# APPENDIX TO CHAPTER 5

| $A$ | constant |
|-----|----------|
| $B$ | constant |
| $Q_s$ | volumetric flow rate of solids |
| $Q_m$ | volumetric flow rate of mixture |
| $C_{vt}$ | transport concentration |
| $C_{vd}$ | delivered volumetric concentration |

| | |
|---|---|
| $t$ | time in hours |
| $V_m$ | mean velocity of mixture |
| $w$ | weight loss in grammes |

# APPENDIX TO CHAPTER 6

| | |
|---|---|
| $a$ | empirical constant |
| $B$ | a function of impeller geometry |
| $b$ | empirical constant |
| $c$ | empirical constant |
| $D$ | impeller diameter/distance travelled by slurry |
| $i$ | counting integer |
| $j$ | counting integer |
| $L$ | distance pumped/mass of particles below specified size |
| $m$ | empirical constant |
| $N_p$ | power number |
| $p$ | power |
| $R_L$ | quantity of solids retained on screen |
| $S_i$ | selection rate constant |
| $\tau$ | shear stress near impeller |
| $\Delta W$ | specific energy consumption |

# APPENDIX TO CHAPTER 7

| | |
|---|---|
| Nu | Nusselt number |
| $Nu_s$ | Nusselt number of slurry |
| $\phi_{30}$ | volumetric concentration of solids less than 30 $\mu$m |

# APPENDIX TO CHAPTER 8

**Table 1.** Slurry pipeline installation
(Some have now ceased operation)

| Location | Length (km) | Diameter (m) | Throughput (Mta) | Velocity (m/s) |
|---|---|---|---|---|
| BORAX PLANT WASTE | | | | |
| Boron, USA | 1.325 | 0.127 | | 1.45–3.64 |
| COAL | | | | |
| Black Mesa, USA | 439 | 0.457 | 4.8 | |
| Belovo-Novosibirsk, USSR | 256 | 0.50 | 3.0 | |
| Poland | 200 | 0.256 | | |
| Cadiz, USA | 174 | 0.254 | 1.3 | |
| Belovsk, USSR | 10 | | | |
| France | 8.8 | 0.381 | 1.5 | |
| Gneisenau, Germany | 0.7 | 0.2 | 150 t/hr | |
| Safe Harbour, USA | 0.49 | 0.324 | | 4.62 |
| Markham Colliery, UK | 0.09 | 0.15 | | |

**Table 1.**   continued

| Location | Length (km) | Diameter (m) | Throughput (Mta) | Velocity (m/s) |
|---|---|---|---|---|
| **COAL WASTE** | | | | |
| Horden Colliery, UK | 1.2 | 0.25 | | |
| Fairmount, USA | 0.82 | 0.1 | | 2.12 |
| Chesnick, USA | 0.79 | 0.155 | | 2.43 |
| Egeresehi, Hungary | 0.45 | 0.15 | 80 t/hr | |
| **COPPER CONCENTRATE** | | | | |
| Irian Jaya, Indonesia | 112 | 0.114 | 0.3 | |
| KBI, Turkey | 61 | 0.127 | 1.0 | |
| Bougainville, Papua New Guinea | 27.4 | 0.152 | 1.0 | |
| Pinto Valley, USA | 17.1 | | 190 gal/min | |
| Ouganda, South Africa | 13.5 | 0.076 | | |
| Haydon, USA | 0.85 | 0.152 | | 1.03–1.52 |
| **COPPER TAILINGS** | | | | |
| Japan | 64 | 0.2 | 1.0 | |
| Cebu, Philippines | 19 | 0.508 | 24.0 | |
| Chingola, Zambia | 8.2 | 0.600 | 1.1 | |
| Bougainville, Papau New Guinea | 4.54 | 0.914 | 2268 l/sec | |
| White Pine, Michigan, USA | 4.0 | 0.74 | | 3.22 |
| Ajo, Arizona, USA | 3.4 | 0.406 | | |
| **FILTER EFFLUENT** | | | | |
| Safe Harbour, Penns., USA | 1.07 | 0.254 | | 1.85 |
| **FLY ASH** | | | | |
| Sunbury, Penns., USA | 2.41 | 0.247 | | 4.37 |
| Safe Harbour, Penns., USA | 1.36 | 0.152 | | 1.61 |
| **FUEL ASH** | | | | |
| Ferrybridge, UK | 10.0 | 0.533 | 1.0 | 1.87 |
| Eggborough, UK | 5.4 | 0.533 | 1.0 | 1.87 |
| Longannet, UK | 5.0 | 0.533 | 1.0 | |
| **GILSONITE** | | | | |
| Bonanza, Utah, USA | 116 | 0.152 | 0.4 | 1.18 |
| **GOLD ORE** | | | | |
| Vaal Reef, South Africa | 2.19 | | | |
| **GOLD SLIME** | | | | |
| Freddie North, South Africa | 9.45 | 0.228 | | 1.32 |
| **GOLD TAILINGS** | | | | |
| South Africa | 35.4 | 0.228 | 1.05 | |
| **HEAVY MINERAL CONCENTRATE** | | | | |
| Lawtey, Florida, USA | 4.83 | 0.102 | | 2.6 |
| **IRON ORE CONCENTRATE** | | | | |
| Samarco, Brazil | 396 | 0.508,0.457 | 6.2 | |
| Chongin, North Korea | 98 | – | 4.5 | |
| La Perlat Hercules, Mexico | 85/293 | 0.203,0.356 | 4.5 | |
| Savage River, Tasmania | 85 | 0.228 | 2.3 | |
| Kudremukh, India | 67.7 | 0.457,0.406 | 7.5 | |

Table 1. continued

| Location | Length (km) | Diameter (m) | Throughput (Mta) | Velocity (m/s) |
|---|---|---|---|---|
| Pena Colorada, Mexico | 48 | 0.219 | 1.8 | |
| Sierra Grande, Argentina | 32 | 0.219 | 2.0 | |
| Las Truchas, Mexico | 27 | 0.273 | 1.5 | |
| Waipipi, New Zealand | 6.4/3.0 | 0.203,0.3 | 1.0 | |
| IRON ORE TAILINGS | | | | |
| Hibbing, Minnesota, USA | 5.5 | 0.490 | | 2.36 |
| Honner Plant, USA | 5.3 | 0.406 | | 2.92 |
| Morgantown, Penns., USA | 2.86 | 0.254 | | 2.26 |
| Calumet, Minnesota, USA | 2.62 | 0.406 | | 2.7 |
| Calumet, Minnesota, USA | 1.74 | 0.610 | | 2.38 |
| Grand Rapids, Minnesota, USA | 1.63 | 0.614 | | 2.52–2.74 |
| Star Lake, New York, USA | 1.52 | 0.298 | | 3.65 |
| Canisteo, Minnesota, USA | 0.78 | 0.256 | | 2.68 |
| Sellwood, Ontario, Canada | 0.48 | 0.356 | | 3.2 |
| KAOLIN | | | | |
| Georgia, USA | 25.7 | 0.203 | | 1.3 |
| Georgia, USA | 17.7 | 0.203 | | 1.3 |
| Georgia, USA | 8.04 | 0.300 | | 1.3 |
| LIMESTONE | | | | |
| Kensworth, Beds, UK | 92.2 | 0.254 | 1.7 | |
| Calaveras, USA | 28.0 | 0.194 | 1.5 | |
| Columbia,USA | 27.4 | 0.178 | 0.4 | |
| Gladstone, Australia | 24.0 | 0.203 | 1.8 | |
| Rugby, UK | 14.5 | 0.254 | | |
| Ocenden, UK | 13.0 | 0.330 | | |
| Trinidad | 9.65 | 0.203 | 0.6 | |
| Swanscombe, UK | 5 | 0.330 | | |
| Sewell, Beds, UK | 2.58 | 0.203 | | |
| MAGNETITE CONCENTRATE | | | | |
| Waipipi, New Zealand | 6.43 | 0.203 | 1 | |
| Waipipi, New Zealand | 2.93 | 0.304 | 1 | |
| MILL TAILINGS | | | | |
| Odate City, Japan | 68.14 | 0.300 | 0.6 | |
| Mesabi, Minnesota, USA | 12 | 0.304 | 14 | |
| Sunbright, Virginia, USA | 0.36 | 0.097 | | 1.16 |
| NICKEL REFINERY TAILINGS | | | | |
| Western Mining, USA | 6.92 | 0.101 | 0.1 | |
| PHOSPHATE SLURRY | | | | |
| Valep, Brazil | 120 | 0.244 | 2 | |
| Golasfertil, Brazil | 14.4 | 0.152 | 0.9 | |
| Michols, Florida, USA | 7.72 | 0.356 | | 2.9 |
| Achan, Florida, USA | 6.43 | 0.508 | | 4.57 |
| Michols, Florida, USA | 6.29 | 0.406 | | 3.63 |
| Sydney Mine, Florida, USA | 6 | 0.406 | | 3.63 |
| Tenaroc, Florida, USA | 4 | 0.406 | | 3.7 |
| Barton, Florida, USA | 3.2 | 0.406 | | 3.66 |

**Table 1.** continued

| Location | Length (km) | Diameter (m) | Throughput (Mta) | Velocity (m/s) |
|---|---|---|---|---|
| PLANT REFUSE | | | | |
| Dethue, West Virginia, USA | 0.102 | | | 1.23 |
| SAND | | | | |
| Tatabanya, Hungary | 8.01 | 0.200 | 300–360 t/hr | |
| Kellogg, Idaho, USA | 2.87 | 0.127 | | 1.66 |
| Blackbird, Idaho, USA | 2.26 | 0.150 | | 1.39 |
| Bralorne, B.C., Canada | 1.34 | 0.075 | | 1.65 |
| SMELTER SLAG | | | | |
| Thompson, Manitoba, Canada | 0.30 | 0.127 | | 3.35 |
| TAILINGS | | | | |
| Creighton, Penns., USA | 1.70 | 0.097 | | 1.95 |
| URANIUM BEARING GOLD SLIME | | | | |
| Barbrosco, South Africa | 19 | 0.152 | | 0.97 |
| Ellaton, South Africa | 15.26 | 0.228 | | 1.03 |
| Freddie South, South Africa | 11.13 | 0.406 | | 1.2 |
| Doornfontein, South Africa | 10.66 | 0.226 | | 0.87 |
| East Champ D'or, South Africa | 10.2 | 0.152 | | 0.75 |
| Freddie North, South Africa | 8.14 | 0.304 | | 1.75 |
| Welcom, South Africa | 6.4 | 0.228 | | 1.13 |
| President Brand, South Africa | 3.67 | 0.254 | | 1.39 |
| Welcom, South Africa | 2.74 | 0.228 | | 1.13 |
| URANIUM BEARING PYRITE CONCENTRATE | | | | |
| Merriespruit, South Africa | 5.22 | 0.127 | | 1.37 |
| Daggefontein, South Africa | 2.85 | 0.127 | | 1.35 |

# INDEX